Philosophical Mechanics
in the Age of Reason

Philosophical Mechanics
in the
Age of Reason

KATHERINE BRADING AND MARIUS STAN

OXFORD
UNIVERSITY PRESS

OXFORD
UNIVERSITY PRESS

Oxford University Press is a department of the University of Oxford. It furthers
the University's objective of excellence in research, scholarship, and education
by publishing worldwide. Oxford is a registered trade mark of Oxford University
Press in the UK and certain other countries.

Published in the United States of America by Oxford University Press
198 Madison Avenue, New York, NY 10016, United States of America.

© Oxford University Press 2023

Library of Congress Control Number: 2023952439

ISBN 978-0-19-767895-4

DOI: 10.1093/oso/9780197678954.001.0001

Printed by Integrated Books International, United States of America

For Cherie & Lomey & Momo
For Mark & Thomas & Matthew & Toby & Max
Your love sustains us, and makes it all possible

Contents

Preface   xi

1. A golden era   1
   1.1 Introduction   1
   1.2 The problem of bodies   2
   1.3 Philosophical mechanics   3
   1.4 Constructive and principle approaches   8
   1.5 The unity of physical theory   12
   1.6 Collisions and constraints: PCOL and PCON   13
   1.7 Methods   17
   1.8 Audience   19
   1.9 Overview   22
   1.10 Conclusions   25

2. Malebranche and the Cartesian foundations of natural
   philosophy   27
   2.1 Introduction   27
   2.2 Correcting Descartes: Malebranche's early theory of collisions   29
   2.3 Leibniz's objections to Malebranche's early collision theory   34
   2.4 Malebranche's mature theory of collisions   37
   2.5 After Malebranche: hard bodies in the Paris Academy prize
        competition of 1724 and beyond   48
   2.6 After Malebranche: elastic rebound and the Paris Academy prize
        competition of 1726   52
   2.7 Open questions, hidden problems   57
   2.8 Conclusions   61

3. Beyond Newton and Leibniz: Bodies in collision   63
   3.1 Introduction   63
   3.2 Newtonian collisions   64
   3.3 Leibniz on collisions   74
   3.4 Leibnizian collisions in Hermann and Wolff   82
   3.5 The problem of collisions   95

4. The Problem of Bodies   97
   4.1 Introduction   97
   4.2 The scope and remit of physics   99

4.3 The problem of bodies: nature and action   103
4.4 The problem of bodies: evidence and principle   105
4.5 The methods of Newtonian physics   106
4.6 Substance and causation   124
4.7 The goal: a philosophical mechanics   129

5. Body and force in the physics of collisions: Du Châtelet
   and Euler   132
5.1 Introduction   132
5.2 Nature: extension as a property of bodies   133
5.3 Action   143
5.4 Du Châtelet and Action   149
5.5 Euler and Action   154
5.6 Conclusions   167

6. Searching for a new physics: Kant and Boscovich   169
6.1 Introduction   169
6.2 The physics of bodies in Kant and Boscovich   170
6.3 Kant's philosophical mechanics   180
6.4 Boscovich's philosophical mechanics   184
6.5 Conclusions   198

7. Shifting sands in philosophical mechanics   201
7.1 Introduction   201
7.2 Methodology   202
7.3 Elusive mass   204
7.4 Contact action   210
7.5 A general theory of bodies in motion   213
7.6 Shifting sands   217
7.7 From rational to philosophical mechanics   223
7.8 Rational mechanics ascendant   227
7.9 Conclusions   228

8. Early work in the rational mechanics of constrained motion   231
8.1 Introduction   231
8.2 Personnel and work sites   232
8.3 New territory: oscillating systems   234
8.4 The compound pendulum   238
8.5 From special problems to general principles   245
8.6 Implications for philosophical mechanics   249
8.7 Conclusions   261

9. Constructive and principle approaches in d'Almbert's *Treatise* 263
9.1 Introduction   263
9.2 Constructive and principle approaches   265
9.3 D'Almbert's *Treatise on Dynamics*: its structure and contents 268

9.4 D'Alembert's *Treatise* as a contribution to rational mechanics 285
9.5 D'Alembert's *Treatise* as a contribution to philosophical
    mechanics: a constructive approach 287
9.6 D'Alembert's *Treatise* as a contribution to philosophical
    mechanics: a principle approach 289
9.7 Ontic and nomic unity 295
9.8 Nature, Action, Evidence, and Principle 296
9.9 Conclusions 297

10. Building bodies: Euler and impressed force mechanics 299
    10.1 Introduction 299
    10.2 Solving MCON 300
    10.3 Newton's *Lex Secunda*, Euler's principles, Cauchy's laws
         of motion 303
    10.4 Solving MCON1 307
    10.5 Assessment 324
    10.6 Conclusions 331

11. External obstacles: Lagrange and the mechanics of constraints 333
    11.1 Introduction 333
    11.2 The Principle of Virtual Velocities and "Lagrange's Principle" 334
    11.3 Constraints: equations of condition 341
    11.4 Lagrange's Relaxation Postulate: the kinematics and dynamics of
         constraints 345
    11.5 Philosophical mechanics and Lagrange's *Mechanique* 349
    11.6 Action 356
    11.7 Evidence 358
    11.8 Assessment 362
    11.9 Conclusions 364

12. Philosophical mechanics in the Late Enlightenment 366
    12.1 Introduction 366
    12.2 Makers and spaces 367
    12.3 Lagrangian nomic unification 369
    12.4 Molecular ontic unification 377
    12.5 The Cauchy package 385
    12.6 Conclusions 394

Conclusions 396

Bibliography 401
Index 415

Preface

The enormous project in this book resulted from two strands of interest that converged, in a meeting of the minds, at a 2014 workshop that Michela Massimi and Angela Breitenbach organized in Cambridge, UK. Katherine Brading is a philosopher of science, and of physics in particular, who had recently begun work on foundational problems left unsolved in the wake of Newton's *Principia*. She had long been curious about the question of how and when philosophy and physics came apart—because it was clear they had not separated in the decades after the *Principia* first appeared. Marius Stan is a philosopher and historian of science whose work on mechanics has taken him deep into the 18th century. He relies on linguistic, technical, and philosophical tools to cast light on the deep structure of classical mechanics, investigating its enormous growth, and explicating its philosophical foundations. This book lies at the happy crossroads of their inclinations, interests, and expertise.

It has been many years in the making; during them all sorts of upheavals have taken place, from family illness to a global pandemic. Our work together has been a rock in troubled times, and speaks to the power of collaborative scholarship. But even with each other, we could not have done it alone. We owe a great debt of gratitude to many colleagues, friends, mentors, and students; it is our great pleasure to acknowledge them here.

We humbly thank those who have shaped and inspired us as scholars. KB's interest in the entwined history of physics and philosophy developed slowly from seeds sown and nurtured by John Roche, Dan Garber, Bill Harper, and John Henry. She is forever grateful to Dana Jalobeanu and Hasok Chang for the many hours spent pondering the relationships between Descartes' and Newton's physics and between history, philosophy, and science, respectively. Her appreciation for the philosophical importance of the details in the physics owes most to Julian Barbour, Harvey Brown, Tom Ryckman, and the philosophy of physics group at Oxford. She thanks the students and faculty of the Notre Dame History and Philosophy of Science Graduate Program for creating an environment in which integrated history and philosophy of science is the norm.

MS learned the history of classical physics from Jed Buchwald, a gift for which he is forever grateful. For the way he looks at early modern matter theories, he is greatly indebted to Sheldon Smith. Dan Garber and Vincenzo De Risi inspired him to wonder about the relation of early modern philosophy to science. His broader grasp of classical mechanics has been shaped by Mark Wilson, Iain Stewart at MIT, and the late Clifford Truesdell. For their generous mentoring in the history and philosophy of science, MS is grateful to Moti Feingold, Gideon Manning, and Chris Hitchcock.

This book is not about Newton, but his legacy and impact on the Enlightenment is our backdrop. In grappling with that legacy, we have benefited enormously from long and illuminating conversations with Zvi Biener, Mary Domski, Geoff Gorham, Dana Jalobeanu, Andrew Janiak, Oliver Pooley, Eric Schliesser, Ed Slowik, Chris Smeenk, and Monica Solomon. For us both, the influence of George E. Smith as scholar, mentor, and friend is beyond measure.

MS's path to Enlightenment mechanics went through Kant. For guidance in that difficult area, we wish to thank Eric Watkins and Michael Friedman. We thank warmly Marij van Strien, Josh Eisenthal, Sandro Caparrini, and Andreas Verdun for many enlightening discussions about the history of classical mechanics and its foundational subtleties. And we thank Alison Peterman, Jeff McDonough, and David Marshall Miller, who have been our thoughtful, generous interlocutors about early modern natural philosophy for many years.

Our students, past and present, have kindly given us feedback on the various parts of this book as it grew. For their comments, we thank especially Qiu Lin, Michael Veldman, and the students in KB's research seminars at Duke University.

For penetrating questions, comments, and advice, we thank audiences at the Oxford Philosophy of Physics Seminar, the Division of Humanities at Caltech, the philosophy departments at Princeton and Toronto, the Logic and Philosophy of Science Department at the University of California Irvine, the HOPOS 2022 Conference, and the participants at the 2018 workshop Mechanics and Matter Theory in the Enlightenment at Duke.

We thank the American Council of Learned Societies, whose generous support allowed us to take time off from teaching in 2017–18 and do a great deal of primary research and sustained reflection on the problems treated here.

We are grateful to Oxford University Press; to Peter Ohlin for support and encouragement; and to his two anonymous reviewers who gave us invaluable feedback earlier in the project.

Last, and most importantly, we thank our families and our close personal friends, who have endured our dedication to this project with good humor and unfailing support. Thank you.

1
A golden era

1.1 Introduction

This is a book about philosophy, physics, and mechanics in the 18th century, and the struggle for a theory of bodies. Bodies are everywhere, or so it seems: from pebbles to planets, tigers to tables, pine trees to people; animate and inanimate, natural and artificial, they populate the world, acting and interacting with one another. And they are the subject-matter of Newton's laws of motion. At the beginning of the 18th century, physics was that branch of philosophy tasked with the study of body in general. With such an account in hand, the special areas of philosophy (whether natural, moral, or political) that presuppose special kinds of bodies (such as plants, animals, and human beings) could proceed assured of the viability of their objects and the unity of their shared enquiries. For all had "bodies" in common. So: What is a body? And how can we know? This is the *Problem of Bodies*, and the quest for a solution animated natural philosophy throughout the Age of Reason.

How so? Because, inherited from the 17th century as a foundational concern, the *Problem of Bodies* proved surprisingly resistant to solution. Consequently, it ensnared a wide range of 18th century figures who brought to bear a diverse assortment of resources. At the forefront we find familiar characters from the received philosophical canon, such as Leibniz, Malebranche, Wolff, Hume, and Kant, wrestling with BODY alongside others of equal or greater import, such as Maupertuis, Musschenbroek, Du Châtelet, Euler, and d'Alembert. Their attempted solutions drew on matter theory, metaphysics, physics, and mechanics; they appealed to a variety of principles metaphysical, epistemological, and methodological; and they simultaneously disputed the appropriate criteria for success. At stake were two central issues of philosophy from the period: material substance and causation. Upshot: the contours and depths of the problem are a philosophical treasure trove.

Philosophical Mechanics in the Age of Reason. Katherine Brading and Marius Stan, Oxford University Press.
© Oxford University Press 2023. DOI: 10.1093/oso/9780197678954.003.0001

In this chapter, we introduce *the Problem of Bodies*, along with the main analytical tools we use for its investigation (sections 1.2–1.6). We outline our methods (section 1.7) and our intended audiences (section 1.8). Finally, we provide a chapter-by-chapter guide for what is to come in the remainder of this book (section 1.9), and a preview of our main conclusions (section 1.10). Inevitably, some of what we say in this chapter is compressed and may seem somewhat cryptic at first sight, but when read in conjunction with the later chapters it is, we hope, sufficient to anchor the main points of each chapter within the argument of the whole.

1.2   The problem of bodies

The *Problem of Bodies* (hereafter *Body*) is large and unwieldy, as we will see. Nevertheless, we argue that it has a structure that makes it amenable to analysis, in the form of a goal and four criteria for success.[1]

> **Goal:** a single, well-defined concept of body that is simultaneously (i) consistent with an intelligible theory of matter, (ii) adequate for a causal-explanatory account of the behaviors of bodies, and (iii) sufficient for the purposes of mechanics.

Any satisfactory solution to *Body* was expected to meet this goal. To do so, it needed to satisfy the following criteria for success: Nature, Action, Evidence, and Principle (NAEP).

> **Nature:** Determine the nature of bodies. Ascertain their essential properties, causal powers, and generic behaviors.

> **Action:** Explain how bodies act on one another. Give an explanation of how, if at all, one body changes another's state (where specifying the "state" of a body is addressed by Nature).

These first two are metaphysical. The next two are epistemological: they seek to uncover the justificatory reasoning behind Nature and Action.

---

[1] We arrived at this structure for *Body* by examining the arguments of the participants in the debate. We then made the elements explicit and used them to assess purported solutions.

Evidence: Elucidate the evidential reasoning behind Nature and Action. Spell out what counts as evidence for these claims, and what patterns of inference take us to them as conclusions.

Principle: Elucidate the constraining principles appealed to in attempting a solution to BODY, and check that proposed solutions conform.

Such principles include the Principle of Sufficient Reason, the Law of Continuity, the restriction to contact action, and the criterion of clear and distinct ideas. Our protagonists understood their principles in different ways—as a priori philosophical requirements, defeasible heuristics, and so forth—but such principles were always in play, whether implicitly or explicitly.

The four criteria (NAEP) may be variously interpreted and implemented. Their more precise specification differs from one philosopher to another, and this comprises one element of the debate over BODY. Excavating and investigating them is a part of what we do in this book. The diversity of views on offer is proof that BODY was neither easy to state nor straightforward to solve.

1.3 Philosophical mechanics

"Philosophical mechanics" is a term of art.[2] We use it to label the framework within which we use the resources described above ("Goal" and "NAEP") to analyze BODY in the 18th century. Such a framework is justified by its utility: it stands or falls by the work that it does for us in this book, and that assessment can be made only when we reach the end. However, there are some remarks we can make here, at the beginning, that we hope will be helpful.

Simply put, a philosophical mechanics is any project that integrates matter theory with rational mechanics. To motivate the idea, we offer some 17th century background in Descartes' *Principles of Philosophy* and in Newton's *Principia*.[3] In Descartes, we see the connection of BODY to collision theory, and from there to philosophical mechanics. In Newton, we find an explicit

[2] While the term was first used (to our knowledge) in 1800, in Gaspard Riche de Prony's *Mécanique philosophique, ou, Analyse raisonnée des diverses parties de la science de l'équilibre et du mouvement*, we adopt it for our own purposes.
[3] Descartes 1991; Newton 1999.

example of a philosophical mechanics. Reflected in each is an important disciplinary distinction between physics—as a sub-discipline of philosophy; and rational mechanics—as a sub-discipline of mathematics. Philosophical mechanics draws on both, as we will see.

(i) Cartesian origins

In his 1644 *Principles*, Descartes set out to explain all the rich variety of the natural world around us: he sought to provide a complete physics that included everything from planetary motions to the creation of comets, from the formation of mountains to the behavior of the tides, and from earthquakes to magnetism and beyond. Descartes' physics begins with his theory of matter; the principal attribute of Cartesian matter is extension, and the parts of matter have shape, size and motion. All change comes about through matter moving in accordance with the laws of nature.

BODY, as it occurs in *Principles*, concerns the parts of matter, for these are Descartes' bodies. His laws of nature take parts of matter—or bodies—as their subject-matter. The first issue is whether (and if so, how) he succeeds in giving a viable account of bodies prior to his introduction of the laws of nature. For, he suggests that matter is divided into parts by means of motion, but he also defines motion by appeal to the parts of matter, generating an undesirable circle. This issue was widely appreciated at the time, and has been much discussed since.[4] If we set it aside, and presume that Descartes has "parts of matter" available, a second issue then comes to the fore. While the laws of nature supposedly take these bodies as their subject-matter, Descartes' bodies seem not to have the properties and qualities demanded by the laws. The only properties that his matter theory secures for bodies are shape, size, and motion. However, as we move through the exposition of his laws, we find Descartes appealing to "stronger" and "weaker" bodies; "hardness"; "yielding" and "unyielding" bodies; the "tendency" of bodies to move in a straight line; and the "force" of a body, which is nothing other than the

---

[4] See Garber 1992, 181; Brading 2012; and references therein. One upshot is a tendency among Cartesian philosophers toward mind-dependent bodies, as seen in both Desgabets and Régis, for example. Lennon (1993, 25) writes of Régis: "Individual things result from our projection of sensations on otherwise homogeneous and undifferentiated extension. On this view individual things are what Malebranche and Arnauld took to be the representations of things."

"power" of a body to remain in the same state. But it is far from clear that these are reducible to shape, size, and motion.[5]

The issue is pressing because of the role of bodies in Descartes' physics. It is by means of bodies moving in accordance with the laws of nature that he aims to explain all of the material world. If his philosophy lacks the resources for his matter theory to yield bodies, his physics cannot get off the ground.

A necessary condition on a viable solution to BODY, within Descartes' system, is that the resulting bodies are capable of undergoing collision. This is because all change takes place through *impact* among the parts of matter. Collisions therefore lie at the heart of his physics, and Descartes supplements his laws of nature with seven rules of collision. The rules, like the laws, appeal to his prior theory of matter: to the essential attribute of matter (extension), and its modes (shape, size, and motion). To sum up: collision theory is foundational for Descartes' physics, and it combines two elements: matter theory and rules of impact.

The centrality of collisions is not confined to Descartes' philosophy. For anyone pursuing "mechanical philosophy" in the 17th century, impact was the *only* kind of causal process by which change comes about in the material world. Moreover, even for philosophers who, in the wake of Newton, sought to move beyond "mechanical philosophy" by endowing bodies with additional "forces," collisions remained an important means of action and interaction among bodies. As a result, collision theory was foundational for natural philosophy in the late 17th century. Moreover, following Descartes, any adequate collision theory was required to combine a theory of matter with rules of collision: we call this a *philosophical mechanics* of collisions.

(ii) The integration of philosophical physics with rational mechanics

Projects in philosophical mechanics seek to meet the demands of both physics and rational mechanics. What do we mean by this?

---

[5] Impenetrability, solidity, and hardness are among the properties of body that some philosophers thought Descartes was not entitled to (as in Locke's discussion of body, for example). The question of whether a notion of force must be added (cf. Leibniz) persists long into the 18th century, as we shall see in later chapters of this book.

Early modern physics retained the Aristotelian aim of seeking the most general principles and causes of natural things, and of their changes. The primary subject-matter of physics was *bodies*: the role of physics was to provide a causal account of the nature, properties, and behaviors of bodies in general. Frequently, the term "physics" was used interchangeably with "natural philosophy." This reflects the fact that early modern physics was a sub-discipline of philosophy, practiced by self-professed philosophers who retained responsibility for and authority over the account of body in general. When other areas of philosophy (such as those treating specific kinds of bodies) and other disciplines (such as mechanics) presupposed bodies, they did so with the presumption that physics succeeds in providing an account of bodies in general.

The term "mechanics," on the other hand, had several senses, ranging from the science of machines to the various strands of "mechanical philosophy," but here we use it with one particular connotation, current at the time and broadly familiar from present-day usage. Specifically, we are interested in *rational mechanics*, namely, the mathematical study of patterns of local motion and mutual rest. It was a descriptive approach that represented mechanical attributes (mass, speed, force, and the like) as measurable quantities. Its inferences were subject to laws of motion and equilibrium conditions, functioning as constraints on admissible conclusions. Put modernly, rational mechanics pursued deductive schemas for moving from values of relevant parameters to integrals of motion or to differential equations relating these parameters. At first, its representational framework was heavily geometric, but through the 1700s algebraic methods increasingly supplanted the earlier reliance on synthetic geometry. Our use of the term "rational mechanics" is one that came to dominate by the end of the 18th century, and it can be found explicitly one hundred years earlier, in the Preface to Newton's *Principia*.

By "mechanics" we mean, from here on (unless stated otherwise), rational mechanics. In his Preface, Newton offered a taxonomy of mechanics in which he divided "universal mechanics" into three: practical mechanics, rational mechanics, and geometry. For our purposes, the key points are as follows. Like geometry, rational mechanics is mathematical and exact: it suffers from none of the imperfections of practical mechanics. Unlike geometry, however, rational mechanics goes beyond the treatment of magnitudes to include motions and forces. Rational mechanics, Newton says, is the "science, expressed in exact propositions and demonstrations, of the motions that result from any forces whatever and of the forces that are required for any motions whatever."[6]

The term "mechanics" today typically denotes some *branch* of physics (e.g., classical mechanics, quantum mechanics, statistical mechanics, and so forth). At the beginning of the 18th century, this was not the case: physics and mechanics were distinct fields.[7] Unlike mechanics, physics was largely qualitative and, as we have said, practiced by philosophers. Mechanics, on the other hand, fell under the authority of mathematicians. While some people at the time, including some of the most influential figures of the period, were both philosophers and mathematicians, the two disciplines were distinct. They had distinct methods, distinct goals, and distinct domains of authority.

With this in mind, we see that physics in the early 18th century is importantly different from physics today, in its goal (of providing a causal account of the nature, properties, and behaviors of bodies in general), methods (which were qualitative), and disciplinary relations (within philosophy, and distinct from mechanics). Physics thus understood is central to the arguments of our book, and it is this 18th century conception that the term "physics" denotes. Sometimes, we will use the term "philosophical physics" as a reminder that in the 1700s physics was a non-mathematical branch of philosophy.

Descartes' account of bodies falls within his philosophical physics. His rules of collision, insofar as they are mathematical and exact, fall under the remit of rational mechanics. According to the analysis that we offer, Descartes' theory of collisions integrates resources from physics and rational mechanics in order to provide a philosophical mechanics of impact. By itself, this makes overly hard work of Descartes on collisions, with a superfluity of terminology for little philosophical gain. The payoff comes from the application of the same analytical tools over the next 150 years of developments.

(iii) Newton's *Principia* as a project in philosophical mechanics

Newton's *Principia* contains a rational mechanics, but it is not *merely* a text in rational mechanics. His choice of title, *Mathematical Principles of Natural Philosophy*, is revealing. Newton declares that the forces to be treated

---

[6] Newton 1999, 382.
[7] Guicciardini 2009.

mathematically include natural forces, such as gravity. In this way, rational mechanics becomes a tool for the pursuit of natural philosophy. In Book III, Newton applies the results of his rational mechanics from Books I and II to the particular case of gravity, and thereby provides a causal account of the motions of material bodies under the force of gravity: he offers a contribution to philosophical physics.

The *Principia* therefore contains both rational mechanics and physics. Newton explicitly set out the relationship between the two, as he understood it. Together, the overall project forms a framework for pursuing the science of bodies in motion, in which a rational mechanics (an exact mathematical treatment of bodies and the forces that act upon them) is to be integrated with a physics (a treatment of the causes of the motions of bodies). Newton's physics is incomplete, but his intention to contribute to both rational mechanics and physics is clear. Indeed, Newton's "Axioms, or Laws of Motion" belong to both. The 17th century, prior to Newton, had seen discussions over whether the laws of nature (such as those found in Descartes' *Principles*) might also serve as axioms of mechanics. Up until Newton, books of physics and books of mechanics were distinct, and the principles of each differed. The *Principia*, in attempting to combine rational mechanics with physics, is an important example of a philosophical mechanics.[8]

1.4 Constructive and principle approaches

We have seen the centrality of collisions within Descartes' natural philosophy. Yet, as is well known, his rules of collision were rejected for their inadequacy with respect to observation.[9] The ensuing 17th century discussions are an important background for our book. First, they reveal hints of two distinct heuristics for tackling BODY—constructive and principle—and we discuss these here. Second, they preview the problems with collision theory that 18th century natural philosophy was to inherit.

In October 1668, Henry Oldenburg, the secretary of the Royal Society, wrote to Huygens and Wren asking for their theories of motion and collision.

---

[8] For a detailed discussion of Newton's philosophical mechanics see Brading, forthcoming.

[9] Descartes himself maintained that his rules applied only to microscopic (and therefore unobservable) collisions, not to the bodies of our experience. The rejection of this defense of his rules speaks to the question of epistemology: of the means by which we are to determine whether or not the proposed rules are to be accepted.

Soon, Huygens, Wren, and Wallis submitted their proposals. Those of Huygens and Wren cover the case of perfectly elastic bodies while Wallis' pertain to perfectly inelastic collisions.[10] One might think that this is where the story should end: we have the correct rules of collision, so what else is there?

As Jalobeanu describes, the issues were far from resolved by the arrival of the rules from Huygens, Wren, and Wallis.[11] On December 1, 1668, Oldenburg wrote to Wallis (and others) asking about the physical causes of rebound; whether resting matter resists motion; whether motion is conserved; whether motion is transferred from one body to another when they collide; and so forth. The questions concerned the material nature of bodies and the physical causes of their behaviors during impact.

In their submissions, Huygens, Wren, and Wallis had not discussed the material constitution of bodies, let alone used such considerations as pertinent to the problem. Wallis responded by claiming that the rules themselves provide an account of the physical causes:

> I have this to adde ... you tell mee yt *ye Society in their present disquisitions have rather an Eye to the Physical causes of Motion, & the Principles thereof, than ye Mathematical Rules of it*. It is this, That ye Hypothesis I sent, is indeed of ye *Physical* Laws of Motion, but *Mathematically* demonstrated. (Oldenburg 1968, 220–2)

Huygens explicitly set aside causes:[12]

> Whatever may be the cause of hard bodies rebounding from mutual contact when they collide with one another, let us suppose that when two bodies, equal to each other and having equal speed, directly collide with one another, each rebounds with the same speed which it had before the collision. (Huygens 1977, 574)

---

[10] For simplicity of exposition, we use today's terminology here. The problem of how to categorize bodies—as hard, soft, elastic, inelastic, rigid, malleable, unbreakable, infinitely divisible, etc.—and how to correlate these terms with the various behaviors of bodies (when pressed upon, during impacts, etc.) persisted into the 18th century, as we shall see. For discussion of the 17th century struggles with "hardness" in the context of collisions, and in relation to the Royal Society debates, see Scott (1970, 12ff.).

[11] See Jalobeanu 2011.

[12] This excerpt is from a paper published posthumously in 1703. See also Murray, Harper, & Wilson (2011, 189, footnote 8) who note that this phrase does not appear in the original letter sent to Oldenburg but was added prior to publication.

But, their quietism about material properties and causes met with resistance at the Royal Society. Another member, William Neile, argued that the collision rules should be supplemented by an underlying matter theory so as to account for the "physical causes" of the observed phenomena, such as rebound.[13]

The problem to be solved arises in the following way. If the properties of bodies (such as hardness and "springyness") and the principles concerning the behavior of bodies (such as conservation of quantity of motion) appealed to in the rules of collision arise from the nature of matter (as they do in Cartesian physics), then a problem in mechanics—finding the correct rules of collision—is inevitably entangled with matter theory. More generally, Neile's objection signals a theme that persists late into the 18th century: the search for a causal-explanatory account of the properties of bodies, and of the collision process, that integrates the rules of impact into a theory of matter: a philosophical mechanics of collisions.

The Royal Society dispute can be analyzed as offering two general approaches for tackling problems within philosophical mechanics. Following Neile, we may decide to begin with a theory of matter, and develop our collision theory from there. We call this the *constructive* approach. Following Wallis, we may decide to begin with the rules, and seek to build our matter theory from them. We call this the *principle* approach. By "approach" we here mean a general strategy consisting of a broad heuristic along with a reservoir of initial evidence and explanatory premises. We will argue that 18th century attempts to solve BODY are best understood as pursuing a philosophical mechanics of bodies by means of these two general approaches. We further specify them as follows.

Constructive approach (bodies): The qualities and properties of matter are the primary resource for solving BODY.

From the properties and powers of matter, we construct concepts of body (Nature) and bodily action (Action) consistent with Principle and Evidence (see section 1.2) to arrive at a philosophical mechanics in which the resulting account of bodies yields, or is at least consistent with, the notion of body that rational mechanics presupposes.

---

[13] For context and discussion, see Jalobeanu 2011 and Stan 2009.

This approach comes in two varieties, a stronger and a weaker. The stronger begins from an explicit theory of matter, and from there constructs bodies. The weaker eschews a foundation in matter theory, working directly with bodies instead; it presumes that the methods and resources of philosophical physics itself are sufficient for determining their qualities and properties.

Principle approach (bodies): Theoretical principles, such as the laws of motion, are the primary resource for solving BODY.

In the case of laws of motion, the principle approach means drawing the concept of body, and of bodily action, from the laws themselves, without appeal to any prior theory of the material constitution of bodies. From there, we arrive at a philosophical mechanics by showing that the resulting account coheres with a philosophically viable theory of matter in meeting the demands of NAEP. By "coheres with," we likewise include two different possibilities, a stronger and a weaker once again. The weaker is that the laws are deemed *necessary* in constructing a body concept, but that extra-legal ingredients are also needed, drawn perhaps from an independent matter theory. The second, stronger, position is that the laws are both necessary *and* sufficient for the construction of an adequate body concept.

Insofar as matter theory and physics fall under the authority of philosophy, and laws of motion under rational mechanics, the constructive and principle approaches align with two distinct routes to a philosophical mechanics: one which prioritizes philosophy—including matter theory and physics—and the other which prioritizes rational mechanics. To see this play out requires the rest of our book.

The constructive and principle approaches generalize beyond bodies. At the beginning of the 18th century, bodies were presumed to be the objects of study in both physics and rational mechanics. As the century wore on, this presumption came under increasing pressure. From the perspective of mechanics, candidates for the objects of study in the 1700s included point particles, flexible and elastic solids, inviscid fluids, and mass volumes in equilibrium configurations.[14] If we relax the assumption that the objects of physical theorizing are bodies, then it becomes an open question what those objects might be, and BODY becomes a more general problem, the *Problem of*

---

[14] If we move beyond the 18th century we soon add classical fields, quantum particles, quantum fields, and so forth, as objects of theorizing. The problem persists of how best to specify these objects.

Objects: What are the objects of physical theorizing? And how can we know? As our analysis of the 18th century unfolds in later chapters, we see the constructive and principle approaches tracking this generalization.

Constructive approach (objects): Matter theory is the primary resource for constituting the objects of physical theorizing.

Principle approach (objects): Theoretical principles, such as laws, are the primary resource for constituting the objects of physical theorizing.

As in the case of bodies, these approaches come in weaker and stronger forms.

1.5 The unity of physical theory

At the beginning of the 18th century, the role of physics was to provide the general account of bodies. Other areas of enquiry, both within philosophy and beyond, took such bodies for granted, as a given in theorizing. Special areas of natural philosophy (e.g., botany) took the general account of body and then studied the additional specifics (e.g., as appropriate to plants). The general concept of body provided by physics thus played a unifying role as the common object of philosophy, mechanics, and so forth. This is an example of ontic unity in physical theory.

Ontic unity: a single type of object unifies physical theory.

In order for bodies to play such a unifying role, solutions to BODY must yield a single body concept adequate for all areas of theorizing. To achieve this, one may adopt either a constructive or a principle approach.

There is an alternative to ontic unity, one that arises from the principle approach alone. Rather than unifying our theorizing through a shared *object* (such as a shared account of body), we locate unity directly in the *principles* (such as the laws). We call this "nomic unity".

Nomic unity: a single set of principles, such as a set of dynamical laws, unifies physical theory.

Unity comes from a (small) set of principles that entail the properties and behaviors of all physical systems (such as when the laws entail equations of motion)—or at least all those regarded as tractable at the time. Such an approach makes no explicit commitments about the ontology associated with the principles: it allows for a diverse ontology, for diverse objects, and for cases where there is no explicit specification of objects at all. Rather than unifying our theorizing through a shared *object* (such as a shared account of body), we locate unity directly in the *principles* (such as the laws).

As we will see in the second half of our book, this latter conception of unity emerged in the context of rational mechanics as a consequence of the persistent failure to solve BODY. But it too faced a challenge, viz. to ensure that mechanics is *one* theory, not a patchwork of local accounts joined arbitrarily by blunt juxtaposition in a textbook. Facing up to this challenge is crucial, for if mechanics lacks even this unity then it is unclear whether it has a subject-matter at all. We see examples of this approach to unity, and challenges to it, in the second half of this book.

1.6 Collisions and constraints: PCOL and PCON

We cannot hope to cover all the many aspects of BODY in one book. However, when viewing the 18th century through the lens of philosophical mechanics, we see that two somewhat better defined problems provide the main loci of investigation: the problem of collisions (PCOL) and the problem of constrained motion (PCON). We explain these at length in our book, for they are the focus of the first and second halves, respectively. Here, we state them for future reference, and attempt to give the gist of their significance.

We have already noted that collisions became central to natural philosophy in the wake of Descartes' *Principles*. As a result, the following question became pressing:

PCOL: What is the nature of bodies such that they can undergo collisions?

We argue that solving PCOL became a necessary—but not sufficient—condition on solving BODY. The task was to provide a causal-explanatory account of collisions by integrating rules of impact into a theory of matter. In other words, to provide a philosophical mechanics of collisions.

There are two routes to this: one prioritizing philosophical physics and the other rational mechanics. Within the former, we find two versions of a constructive approach, consistent with the stronger and weaker versions described above. The stronger is matter-theoretic: it starts with an overt, philosophical theory of matter. From its resources, this approach articulates a physics of the bodies that undergo impact, including their properties relevant to the collision process, and their behaviors during and after. The weaker presumes that physics itself has methods and resources sufficient for determining and justifying the qualities and properties of material bodies, without the need to appeal to any explicit theory of matter. The constructive approach to PCOL is the subject of the first half of our book.

Then, we take a new direction. For, despite concerted efforts by a wide range of philosophers, success eluded them as of the mid-1700s. Meanwhile, developments in rational mechanics had begun to change the philosophical space in important ways, so that a new problem supplanted PCOL as the most important locus of research relevant to BODY. This is the focus of the second half of our book, and it is here that the principle approach comes to the fore.

A key assumption in any attempt to address PCOL is that bodies are extended and mobile. Such are the bodies of rational mechanics, and Euler's *Mechanica* of 1736 positioned collisions within a projected general theory of the motions of extended bodies. The 17th century had tackled the motion of an extended body by tracking a representative point.[15] This approach has two serious limitations that yielded two corresponding challenges for 18th century rational mechanics, both of which proved highly consequential for attempts to solve BODY.

First, treating a representative point yields the overall trajectory of a body (e.g., the path of an asteroid), but it does not determine the motions of the parts of the body as it executes that trajectory (e.g., the tumbling of the asteroid as it careens toward the earth). The challenge was to construct a rational mechanics that goes beyond the representative point when treating the motion of the whole. Within mechanics, it falls within the theory of constrained motions, and we call it MCON1.

MCON1: Given an extended body subject to *internal* constraints, how does it move? More specifically, given an extended body whose parts are mutually constrained among themselves (i.e., held together to form *one* body), what is the motion of each of the parts?

---

[15] This proved hard enough for many systems. See Chapters 7 and 8, as well as Stan 2022.

A solution to MCON1 would enable us to determine the motion of *every* part of an extended body, as the whole moves. The simplest case is the hard (or perfectly rigid) body, in which there are no relative motions among the parts of the body. To achieve rigidity, the presumptions are that (i) forces acting on the body produce no change in shape (no compression forces, no torsion); (ii) forces acting on the body through a point other than the centroid (representative point) produce only rotational motion, no torsion; and (iii) any rotational motion has no effect on the shape of the whole. As soon as these assumptions are relaxed, relative motions among the parts of the body, and their consequences for the motion of the whole, must be addressed. As you might imagine, MCON1 is horribly difficult.

The second limitation concerns the motions of bodies that are impeded by other bodies, such as when a ball is prevented from falling by the presence of an inclined plane. One might hope to treat such obstructions in terms of Newton's laws of motions, via the forces at work as one body acts on another. However, such hopes are often ill-founded, especially when the forces are many or when they change at every moment of the motion. As a classic example, consider a bead constrained to move along an arbitrarily curved wire. As the bead moves, the direction and magnitude of the impressed force change at every instant. To overcome this complexity, and the resulting intractability of the problem,[16] we consider the bead as subject to *kinematic constraints*: we treat the wire as restricting the motion of the bead to a particular spatial region, without concern for the forces that bring this about. More generally, we theorize the obstructed motion of the target body as encountering obstacles that render certain regions of space inaccessible. By this means, we can seek to determine the motions of bodies when subject to a variety of external obstacles. We call this MCON2.

> MCON2: Given an extended body subject to *external* constraints, how does it move? More specifically, when the motion of a body is impeded by an obstacle, what is its resulting motion?

[16] It can be tempting to think that intractability concerns the limits of what is practical for us and is therefore unimportant for the claim that "in principle" everything moves in accordance with Newton's second laws. This would be a mistake because of the relationship between Evidence and Nature: our evidence for our claims about Nature depends on showing that bodies do indeed move in accordance with theories of motion that are at least consistent with Nature, so if we cannot solve problems of motion using those theories, then we break the link between Evidence and Nature, and our claims about Nature lose their justification.

A solution to MCON2 would enable us to determine the motion of a body when subject to any kinematic constraints whatsoever.

Both MCON1 and MCON2 belong to the theory of constrained motions within rational mechanics. As this theory developed in the 18th century it provided a new locus of investigation into BODY that we call "the problem of constrained motion," PCON:

> **PCON:** What is the nature of bodies such that they can be the object of a general mechanics?

The parallels with PCOL will be helpful. In tackling PCOL, the rules of collision place important demands on the nature of bodies undergoing impact and thereby play a significant role in determining conditions of adequacy for any solution to BODY. In moving beyond collisions to consider a more general mechanics, we turn our attention to the more general rules for the motions of bodies, as formulated in equations of motion and so forth. In particular, we seek a rational mechanics that provides solutions to MCON1 and MCON2. The resulting theory in relation to PCON is analogous to the rules of collision in relation to PCOL. The crucial difference is that, unlike in the case of PCOL where the rules of collision had been widely agreed since the mid-1660s, as of the early 18th century mechanics did not yet *have* the "rules of motion" for a general mechanics. Rather, it wasn't until the 18th century itself that a general theory became an explicit goal for rational mechanics (MCON1 and MCON2), and that PCON came to the fore as a critical locus of investigation for BODY. This is the subject of the second half of our book.

We argue that addressing PCON became a necessary condition on any adequate solution to BODY. The nature of bodies must be such that they are capable of cohering so as to move in accordance with the demands of MCON1, and of undergoing constrained motion in accordance with the demands of MCON2. As the century progressed, the relationship of BODY—and the vulnerability of proposed solutions thereof—to developments in rational mechanics became increasingly fraught with philosophical and conceptual difficulties. For philosophers, the lesson is this: any attempted solution to BODY must keep up with developments in rational mechanics, especially MCON1 and MCON2.

Historically, we can map the relationships between BODY, PCOL, and PCON in the 18th century as follows. From the 17th century, philosophers inherited collisions as fundamental for natural philosophy, along with an unsolved problem, PCOL. The failure of philosophical physics to solve this during the early decades of the 18th century (see Chapters 2–6) coincided with the independent rise of rational mechanics (see Chapter 7). Following this, PCON emerged as critical for solving BODY, and impact now lay downstream, far from the foundational problems of the newly developing generalized theories of rational mechanics (see Chapters 8–11). What then of BODY come the end of the 18th century (see Chapter 12)?

1.7 Methods

BODY belongs to philosophy, and remains a problem in contemporary philosophy today.[17] Yet, in this book, we approach it from a historical vantage point, and limit our attention to the 18th century. Why?

Three reasons lead us to take a historical approach. First, BODY is more than three hundred years old, as are attempts to solve it. If we study just its current version, we risk working in an impoverished problem-space, bound by a thin, narrow slice of a philosophical picture that is both bigger and richer. In consequence, even those with strictly contemporary interests stand to benefit from taking a longer temporal view. We are familiar with this from work by philosophers of physics on space, time, and motion, where not just early 20th century, but also 17th, 18th, and 19th century considerations deepen our understanding of the philosophical issues at stake. The same is true for BODY.

Second, the contours of BODY are historically sensitive: how BODY is formulated, its place in the system of knowledge, the preferred heuristics for solving it, and—most importantly—the criteria for an acceptable solution, vary with time as the philosophical context for addressing it shifts and changes. Therefore, BODY raises different philosophical challenges and questions at different moments in the history of philosophy. These are of philosophical interest in their own right.

Finally, this diachronic dimension is developmental and interactive. Philosophers' understanding of BODY changed and developed over time not

---

[17] There are two main strands of BODY in contemporary philosophy. The first concerns macroscopic bodies, their metaphysical status, and their relationship to "fundamental" objects. The second is the generalization of BODY to the *Problem of Object*: that is, the problem of specifying the object of a given theory, whether that be a body, particle, field, gene, or whatever it may be. See van Inwagen 1990, and the subsequent literature in metaphysics; the vast literature on reductionism in philosophy of science; discussions of the appropriate ontology for quantum mechanics (see, e.g., Ney 2020) and quantum field theory (see, e.g., Fraser 2008); and so forth.

only in response to, but also—and crucially—*contributing to* the evolving philosophical context. To study the unfolding of BODY in the history of philosophy is, in our view, the best path to understanding both BODY itself and its significance for philosophy.

A contrastive characterization of our project may be helpful, for our book is situated between two alternative historical approaches. On the one hand, it is not a work in intellectual history: we do not set out to track the "emergence" of a concept or idea, or the semantic shifts undergone by a word or concept. On the other hand, neither is this a work in history of material culture. We do not seek to map chains of belief transmission through networks of patronage, mentorship, correspondence, and the like, or to study the means by which such transmission takes place. Rather, our book studies a philosophical problem as a historically situated object whose characteristics are: *determined by* the historical figures who formulated it and struggled with it; *revealed by* the argumentative and evidential resources those figures employed; and *presented in* the material books, papers, letters, manuscripts, and notes those figures left behind. As a result, conceptual developments and material circumstances have an important role to play, but only when and where they make a *philosophical* difference to the argumentative, explanatory, or evidential elements of BODY.

To pursue our goal, we employ three methodological heuristics. First, we seek to recover meaning from use, where by "use" we mean: in philosophical argumentation and in theoretical problem-solving. We do not confine ourselves to prefaces, manifestos, and programmatic declarations; while these are important, we give greater evidential weight to the details that come later: the places in the texts where the opening declarations are tamed and re-shaped by the argumentative and evidential constraints of the problems at hand. It is here that most of the philosophical action takes place, in our view.

Second, we use anachronism judiciously. Situated as we are in the 21st century, later developments (in both philosophy and classical mechanics) provide us with resources unavailable to our protagonists. We use the resulting insights only where they can be translated without remainder into the concepts of the historical period at issue. We aim to state and shape our explanations or objections just as a leading authority, fully au courant with the state of the art then, could have done, with the proviso that some of the words we use have changed their meaning since that time (and where this is of philosophical import for our project, we say so). We neither state nor assess any historically given answers to BODY in terms or by standards that greatly post-date the context of the answer at issue. In this way, we seek to preserve the diachronic dimensions of BODY described above.

Finally, we explicitly recognize our own authorship. BODY is both *our* problem and *their* problem, and this book is the product of an engagement between the two, of course. We chose BODY as our central theme because it interests us. As philosophers living in the 21st century, our philosophical backgrounds, sensibilities, and motivations for embarking on this project frame and guide the work that we do in this book. We do not offer a history that pretends to wash out our own presence: rather, we offer a philosophy of a problem that has a long history, and we seek to explain what interests us about it and why.

1.8 Audience

We offer three strands of argument, of interest to the three groups of people for whom we wrote this book: philosophers of physics; historians of modern philosophy; and philosophers of science and metaphysicians interested in the epistemology and metaphysics of science.

(i) Philosophy of physics

From Aristotle to Newton, physics was the study of bodies. If we turn our attention to modern physics, however, we find that bodies are no longer its principal object, and indeed "body" is not even among its central concepts. This observation, mundane though it may seem, turns out to hide an abundance of interesting philosophical problems. When we ask: "Why did bodies get displaced from their privileged position, how did that come to be, and with what consequences?," the answers that we demand—and that we offer in this book—are metaphysical, epistemological, and conceptual. One upshot is this: the 18th century becomes a period of focal interest for philosophers of physics, equal to the 17th and early 20th centuries in its import. This is because it was then that physics proved unable to articulate a satisfactory account of body, and rational mechanics (then a separate discipline) attempted to fill the void—also unsuccessfully, as it will turn out. In the process, the conceptual foundations of physics and mechanics received profound scrutiny and reformation. The consequences shape contemporary physics today, as we shall see.

Most philosophers of physics will be familiar with 18th century debates over space, time, motion, and gravity. These will not be our focus. Instead, our goal is to open up a new area of enquiry for philosophers of physics, one that has yet to receive detailed scrutiny. If we consider Newton's *Principia* as a text in philosophical mechanics, we see that the scope of his rational mechanics contrasts sharply with that of his physics: while Books I and II are intended to be general, Book III concerns one force only, gravitation. Against this foil, we restrict our attention to philosophical mechanics where the gravitational behavior of bodies is *not* the target of investigation: namely, to non-gravitational mechanics, and to terrestrial physics. It is here that many foundational issues in classical mechanics were worked out, for it is here that the pressing need to treat constrained systems, and the limitations of Newton's laws of motion for this purpose, come to the fore.

The comparison with gravitation is useful in one further respect. Recent work has done much to elucidate the evidential support for Newton's theory of universal gravitation, as it was developed and accrued during the 18th century (Smith 2014). Unlike celestial mechanics, non-gravitational mechanics in the 18th century did not have centuries of observational data and mathematical theorizing to work with. The only potential analogue of positional astronomy for terrestrial mechanics is the set of terrestrial machines studied in ancient mechanics, but the new mechanics of the 18th century explicitly sought to move beyond this limited set. The contrast with gravitation brings the following question into focus: what counts as evidence in those parts of philosophical mechanics *not* primarily concerned with gravitation?

(ii) History of modern philosophy

Philosophers have long read the 18th century as grappling with problems inherited from Descartes: Cartesian skepticism, Cartesian dualism, and the Cartesian circle, to name but a few. Most often, these are cast in an epistemological vein. However, there is a further problem, also originating with Descartes and equally evident in the collective philosophical struggles of the 18th century: BODY.

BODY directly confronts core topics of the period: the metaphysics of substance and causation, and the associated epistemological issue of how, if at all, we can arrive at knowledge of either. All created substances were presumed (by almost everyone at the time) to depend on God, the primary substance and primary cause of all things. As a result, discussions of substance and causation divide into two: primary and secondary. Our concern is exclusively with secondary substances (bodies) and secondary causation (agency among bodies). When approached with primary substance and causation as the entry point, we find the familiar range of opinions on secondary substance and causation—from Leibniz's monads and pre-established harmony through Malebranche's occasionalism and physical influx—and rehashing these debates is not our goal. If we focus our attention exclusively on secondary substances and causation instead, a different picture emerges, one in which there is—perhaps surprisingly—widely shared broad-brush agreement on what would count as an adequate theory of secondary substance and causation. The disagreements are over how and whether any such theory can be developed. Detailing *this* debate, and its philosophical consequences, is our concern.

Taming the problems that 18th century natural philosophers engaged with, so as to render them soluble, required developments in not just the appropriate technologies (experimental, mathematical, conceptual) but also the appropriate epistemologies and accompanying methodologies. Canonical figures such as Locke, Berkeley, Hume, and Kant, like Descartes before them, presumed a starting point for epistemology in our ideas. An individual is assumed to have ideas whose contents they may inspect, with at least some of those contents being sufficiently determinate and accessible that they may serve as the basis of a viable epistemology. We are familiar with reading Hume's views as the terminus of this line of enquiry into causation, and conceding the point lies at the heart of Kant's Critical turn.[18] However, not all who contributed to BODY take ideas as their epistemological fountainhead, and BODY therefore requires us to widen our epistemological purview. For example, many of the figures we discuss approached questions of justification, certainty, and truth by way of criteria of success (theoretical and empirical) in solving problems.

Whereas Enlightenment natural philosophy has predominantly been cast as grappling with the world according to Newton, a different picture emerges when BODY provides our lens. Though he engaged with BODY, Newton was neither the first nor the last to do so. For the most part, philosophers of the

---

[18] See Clatterbaugh 1999.

18th century attacked BODY within parameters set largely by Descartes, or they recast it in new terms that had no precedent in Newton. As a result, our book provides a new perspective on the relationship between philosophy and the exact science of nature then. More generally, it is an invitation to historians of philosophy to revisit the epistemological and metaphysical commitments, arguments, and methodologies of Enlightenment philosophers engaged with understanding the material world, and the philosophical consequences of these that we inherit today.

(iii) Philosophy of science and metaphysics of science

BODY lies at the intersection of metaphysics, physics, and mechanics during the 18th century. Attempts to address it involve inferences from one domain to another, disputes over the authority of one domain with respect to another, and indeed problematize where the boundaries between domains might be drawn and on what basis. Moreover, these attempts lead directly to questions of appropriate epistemology and methodology for solving BODY, including questions of what principles might be used to guide, constrain, or evaluate a solution. Enlightenment attempts to grapple with these issues are interesting in themselves, as well as for the light they shed on their contemporary counterparts. We offer our book as an invitation to philosophers of science, metaphysicians of science, and anyone interested in "scientific metaphysics," to engage with either or both.

1.9 Overview

In this introductory chapter we have presented several technical terms (including BODY; philosophical mechanics; Goal; Nature, Action, Evidence, and Principle; PCOL and PCON). In the chapters that follow, these terms will be re-introduced slowly as the need for each arises naturally in the argument of the book. We have collected them together here as a guide to the overall structure of our analysis, and for ease of reference going forward. Whether all are necessary will be clear only by examining the work that they do in the remainder of this book. We proceed as follows.

In the first half of our book, our primary focus is collisions. Accordingly, Chapter 2 is a genetic account of collision theory in France after Descartes.

It documents and explains protracted efforts, by Malebranche and his posterity, to integrate coherently a broadly Cartesian matter theory and the rules of impact.

Chapter 3 is a sequel that uncovers analogous efforts to build a philosophical mechanics of collision from resources bequeathed by Leibniz and Newton. The joint upshot of these two chapters is that, by 1730, European natural philosophy regarded collision theory as the main locus for solving BODY, and no satisfactory solution was available.

In Chapter 4, we broaden our scope to BODY. We introduce physics as a sub-discipline of philosophy, and show that BODY was its central problem We articulate NAEP within this context, and illustrate some of the difficulties facing philosophers at that time in their attempts to find appropriate resources—metaphysical and epistemological—by which to tackle BODY. We show the relationship of BODY to familiar debates over substance and causation in the period. Finally, we show that PCOL arises naturally within the context of BODY, and that its philosophical significance is best understood against this backdrop. We conclude that BODY is a problem to be solved not just within philosophy, but within philosophical mechanics.

This sets the scene for Chapter 5, which examines mid-century attempts to address BODY. We argue that the two most promising proposals, defended by Du Châtelet and Euler, respectively, face insurmountable obstacles. Both proposals begin with physics, and seek to integrate relevant resources from the mechanics of collision. We claim that success demands meeting Goal: providing a single, well-defined concept of body that is simultaneously (i) consistent with an intelligible theory of matter, (ii) adequate for a causal-explanatory account of the behaviors of bodies, and (iii) sufficient for the purposes of mechanics. And we argue that neither succeeds.

Chapter 6 studies two radically new ways of constructing spatially extended, impenetrable, mobile, and causally interacting bodies, found in Kant and Boscovich. We explain how their theories transform the goals of physics while simultaneously falling short when it comes to philosophical mechanics.

In Chapter 7, we argue that, around 1750, the locus for grappling with BODY moved from philosophy to rational mechanics. We explain the reasons for this watershed transition. First, we identify conceptual difficulties with three notions that philosophers had relied on—mass, contact action, and extended-body motion. Then, we argue that professional philosophers at the time failed to incorporate pertinent advances in mechanics into their accounts of BODY, opening a rift between philosophical physics and rational mechanics. Finally, we uncover the key challenge within mechanics going forward; namely, PCON. This problem is our lens, throughout the subsequent chapters, for analyzing developments in rational mechanics most relevant to BODY.

In Chapter 8 we return to the early 1700s to review developments in the theory of constrained motion at that time, as they pertain to the goals of this book. We survey a wealth of work on the vibrating string and the compound pendulum. Theorists at this time sought general principles and uniform methods for treating constraints. But, they generally fell short of these desiderata. That insufficient outcome would shape the agenda for rational mechanics through the latter half of the century.

Building on this, in Chapter 9 we turn to d'Alembert's *Treatise on Dynamics*. D'Alembert offered the first systematic attempt at a general treatment of the mechanics of constrained motions. We show that his *Treatise* exemplifies the enormous difficulties involved in PCON, and argue that it is pivotal for the development of philosophical mechanics in the latter half of the 18th century.

Chapters 10 and 11 assess two different strategies for solving MCON. In Chapter 10, we examine how the 18th century dealt with MCON1. After 1730, rational mechanics learned how to tackle the motion of extended bodies with internal constraints (e.g., rigidity and incompressibility). The greatest advances were due to the Bernoullis, d'Alembert, and especially Euler. Accordingly, we focus on their key breakthroughs in pursuit of a rational mechanics of extended bodies.

In Chapter 11 our focus is Lagrange. The problem of external constraints found a wholesale solution in his analytic mechanics of 1788. Two ingredients were key to his solution: a dynamical law, viz. the Principle of Virtual Velocities; and the method of Lagrange multipliers. This combination allowed him to unify all rational mechanics then available. We assess Lagrange's achievement, via a constructive and then principle reading of his theory.

By now, we have reached the end of the 18th century. What, then, is the state of philosophical mechanics? To see this, we follow two strands of development, one seeking nomic unity (following Lagrange), and the other ontic unity (in the physics of the Laplacian School). Work by Cauchy spawned alternatives to both: a new approach to nomic unity, by means of balance laws of force and torque; and a new approach to ontic unity, based in deformable-continuous matter. The upshot is pluralism in philosophical mechanics, with consequences for BODY and for the relationships among philosophy, physics, and mechanics. We end with a brief review of the main conclusions of our book.

1.10 Conclusions

To sum up. The 18th century was a golden era for philosophical mechanics. As the century began, physics was a subdiscipline of philosophy, and its primary task was BODY. By 1800, this was no longer the case. Physics had become an independent discipline, and BODY was not its driving concern anymore. In this book, we argue that the philosophical reasons for this transformation—and thus its consequences—come into view if we analyze the century as an era of philosophical mechanics. That is, as an age of wide-spread, long-lasting, and concerted efforts to address BODY by integrating rational mechanics with philosophical physics.

This is an entirely new way of thinking about philosophy, physics, and mechanics in the 18th century, diverging sharply from prior accounts. According to Mach, with Newton's *Principia* classical mechanics is complete as regards its principles; all that remains is the technical challenge of *using* these principles to treat ever more complex and difficult phenomena. This view of post-Newtonian mechanics is perpetuated in Kuhn. For him, the *Principia* is the culmination of a scientific revolution, after which all "classical mechanics" becomes normal science within the Newtonian paradigm. The principles, methods, and basic ontological commitments are secure; all we need to do now is solve puzzles.[19]

But it is simply not true that 18th century physics is "normal science," nor that 18th century mechanics has settled foundations and is philosophically uninteresting. This is not a new point to make, yet for its significance to shine through, for it to be something we can use in our research and teach in our classrooms, we need an alternative way of framing the history, one that is different from Mach's or Kuhn's.

Our proposal, philosophical mechanics, is built on evidence from the books and papers of the time, and it enables us to do a wide variety of new and interesting work. Books whose philosophical importance is largely

---

[19] See, respectively, Mach (1883, 239) and Kuhn (1962, 10–14).

invisible under old framings (such as Du Châtelet's *Foundations of Physics* and d'Alembert's *Treatise on Dynamics*) come to the fore. New questions arise about more familiar works (such as Boscovich's *Theory of Natural Philosophy* and its relation to constrained mechanics). A broad range of first-rank physicists are seen to be doing work with *philosophical* import. This latter is in stark contrast with the Mach–Kuhn picture, where the work of some physicists (such as Newton and Einstein) is philosophically important, while the work of others (such as Euler and Lagrange) is philosophically inert—because all they did was use Newton's principles to solve problems within the already-existing Newtonian paradigm. The work on BODY that we discuss makes little sense from a Mach–Kuhn perspective. Yet, it was important. From our vantage point it appears as a widely shared and long-lasting project—of integrating philosophical physics with rational mechanics.

We intend our book to be an example of the importance of telling and re-telling our history, keeping it alive over and again with every new generation of students and scholars. We hope you will find it worthwhile.

2
Malebranche and the Cartesian foundations of natural philosophy

2.1 Introduction

In 1668, Huygens, Wren, and Wallis sent rules of collision to the Royal Society of London. We recognize these today as the rules of elastic and inelastic collisions (Huygens and Wren for the former, Wallis the latter).[1] Yet almost sixty years later, in 1724 and 1726, the Paris Academy of Sciences set up prize competitions concerning the rules of collision. Why? What was left unsolved by Huygens, Wren, and Wallis, and why was it so important as to demand not one but two prize competitions? These are the questions addressed in this chapter.

If we focus our attention on early 18th century France, it is not hard to see why collisions were foundational to natural philosophy. The towering figure in French philosophy was Malebranche, and in natural philosophy Malebranche was broadly Cartesian (modulo his explicit occasionalism, and with further differences that, when important, we highlight below). For Cartesian philosophers, the material world is a plenum, and all change comes about through collisions. The explanation for all that we see around us in the material world is to be given in terms of the parts of matter in motion, colliding with one another so as to change direction and speed in accordance with the rules of collision. Without collisions, no change in the natural world; without the rules of collision, no theory of such change: no natural philosophy. So we begin our discussion of collisions in France, before widening our purview in later chapters.

The 18th century opens with the fifth edition of Malebranche's *Search after Truth* (published in 1700), which contains his mature theory of collisions.[2]

---

[1] For several discussions of this episode, see papers in Jalobeanu & Anstey 2011 as well as Chapter 1.
[2] During his lifetime, Malebranche reissued *Search after Truth* six times, every time with some corrections and additions. For convenience, we shall denote them by means of Roman capitals, as follows: edition A (1674–5), B (1676), C (1678), D (1688), E (1700), and F (1712). Hereafter, we cite

This theory sets the agenda for collision theory in France and is (or so we will argue) the impetus for the 1724 Paris Academy prize competition. Malebranche's theory evolved through multiple versions, not least due to Leibniz's repeated pressing and prodding. If we are to appreciate the goals of Malebranche's mature theory, and the problems that it faced, we must go back to his earlier discussions of collisions, beginning with the first edition of the *Search after Truth*, published in 1674–5. There, Malebranche set out to correct the collision theory inherited from Descartes' *Principles of Philosophy*.[3] Leibniz then pounced on what he claimed to be Malebranche's perpetuation of Cartesian mistakes, and this led the Frenchman to revise his account twice. We describe these developments below. But to jump ahead: Malebranche's mature theory of collisions created more problems than it solved. Some of these soon became apparent and turned into a research agenda for his disciples into the late 1730s, while others were less immediately visible; all spelled trouble for the project of founding natural philosophy on collisions.

In what follows, we argue that the goal for an adequate theory of collisions, as bequeathed to natural philosophers in France by Malebranche, was as follows: to provide a causal-explanatory account of impact within a single theory that unifies the matter-theoretic account of body with rules of collision. We argue that the Paris Academy prize competitions of the 1720s, on the topic of collisions, were among the efforts directed at finding such an account. Finally, we argue that between 1700 (when Malebranche published his mature collision theory) and the 1730s no satisfactory account was found. On the contrary, the depth of the difficulties involved in developing such an account began to make itself visible.

To that aim, we begin with a synopsis of Malebranche's first collision theory (section 2.2). Then we explain Leibniz's protracted objections to it (section 2.3) and Malebranche's ensuing responses and improvements (section 2.4). We will argue that, despite his later efforts, Malebranche's theory had failed to solve two key problems, which the French Academy took up in search of solutions (sections 2.5 and 2.6). We end with a forward-looking account of some open questions and hidden problems in the natural philosophy of collisions (section 2.7).

Malebranche from the canonical edition Malebranche 1958–70, which we abbreviate as *OM* followed by volume number (e.g., *OM* xvii.2, which is in two parts).

[3] See Descartes 1991.

2.2 Correcting Descartes: Malebranche's early theory of collisions

We begin, then, with Malebranche's early theory of collisions. Its origin lies in Descartes' philosophy. In 1664, the young Malebranche came across a copy of Descartes' *L'homme*, and its mechanistic program for physiology quickly seduced him with its seeming power to explain so much from so little. Destined to an exclusive career in the priesthood at the time, he re-trained himself in mathematics and mechanics—a "Cartesian novitiate," so to speak—such that within a decade Malebranche became a leading authority in the new French physics.[4] In the 1690s, to this knowledge he added a solid initiation in the emerging differential calculus, taught to his friend, the marquis de l'Hôpital, by Johann Bernoulli (who, after a long visit to Paris in 1691, was hired by l'Hôpital as private coach in mathematics).

Malebranche took Descartes' chief legacy to be his doctrine of method, which he adopted. So taken was he with the requirement of "clear ideas" that he advertised his forays into collision theory as applications of that Cartesian method. This explains their place in *Search after Truth* as a quantitative coda to the last section of his treatise. In addition, Malebranche took from Descartes the notion that matter has just one essential attribute, viz. extension.

What is novel in Malebranche's theory of collisions, and in his matter theory more generally? Malebranche's greatest departure—heavy with implications, it turns out—was his decision to discard Descartes' "force of rest." While Descartes had claimed to remove any notion of force from bodies, his words belied his aim, especially in his statements and explanations of his rules of collision. Already in the first edition of his book (*Search* A), Malebranche sought to improve the situation, claiming that "rest is a pure privation that does not suppose any positive will on God's part."[5]

Purging the "force of rest" from the metaphysics of body creates two problems. One is in matter theory: what causes cohesion? Specifically, what prevents the parts of certain matter-volumes (namely, solids) from dispersing away from each other? More importantly for the project, what is the reason

---

[4] Gouhier 1926, 68.
[5] Malebranche's early argument seems to be that corporeal rest obtains by the very same (divine) act of willing that brings that body into being, or, alternately, by God ceasing to will that the body move (cf. Mouy 1927, 35). Because a force (in some body) requires a distinct, or "positive," act of willing, he concludes that rest is not associated with any sui generis force.

that, from before to after collision, the parts of a solid body move together, as a whole?

The second problem arises directly in the rules of collision. In the *Principles*, Descartes had supposed one body to exert a "force to remain at rest" and had used this in deriving his rules. Malebranche needed to repair the rules somehow, a problem whose depth is disguised by the humble place he gave it in the presentation structure of the *Search*, yet whose attempted resolution soon led to trouble.

Descartes' paradigm object is his "hard body," which is rigid: the parts of the body remain at rest with respect to one another, throughout the interaction with any other body. But Cartesian rigidity arises from the body parts having a "force to remain at rest" relative to each other. Malebranche, by discarding the force of rest, leaves himself without the resources in his matter theory to provide the "hard" bodies that are the subject-matter of the Cartesian rules of collision. Nevertheless, he retained "hard" bodies within his theoretical treatment of bodies, and added to them two additional kinds of body: "soft" and "elastic."[6] As we will see, "soft" bodies come to play a foundational role in his matter theory, and his account of "elastic" bodies leads to a research program taken up by others. But that is to get ahead. The important point for now is that Malebranche's collision theory begins from his matter-theoretic account of bodies, and incorporates this three-way taxonomy of bodies into that theory.

The main elements of Malebranche's approach to collision theory are as follows. For Malebranche, extension is an essential property of bodies, on which the further essential properties of divisibility, impenetrability, and figure (size and shape) depend. From these matter-theoretic resources, Malebranche seeks to provide an account of "hard," "soft," and "elastic" bodies (more on this below). To this three-way taxonomy of kinds of body, Malebranche added a threefold collision theory, with rules for each kind of

---

[6] The terminology of "hard," "soft," and "elastic" varied in meaning in the 18th century, from author to author, and so must be read with care, and most especially not through the lens of modern-day physics. Our age has entrenched "hard" and "elastic" in different regimes, viz. analytic and continuum mechanics, respectively. In these fields hard, elastic, and plastic (a lateral descendant of "soft body," now extinct) are defined in terms of stress and strain, by specifying the local (infinitesimal) deformation in response to applied contact forces. Each class of possible response is governed by a so-called constitutive relation, expressed by a (partial) differential equation. For example, simple elastic materials are ruled by the Generalized Hooke's Law. For Malebranche (and for many of his contemporaries), the term "soft" is broadly coextensive with our notion of a flexible body today: a mesoscopic matter-volume that undergoes only irreversible isochoric deformations. Here, and throughout, the term "rigid" denotes that no two points in the body change distance relative to each other; hence, the body remains self-congruent at all times (including throughout impact).

body and a causal account of the collision process for each. This fact is significant. While Malebranche revised his collision mechanics several times, he always ran it along three tracks, one for each basic body kind. His project—to have the matter theory and the rules of collision cooperate toward a unified picture of body–body action—remained a constant heuristic, visible under his opaque account of "communication of motion," or speed exchanges in impact. Moreover, this threefold division and this project turn out to run deep and wide in 18th century Europe. (Just why, and with what consequences, is something we seek to excavate in this book.) For Malebranche, motion is a universal but inessential property of bodies: it "always accompanies" matter.[7] The overarching principle applying to collision processes is conservation of total quantity of "force of motion," defined as the product of size and speed (more on this term below). This completes the foundation for his collision theory, a foundation that persists throughout the later versions of his theory.

We can locate Malebranche's approach to collisions within the taxonomy set out in Chapter 1, where we distinguished between the constructive and principle approaches. The principle approach begins with privileged laws or rules and seeks to develop an adequate theory of collisions from there. The constructive approach, by contrast, sets out from a theory of matter. We can further specify these two approaches to collision theory as follows.

Principle approach to collisions: Start with a privileged principle, such as a law or rule, that involves system-specific speeds. Introduce a purely kinematic condition on the outcome (i.e., one that constrains post-collision speeds). From these two premises derive the two exit speeds as a function of known pre-collision quantities.

Crucially, this heuristic bypasses matter theory entirely: it leaves out of the account any causal processes (either within the interacting bodies or at their common contact surface) and thus any reference to bodily attributes beyond those mentioned in the dynamical law.

This principle approach, which can be found in Newton, Huygens, and Leibniz (among others), is in contrast with the constructive approach:

[7] He claims that we can conceive of extension qua always at rest, so motion does not depend on it. On the other hand, it is universal because matter without motion "would be wholly useless, and incapable of the variety of forms for which it was made; and it is impossible that an intelligent being would have made it so deprived." See *OM* i.382f. and Robinet 1970, 72ff.

Constructive approach to collisions: Give a taxonomy of basic species of body, with a causal-powers mechanism proper to each. Then, for each species, describe a causal mechanism of contact action during impact. Next, use the specific description to infer the kinematic outcome, and hence the post-collision speeds. Finally, derive the system-wide quantities, if any, conserved throughout the interaction.

Malebranche took this second approach and offered a causal-explanatory account of the collision process for each of his three kinds of body, as we shall see in detail below. But *why* did he take this approach? In later years, his justification—though always rather understated—was the need to find the "Physical reason for the laws of motion," echoing the language of William Neile during the Royal Society discussion of collisions (see Chapter 1), and by which both he and Neile meant a *causal explanation* for the rules of collision in terms of an *underlying matter theory*.[8] This is something the principle approach eschews. Still, why did he choose the material approach in 1675? There is not enough direct evidence to answer it, and our conjecture is that, having discarded the "force of rest" from Cartesian matter, Malebranche saw that this had consequences for Descartes' rules of collision and for the "hard" bodies to which they apply; and he sought a means of modifying the rules consistent with the changes he made to the theory of matter. This makes the constructive approach to collisions the natural one to pursue. "Hard" bodies are retained by Malebranche—perhaps out of deference to Descartes—as the theoretical starting point from which to reason about "soft" and "elastic" bodies, these latter being the bodies of our experience.

We can divide the evolution of Malebranche's collision theory into three stages: the early theory (described in this section), a second stage prompted by criticisms from Leibniz, and the mature theory published in 1700.[9] The first stage, the early theory of collisions presented in *Search* A and B, begins with hard-body collisions and uses "force of motion" along with the speeds of approach to describe the collision process and resulting outcome for three possible collision scenarios: catch-up collision (when the bodies move in the same direction and the one following behind moves faster), head-on

---

[8] *Search* E, § 16, *OM* xvii.1, 81. In early editions of *Search after Truth*, Malebranche adhered fairly closely to Descartes' linguistic separation of "laws" and "rules of motion." In the 1692 treatise and later, he is less careful to keep these terms distinct; see Robinet 1970, 137. So, at times it is up to the surrounding context to make clear if by "laws" Malebranche means dynamical principles or just kinematic rules of collision.

[9] For Leibniz's criticisms, see section 2.3, and for his revisions in response to them, section 2.4.

Table 2.1 Taxonomies for Kinds of Body

Kind of body	Geometric behavior	Kinematic behavior
Hard	No shape deformation	No rebound
Soft	Irreversible shape deformation	No rebound
Elastic (or "hard with rebound")	Reversible shape deformation	Rebound

collision (when the bodies approach one another from opposite directions) with equal forces, and head-on collision with unequal forces. In this account, as in Descartes', stronger bodies (defined in terms of their force of motion) defeat weaker ones. Malebranche uses hard-body impact as the basis from which to arrive at his accounts of soft and elastic collision. Soft bodies, he claims, behave exactly like hard ones, with one exception: unlike hard bodies, soft bodies deform on impact, expending "force" to compress as well as to arrest each other. For elastic-body collisions, Malebranche offers a two-stage process. In the first, the two bodies behave like "soft" ones, expending "force" to compress and arrest each other. But then shape restoration begins, and in this second stage the bodies press back on each other, driving them apart from one another, and resulting in rebound.[10]

We summarize Malebranche's threefold taxonomy of bodies in Table 2.1. The kinds of body are distinguished by means of two criteria, one geometric (concerning the shape of the body) and the other kinematic (concerning the motion of the body). Both criteria refer to the body type's *behavior during the collision*.

This taxonomy remained stable throughout the evolution of Malebranche's collision theory and was widespread in the early 18th century. It is immediately obvious from this table that the geometric criterion yields three kinds of body, whereas the kinematic criterion yields only two. The kinematic division, arising from the rules of collision, is that which the principle approach invites: the rules of Huygens, Wren, and Wallis divide collisions (and thereby kinds of body) into elastic and inelastic (to use today's terminology); and it is these rules to which Malebranche appeals. Whether this discrepancy between the geometric and kinematic criteria leaves us with an unsolved puzzle is an issue to which we will return.

[10] For a brief overview, see also Beeson 1992, 19ff. and Pyle 2003, 131–47.

2.3 Leibniz's objections to Malebranche's early collision theory

In January 1686, Leibniz published *A Brief Demonstration of a Notable Error of Descartes and Others Concerning a Natural Law, According to which God Is Said Always to Conserve the Same Quantity of Motion; a Law which They also Misuse in Mechanics*, first in Latin in the *Acta Eruditorum* and then in French translation, already in September 1686, in the *Nouvelles de la République des Lettres*. The provocation was the mechanics of Malebranche's *Search* B (of 1676), and though by title the paper reproached "Descartes and others," no one in France would have missed that Malebranche was the true target. Leibniz's criticisms led Malebranche to revise his account, eventually yielding his mature theory of 1700, which set the agenda for research on collisions in 18th century France.

Leibniz excoriated Malebranche for following Descartes in taking the conserved quantity in collisions to be the "absolute [quantity of] motion," or size times speed.[11] That quantity is not conserved, he argued, and so a collision theory built on it cannot be correct. Instead, he claimed, the conserved quantity is the product of size and speed squared. In other words, it is Leibniz's own *vis viva* that he holds to be conserved. His hope, in pointing out the Cartesian "error," was to convert Malebranche to a reformed notion of body, in which "force" (newly defined as *vis viva*) is, after all, to be admitted:[12]

> I add a remark important for metaphysics. I have shown that the force ought not to be estimated by the product of speed and size, but by its future effect. . . . Hence it follows that we should admit in bodies something else beside size and speed, unless we wish to deny to them all power to act. (Leibniz 1687, 131–5)

Brief and effective in the reasoning, Leibniz's objection and his alternative proposal were well known in France.[13]

In a second line of attack, Leibniz objected that Malebranche's collision rules violate a "general Principle of order, useful in explaining the laws of

---

[11] Speed is, of course, a scalar quantity, unlike velocity, which combines speed and direction.

[12] From Leibniz, "Réplique de M[onsieur] L[eibniz] a Monsieur l'Abbé D[e] C[atelan] . . ., touchant ce qu'on a dit Monsieur Descartes que Dieu conserve toujours dans la nature la même quantité de mouvement." *Nouvelles de la République des Lettres*, February 1687.

[13] Catelan in 1687 had it translated into French and printed in *Nouvelles de la Republique des Lettres*, then edited by Pierre Bayle. See Robinet 1955.

nature by considering divine wisdom." Soon afterward, he and his followers would call this the Law of Continuity. Leibniz's alleged law is multifarious; even in his polemic use against the French, he stated it broadly enough to apply across disciplines. The Law of Continuity—as a guide to (and constraint on) reasoning about the natural world—went on to play a powerful role in the 18th century and beyond. Its initial public statement deserves quotation in full:

> [In responding to my objections] Father de Malebranche gives reasons . . . that contravene a general Principle of order I have noted. I hope he will kindly allow me here to explain this principle, greatly useful in reasoning, and which I see neither applied often enough nor sufficiently known in all its scope. It has its origins in the infinite, it is absolutely necessary in geometry, but it holds in Physics too, because the sovereign wisdom—the source of all things—works as a perfect geometer, and follows a harmony to which nothing can be added. For that reason, this principle often serves me as a test, or touchstone, for detecting prima facie and from the outside the defect in an inadequate view, even before I come to examine it internally.
>
> I could state it thus: *When the difference between two cases can be diminished below any quantity given in the data (or in what has been assumed), the difference must be found to likewise diminish in the sought-after (or in the consequences).* Or, in more familiar terms: *When the cases (or what is given) approach continually, the consequences (or what is sought after) must do so too.* All this depends on a principle even more general, to wit, *If the given is ordered, the required too is ordered.* (Leibniz 1687, 744–5; original emphasis)

With regard to collisions, the law translates as: correct rules of collision must not entail that an infinitesimal change in some initial condition will cause a finite change in the outcome. But with his rules of hard-body impact, Malebranche violates this principle. Specifically, for head-on collisions, his theory makes separate predictions: (1) if the bodies, say C and D, meet with equal force, then they rebound at the same relative speed (to each other) as before impact; but (2) if their forces are unequal, post-impact they move *together* in the direction of the "stronger" (i.e., in the direction of the one which had more "force"). Leibniz's objection amounts to this: there can be two cases, one falling under (1) and the other under (2), that differ "in the given" by an infinitesimal amount and differ in the "sought after" by a finite amount. For example, in (1) the mass of D is m, and in (2) it is $(m + dm)$, where dm is an infinitesimal increment. Malebranche's rules predict that situation (2) results in a mutual speed of 0 units (as C and D move together after impact), whereas (1) produces a finite mutual speed, say v units; the difference in outcomes is thus $(v - 0) = v$, a finite amount.

Leibniz's objections awoke Malebranche rudely; had it not been for them, "I would have never in my life" thought about impact again, he confessed.[14] To appease his critic, Malebranche in 1692 issued a self-standing work, the *Lois de la communication des mouvements*. This treatise of some sixty pages represents the second stage in the evolution of Malebranche's collision theory. Though containing an allegedly revised account, closer inspection reveals that Malebranche's revisions were both superficial and confined to the sub-theory of "hard" bodies, the primary target of Leibniz's attack. The force of Leibniz's critique seems to have eluded Malebranche: a mere tweaking of the rules of collision for hard bodies will not solve the problem that such bodies pose.

Despite this, Malebranche's general approach remains unchanged, and *Lois* reprises his early, constructive approach to collision. Again, he partitions bodies into three kinematic species, he views collision as an asymmetric clash between unequal agents, and he tries to infer the outcome speeds from the "quantity of impact" spent by the "stronger" body so as to displace the "weaker."[15] The Law of Continuity is violated in the same way as before— namely, collision with a body at rest yields a different outcome from when both are in motion. And, *Lois* shows Malebranche a reluctant author; collision theory was not his pressing interest in those years: theological matters had been keeping him busy.[16] His exposition in the treatise is cumbersome and far from transparent.

Unsurprisingly, the 1692 treatise, with its revised Cartesian account, did not satisfy Leibniz much more than Malebranche's first theory. In November 1692, he sent Malebranche an annotated copy of the Oratorian's little treatise. He followed up with more objections in 1693, then again in 1698, when he had disclosed to the public some of his own foundations for dynamics. In all of these missives Leibniz harried Malebranche, accusing him of a lack of neatness and uniformity in his derivation of rules of collision, arguing that the Cartesian "absolute quantity of motion" is not conserved in real-world

---

[14] Cf. [Malebranche] 1692, verso of title page.
[15] [Malebranche] 1692.
[16] Robinet 1970, 129.

collisions and pointing out that Malebranche's revised rules violate the Law of Continuity. Leibniz impressed upon Malebranche the need (as Leibniz saw it) to accept *vis viva*, and moreover to accept this force as an attribute of body, distinct from Cartesian extension.

Father Catelan, then Malebranche's personal secretary, presumed to answer on his behalf and countered that Leibniz's critiques were erroneous, to which the Hanoverian courtier responded, in turn. In December 1698, Malebranche himself stepped in to respond:

> During a leisurely stay in the countryside, I re-read my naughty little *Treatise on the communication of motions*, and—wishing to reassure myself of the third set of laws—I came to see that it is impossible to reconcile experience with Descartes' principle that the absolute [quantity of] motion remains always the same. I have changed this treatise entirely; for now I am convinced that the absolute motion is lost and increased all the time—and that, in collision, only the motion in the same direction is kept always the same. So, I have corrected my treatise entirely, but I don't know when it will appear. (Leibniz 1875, 355f.)

He did not have long to wait. *Search* E, the fifth edition of his opus, came out in 1700, with its collision theory amended as promised. In the final version, *Search* F of 1712, it remained virtually unchanged.

So much for a preamble. Malebranche's treatment of collisions in the fifth edition of *Search* is his mature theory, and it sets the stage for the development of collision theory in 18th century France. It is the details of this that we need to understand.[17]

2.4 Malebranche's mature theory of collisions

The third and final stage in the evolution of Malebranche's collision theory is his mature account, which rests on three pillars: a classification of bodies into three kinds; a conception of the collision process as a clash between bodies of different "strengths;" and an appeal to two quantities: "force of motion" and "quantity of impact." Malebranche offers a general heuristic for how to think

---

[17] For details of that theory, see also Schmit 2020, chapters 2–3, which discusses in rich detail the evolution and context of Malebranche's foundations for his impact mechanics.

about the collision process, and then an account of that process plus a set of collision rules specific to each kind of body. His goal, as it was in his original theory, is to provide a causal-explanatory account of the collision process that integrates the rules of collision into a theory of matter.

As we have seen, Malebranche labels his three kinds of body "hard," "soft," and "elastic." His taxonomic criterion is shape behavior during impact: Malebranchean hard bodies do not change their shape (they remain rigid throughout); soft ones deform irreversibly; and elastic bodies deform and then recover their original shape. The collision process Malebranche envisages as an asymmetric clash of mechanical agencies, with a "stronger" body seeking to displace a "weaker" one that happens to be in its path. Malebranche views all post-collision speeds as kinematic outcomes of such clashes between unequals. The "force of motion" quantifies the inequality in the "strength" of the clashing bodies, whereas the "quantity of impact" is the key by which Malebranche seeks to derive the exit speed of the "weaker" body.

Against this backdrop, Malebranche's general heuristic for the rules of collision is the following. On contact, the "stronger" seeks to defeat the "weaker" so as to free up its own obstructed path. To do so, the "stronger" must spend some force, equal to the "quantity of impact." The "weaker" receives it (modulo occasionalism) and is thereby displaced, hence set in motion. This general heuristic is then worked out in its specifics for each of the three body kinds.

(i) Hard-body collisions

Malebranche's first treatment is of "bodies hard by themselves" (i.e., rigid under applied forces). Moreover, Malebranche supposes them to collide "in the void," such that no subtle matter may interfere with them. He starts from defined notions, viz. "speed," "quantity of motion," and "quantity of impact;" and with some theoretical assumptions:[18]

1. Rest has no force to resist motion.
2. As the bodies are supposed infinitely hard, the force of the impacting acts immediately and in an instant on the impacted body. Hence, it pushes it with all the speed [of the impacting body].

[18] *OM* xvii.1, 67.

3. This force once received, it must distribute itself throughout the mass[19] [of the impacted body], on account of the hardness supposed. So, dividing this force by the mass [of the impacted], we get the outcome speed of that body.

4. The impacting body keeps for itself the motion that it does not give to the impacted. And so, by dividing the remainder motion (that it keeps) by its mass, we get its leftover speed.

With these posits in place, Malebranche finds the rules of direct rigid collision by applying the heuristic above; that is, by viewing impact as a clash of unequals. First, he partitions the possible outcomes in two classes: collisions in which the Cartesian "absolute quantity of motion" must be conserved throughout; and those in which the quantity is allowed to vary. In each case, he obtains a "general rule": an algorithm for computing exit velocities from arbitrary initial masses and speeds.[20] For clarity, we give an algebraic expression of his rule (for the latter case) as derived by Louis Carré, his disciple.[21] Let a body M of mass m and speed v collide head-on with a body N of mass n and speed r, moving against it. Then their exit speeds will be:

$$\text{Of M}: v - \{[2n \times (v+r)]/(m+n)\} \tag{1}$$

$$\text{Of N}: r - \{[2m \times (v+r)]/(m+n)\} \tag{2}$$

In line with his commitment to explain from causes, Malebranche accounts for his rule as follows. In all (but one) possible impacts, one body, W, counts as the weaker. Step 1 is to regard this weaker body as stationary so that the stronger, S, will always count as running into it. Malebranche's justification for this is as follows: either W is already at rest, or it moves against S (head-on), or they both move in the same direction and S is behind W but faster (catch-up). In head-on collisions, contrary motions "destroy each other," hence (at the mid-way stage of collision) W comes to rest,

---

[19] Although Malebranche here uses the term "mass," he is not entitled to help himself to the Newtonian concept of mass; and it is not clear what the term means for him; see below.
[20] Malebranche's rules thus rely on both mass and "speed," but there are difficulties with both; see below.
[21] In a paper for the Paris Academy's proceedings, printed with a motivating account, or *histoire*, by Fontenelle; see Carré 1706, "Proposition générale."

even though just for an instant. That is because, by that stage, *all* of its motion (speed, that is) has been destroyed by the stronger body.

Step 2 is to notice that, qua bodies, S and W are both impenetrable. This means they cannot pass through each other, and so one of them must be displaced. In line with the general heuristic, it is always the weaker that gets displaced. For that, S must spend some force, equal to the quantity of impact q, which gets transferred to W. Thus, S is left with a remainder amount of force, r. So, post-impact, the weaker will have received a speed equal to q divided by its mass; and the stronger will have a speed equal to the remainder r divided by its mass.

Where does "hardness" belong in this explanation? Malebranche is silent on this point, but we can see a place for it, after all. Consider S spending a quantity of impact, q. In principle, q could be spent on three (mutually exclusive) outcomes: to deform W; to deform and move it; or just to move it. Malebranche chooses the third possibility, explaining that "the bodies being supposed perfectly hard, all of their parts proceed or rebound to the same extent; whereas, in an elastic body, the struck part rebounds while the one farthest from the impact surface just continues to move" with the speed it had.[22]

Still, we would like to know, what is the kinematic behavior of "infinitely hard" bodies relative to each other? Although no experiment can yield evidence about such objects, Malebranche feels he ought to say something about them, even obliquely. In essence, he thinks that "infinitely hard" bodies must *rebound* completely after collision.[23] His justification is evasive. He convinces himself that, even though we lack empirical access to them, when a "perfectly hard" body A collides with a body B at rest, A "pushes it with all its force [of motion], hence B must receive all the motion" that A had. "To me, that seems an incontestable principle," he decrees.[24] Hence, in the taxonomy above, his "infinitely hard" bodies count as kinematically elastic (or *k-elastic*, see section 2.5). And yet, right away Malebranche recalls his voluntarism about laws of nature and grants that God "could possibly" have willed that hard bodies do *not* rebound after collision, in a vacuum—hence they may be *k-inelastic* after all.

---

[22] See *OM* xvii.1, 61–2. That is, the elastic body gets deformed as the collision makes its struck part c get closer to the extremal, not-yet-affected-by-collision part f.

[23] This emerges from his tables of predicted exit speeds for particular cases of hard-body impact. For example, when one body strikes an equal one at rest, it comes to a halt, and the resting acquires its entry speed; cf. *OM* xvii.1, 67.

[24] *OM* xvii.1, 63.

Recall that, for Descartes, hard-body impact was the paradigm of physical action at basic scales. With his explanation in place, Malebranche concludes about that view:

> In Descartes' thesis, that God always conserves an equal quantity of motion in the Universe, there is an ambiguity that makes it true in one sense and false in another; in agreement with or contrary to experience. It is true in this sense: the gravity center of two or more bodies, colliding in any possible way, always moves at the same velocity before and after impact. Thus it is true that God conserves an equal quantity of motion in the same direction, or an equal transport of matter. . . . But, his thesis is false and contrary to experience if taken to mean that, after collision, the sum of each body's motion (any way they may collide) is the same as before, viz. that the absolute quantity of motion remains always the same. (*OM* xvii.1, 73)

That is, in Malebranche's mature account, the "quantity of motion" has scalar and vector aspects, but impact conserves just its vector sum while its scalar aggregate varies.

And so, in one respect, Leibniz's critique (in *Brevis demonstratio*) did little to move Malebranche. Even after 1692, he continued to think that hard-body impact is a legitimate thought. Thus, apparently the Law of Continuity did not persuade him, and neither did Leibniz's new force-concept, *vis viva*.

(ii) Soft-body collisions

"Soft" bodies for Malebranche are those that deform irreversibly during impact. Malebranche dispatches the collision theory for these objects briefly: *after* collision they move together just like hard bodies do, and with a common speed inferable from the heuristic above. Their specific difference is the behavior *during* collision: soft bodies compress each other, and remain deformed, because they *lack pores* (that communicate with their exterior). This last feature distinguishes them from Malebranche-elastic bodies, which likewise deform, but regain their shape entirely with help from outside, as it were (see below). As regards the kinematics, the outcome of their encounter is as follows:

General Rule for the Impact of Soft Bodies

When two soft bodies run into each other, their contrary motions (if they have any) destroy each other. The bodies then will move together, with whatever motion is left over in them. Hence, their speed post impact equals the difference of their motions pre-impact, divided by the sum of their masses.

But if they have no contrary motions, after the collision they will move together, with the sum of their [prior] motions. Hence, their exit speed equals the sum of their [initial] motions divided by the sum of their masses. (*Search* F [1712]; *OM* xvii.1, 113)

In his late natural philosophy, this type of encounter stands out through the brief and evasive treatment that Malebranche gives it. He gives no quantitative examples, unlike his profuse accounts of hard and elastic bodies, which he sprinkled generously with particular instances.

Neither does he discuss the geometric behavior of such bodies in collision—the changes of shape that result from it. That is unfortunate because Malebranche faces a dilemma here. On the one hand, he fails to rule out that impact at any speed would flatten two soft spheres into thin films, adjacent to each other.[25] On the other hand, experiment speaks against that possibility: soft spheres deform just partially, to various degrees. By his own admission, Malebranche wants his collision theory to give the "physical reasons" for the outcome of impact, and yet this eludes him entirely for the shape changes in bodies of inelastic, poreless matter.

That is no small problem. At explanatorily fundamental levels, *all* Malebranchean matter is "soft." Stiffness and elasticity are downstream behaviors that arise from interactions between porous "soft" matter and the subtle ether that soaks it. If Malebranche's account of soft-body behavior is incomplete, then a fortiori so is his explanation of elasticity and "hardness with rebound," or stiffness.

---

[25] In plastic bodies, what keeps them from deforming like an incompressible fluid is certain internal forces, between adjacent parts—in particular, shear stresses. Malebranche's blanket denial that bodies have *any* "force of rest" entails inter alia that his bodies neither exert nor undergo any internal stresses. So, his matter theory gives us no reason to suppose that his "soft bodies" are plastic.

(iii) Elastic-body collisions

"Elastic" bodies—which Malebranche confusingly called "hard bodies that rebound"—deform during the collision process (like soft bodies) but recover their shape. For these bodies, Malebranche derives the post-collision speed of each body by means of two "general rules," one for each body. In *Search* F he adopts Carré's 1706 algebraic presentation. Let M and N be two elastic bodies, of masses m and n, respectively. Then their post-impact motions are:

$$\text{Exit speed of M: } v - \{ [2n \times (v + r)] / (m + n) \}$$
$$\text{Exit speed of N: } r - \{ [2m \times (v + r)] / (m + n) \}$$

These recipes closely match the pair of formulas (1) and (2) above. That is because Malebranche argued that "perfectly hard" bodies *behave as k-elastic bodies*, in collision: they rebound. Moreover, this quantitative result was long known. Truly novel was just his picture of the "physical reasons" for it; this picture is also the key to understanding his immediate posterity.

Malebranche offers a two-stage account of the sub-visible mechanism behind the rebound of elastic bodies. Framing the picture is the assumption that such bodies are porous, permeated by microscopic holes, or vesicles, filled with "subtle matter." Moreover, unlike in the analysis of hard-body impact, in this case the bodies collide not in the void but in a plenum: the collision takes place in a medium of subtle matter, the ether.

The impact is a two-stage process. Stage 1 begins when the bodies make kinematic contact, and it amounts to mutual compression. Specifically, as the bodies press on one another, subtle matter is expelled from the pores of each, and each body thereby deforms as its pores are flattened. The expelled subtle matter mixes with the subtle matter surrounding the bodies, adjacent to the compressed body's surface. As a result, that vicinity becomes a region of increased ether pressure.[26] These outcomes mark the mid-point of impact: the colliding bodies' respective "force of contrary movement compresses the subtle matter, and communicates their motion to it."

Stage 2 now begins, with the compressed matter exerting a "reaction," which "makes the rebound."[27] The proximate cause for the reaction is the

---

[26] Put exactly, this compression process results in each body's *apparent* size coinciding with its *true* size (which is never the case, in normal circumstances) and in a net pressure gradient arising on the bounding surface of each body.

[27] *OM* xviii.1, 129.

increased ether pressure on the compressed body's outer surface. This pressure differential endeavors to decrease, Malebranche thinks, so the ether re-enters the pores it had vacated through compression. Its re-entry causes two concurrent effects: the bodies resume their initial shape; and by re-expanding they press on each other, which leads to their mutual rebound. Malebranche now makes a crucial claim: this ether "reaction" induces in the rebounding bodies speeds inversely as their respective masses. These speed increments (call them s and t) must be added to the speeds r and v the bodies had at the end of Stage 1 (where r and v are the post-collision speeds that the bodies would have had were they "hard"). This reasoning leads Malebranche to a final pair of formulas, giving the exit speed of the "elastic" bodies M and N, respectively. He keeps his earlier dichotomic notion of impact being an encounter between a "weaker" and a "stronger" body. Let their masses be m and n and their entry speeds v and r, respectively. Against that background assumption, his rule of collision predicts:[28]

$$\text{Speed of the stronger body: } (mv - nr) / (m + n) \qquad (3)$$
$$\text{Speed of the weaker body: } (-mv + nr) / (m + n) \qquad (4)$$

Before moving on from Malebranche's collision theory, three further aspects of this theory deserve mentioning, for they became relevant soon after his death. First, Malebranche's matter theory gives soft bodies a privileged status. We have seen that Malebranche rested his mechanics on a tripartite division of bodies. Nevertheless, these three kinds of body are not on an equal footing. There are no actual hard bodies for Malebranche: soft and elastic bodies alone are real. Of the two kinds of real body, it is soft bodies that are basic:

> To me it seems clear that, by itself, body is infinitely soft—for rest has no force to resist motion, and so a part (of some body) pushed harder than its neighbor must separate from it. (*OM* xvii.1, 127)

The degrees of softness (and hardness) of bodies as we experience them arise from the pressure of the subtle matter surrounding them. Moreover, elastic rebound is an epiphenomenon, supervening on a certain interaction between "soft" and "subtle" matters, viz. the extrusion and re-entry of ether in body pores. Thus, while hard bodies form the starting point of Malebranche's

---

[28] Again, for the sake of synoptic understanding, we use Louis Carré's algebraic presentation (of 1706), instead of Malebranche's cumbersome verbal account. See *OM* xvii.1, 183.

MALEBRANCHE AND NATURAL PHILOSOPHY   45

argument for his rules of collision, it is soft bodies that lie at the basis of his matter theory.

Second, Malebranche's assertion that there are no hard bodies gives rise to a raft of related difficulties. For example, if there are no hard bodies, then his theory of hard-body collision applies to no actual bodies—not even counter-factual ones, for the reality of hard bodies is ruled out in principle, not just in practice.[29] Such objects count as metaphysically impossible in his system,[30] so his account seemingly applies to nothing. Moreover, it is a mystery how he could infer anything at all about hard bodies, for his very natural philosophy rules out there being any evidence for them, even in principle. On the one hand, his Augustinian tenets led him to voluntarism about God's choice of fundamental laws for the actual world. For him, divine legislation is unconstrained by rationality requirements on God's will, so purely rational methods and knowledge sources yield no insight into the basic laws of our world—only empirical investigation can reveal them, if anything. On the other hand, he admits that no hard body exists in fact, so we can have no empirical evidence from which to infer abductively to claims about the kinematic behavior of hard bodies. Malebranche put all hope in his *intuition* that, when two hard bodies collide, either loses as much motion as the "weaker" had.[31] That is all he has to go on, and he knows it is not quite solid enough: "In the end, it all depends on the Creator's arbitrary volitions, and he may very well have wished that hard bodies without recoil lose their motions in impact."[32] So his theory of hard-body impact succumbs to an epistemic dilemma of his own making: we must discover principles of hard-body action from experience alone, and yet no such experience is ever at hand.

Add to this that Malebranche admits the "inutility for Physics" of hard bodies, and we might wonder about their role in his system: why exactly does he need them, or even bother to theorize about them at all? To his readers, he always insisted that his project in natural philosophy was to correct Cartesian

---

[29] Perhaps we ought to describe hard-body collision (in *Search*) as a "counter-legal" scenario, borrowing a term from Marc Lange.

[30] Again, the reason is that he rejected the Cartesian "force of rest," which secured rigidity in Descartes' system. Without it, Malebranche can at most offer a practical analogue, viz. "relative" rigidity (really, high stiffness relative to a spectrum of applied force). But recall that stiffness, in Malebranche, is an effect induced by ether pressure. In a vacuum, then, all matter is fundamentally "soft." There can be no "hard" bodies in a Malebranche void.

[31] Thereby, any such collision is reduced to impact with a stationary body, and that setup he thinks he can handle unproblematically: through impact, they become one body, as it were, and so the left-over motion permeates the whole "new body" uniformly, etc.

[32] *OM* xvii.1, 65.

physics while changing as little of it as necessary. That minimal change, he thought, was to discard the "force of rest" operative in the *Principles of Philosophy*. "I have given the bodies' laws of collision as Mr. Descartes ought to have given them . . . had he admitted that rest has no force to resist motion," he explained to the end.[33] If it was simply out of deference to Descartes that he kept them in his taxonomy, then the philosophical justification for them remains moot. Hard bodies are treated only in the abstract, as a tool by which to reason about soft and elastic bodies. Could we not arrive at an account of soft and elastic bodies, and the rules of their collision, without going via hard bodies? And indeed, seeing that he is not entitled to the account of hard-body impact that he offers, aren't we required to do just this? Since the rules for soft and elastic bodies are derived as deviations from the rule for hard-body collisions, our epistemic access to and justification of the former are endangered by these problems facing the latter. Malebranche's retention of hard bodies in his theory of collisions, despite his denial that any such bodies exist, seems not to be benign.

Third, Malebranche's views of subtle matter evolved, and in the end they led to a split of sorts. In 1699 he wrote a paper in optics, for admission to the Royal Academy. To solve the problem in it, he postulated the ether to be vortex-full. Namely, subtle matter everywhere moves in spherical patterns around stable but moveable centers; interstices between such touching spheres are filled with smaller vortices; their interstices with other, yet smaller ones; and so on, down to some basic scale. By 1712, this picture had become his preferred foundation for explaining almost everything, as he confessed in Explication XVI:[34]

> The subtle, or ethereal, matter is necessarily composed of small vortices [*tourbillons*]; they are the natural causes of everything that happens to matter. I confirm this Supposition by explaining the most general effects in Physics, such as the hardness, fluidity, weight and lightness of bodies; the light, the refraction and reflection of its rays. The Supposition . . . is not arbitrary, I say. For I am convinced it is the true general principle of Physics, on which particular effects depend. (*OM* iii, 270)

[33] *OM* xvii.1, 79, 75.
[34] Schmit 2020, chapters 4–5, covers exhaustively Malebranche's program of vortex-based explanations coming out of his Explication XVI and its posterity in France after him.

Note what is missing from his explanatory agenda: collision theory, and elastic rebound in particular. That splits his foundation uncomfortably. Specifically, his mechanics of impact assumes a vortex-free ether (that envelops bodies and fills pores), and yet his official doctrine claims that the ether is vortex-full.[35] In particular, the centrifugal forces that obtain in his ether do nothing in his mechanics of contact action. Some disciples strove to address that disconnect, as we will see below.

This concludes our account of Malebranche's mature theory of collisions, which we can summarize as follows. What is sought, it seems to us, is a causal-explanatory account of the collision process that integrates the rules of collision into a theory of matter. The approach that Malebranche offers is constructive: it begins with matter theory, and builds from there. Metaphysically, soft bodies are primary: body in itself is infinitely soft, as we have seen. Malebranche offers a matter-theory account of hard and soft bodies in terms of the pressure of the subtle matter surrounding them, and hard bodies (of which there are no such things, in reality) are the primary bodies from which Malebranche develops his account of collisions. He offers a causal-explanatory account of the rule for hard-body collisions (no deformation, no rebound) in terms of "quantity of force" and "stronger" and "weaker" bodies. And, he accounts for soft-body collisions (irreversible deformation and no rebound) as deviations from hard-body collision due to some "quantity of force" being used up in the deformation process. He also offers a matter-theoretic account of elastic bodies, this time in terms of pores in soft bodies and pressure differences arising in the subtle matter, along with a causal-explanatory account of the rule for elastic-body collisions (reversible deformation and rebound) as deviations from soft-body collision due to the pores and pressure differences.

This, then, is the situation as it stood in France around 1700. Philosophers inherited from Descartes the project of explaining all natural phenomena in terms of matter in motion, with all change occurring via collisions. This placed collisions at the foundation of natural philosophy. They inherited from Malebranche a recognition that Descartes' own rules of collision were in need of revision, and requirements on what any such revisions should achieve. Specifically, the revisions should integrate matter theory with the new rules of collision, so as to yield a causal-explanatory account

[35] A modern analogue is the contrast between a static central-acceleration field and a dynamic (time-dependent) normal pressure field.

of the process and outcomes of collisions: there were to be no remaining inconsistencies between the account of bodies as posited in the rules of impact, and the underlying matter theory from which those bodies and their properties arise. The stakes were clear: no such causal-explanatory account, no foundation for natural philosophy.

In addition to the project just described, French philosophers inherited Malebranche's own attempt to carry out this project, and with it a number of loose ends. These might have turned out to be insignificant, tidied up within a few years and thereby laying the problem of collisions to rest. Natural philosophy would have been free to go its way without further ado. But this was not to be. The loose ends were anchored in deep problems. And tugging on the threads led to an unraveling of the entire project, with disastrous consequences.

2.5 After Malebranche: hard bodies in the Paris Academy prize competition of 1724 and beyond

An area of great disagreement reaching well beyond Malebranche's circle of disciples was his account of "hard" bodies. In hindsight, there were two problems: one regarded their behavior, the other concerned their very existence.

The first is a conceptual problem that transcends Malebranche's specific commitments in metaphysics and theology, as we shall see. Through early modernity, the corporeal attribute "hard" was deeply equivocal (as were its cognates, "soft" and "elastic"). Recall the taxonomy of body kinds set out in Table 2.1. As the table makes vivid, there were two descriptive uses for these terms. Each term was used in a geometric regime, to describe the behavior of shape in impact; and in a kinematic regime, to denote the bodies' post-collision speed relative to each other. From this vantage point, the three attributes above translate thus:[36]

> g-elastic: reversibly deformable: shape returns to its initial configuration
> g-soft: flexible, viz. irreversibly deformed; impact literally bends the body out of shape
> g-hard: rigid, viz. body shape remains self-congruent throughout

---

[36] Evidently, the prefix "g-" stands for "geometric" and "k-" for "kinematic."

k-elastic: the relative exit speed equals their pre-collision mutual speed; the outcome is perfect rebound

k-soft: their relative exit speed is zero (i.e., they proceed together)

k-hard: ???

In Table 2.1, representing Malebranche's taxonomy, *k-hard* bodies do not rebound. But, as we noted above, Malebranche had little justification for this claim, and it was controversial (not least because, for Descartes, *g-hard* bodies *do* rebound, they are *k-elastic*). The first problem of hard bodies was to determine the content of the attribute *k-hard*; and to do so from premises about *g-hard* bodies. In essence, a theoretical explanation was needed to take us from facts about *g-hardness* (shape behavior) to empirical claims about *k-hardness* (motion behavior). Of course, the argument must be sound; in particular, the evidence for it must be as strong as the (specific author's) epistemology of mechanics allows for the rest of their fundamental principles. Malebranche faltered on this count, as we have seen (section 2.4).

This perceived failure in Malebranche's account of collisions is, we submit, the main driving force behind the Paris Academy's 1724 decision to admit that hard-body impact was an open problem, and to invite proposed solutions to it: "Which are the laws whereby a perfectly hard body in motion will move another body of the same nature through collision, be it in a vacuum or in a plenum?"[37] The prize went to a Newtonian of sorts, Colin MacLaurin, who was then in France; Johann Bernoulli's submission received special mention.

In the winning paper, MacLaurin divides his official foundation into two parts; one is a set of seven "Axioms and Principles." Three of these evoke Newton's laws, though they really differ from them; one is a problematic relativity principle for collisions.[38] The other part is three "definitions" of body kinds. The reader will recognize them as the three species of geometric attributes above, viz. *g-hard*, *g-soft*, and *g-elastic*; for example, "we call perfectly hard the Bodies whose parts yield not at all in impact," he explains.[39]

---

[37] Bernoulli 1727, 8. Scott (1970, 17–8) emphasizes the relationship between hardness and atomism and a line of arguments concerning this issue. He claims that the arguments between Leibniz and Hartsoecker precipitated the 1724 competition. We locate these discussions within the broader context of the nature of bodies and how one body acts on another, rather than in the specific—though important—debate over atomism.

[38] "Axiom VII" claims that impact "takes place the same way" in two Galilean frames—which is true; and also "whether the Earth turns on its axis or is immobile, as in the Ptolemaic" system—which is egregiously false, and shows a deficient grasp of Newton's theory.

[39] MacLaurin 1724, 15.

This shows him perfectly aware of what the real problem is: to infer the *k*-behavior of *g-hard* solids. What is his solution? He answers:

> As these Bodies have no recoil [*ressort*], they will not rebound after the collision. Instead, they will keep moving in the same direction at the same speed, as if they made up a single mass. (MacLaurin 1724, 16)

In our terms, he claims that *g-hards* are *k-soft*, just as Malebranche had decreed; that is all he deemed necessary for a justification. There is a way to unfold MacLaurin's reasoning, but no new insight emerges. Like Malebranche, MacLaurin faces a clash between his assertion that there are no genuine hard bodies, and his epistemology. MacLaurin was broadly an empiricist in regard to foundations, but since hard bodies do not exist, empiricism about their *k*-behavior is not an option; and yet he thinks that, like ideal frictionless fluids, true rigids are a legitimate object for physical theory.[40] So MacLaurin's reasoning about them becomes strongly aprioristic, befitting a rationalist of sorts. In essence his argument for his predicted exit speeds in hard-body impact is:

1. Bodies are either *g-hard* or *g-soft* or *g-elastic*.
2. Only *g-elastics* rebound (i.e., are *k-elastic*).
3. Hard bodies are not *g-elastic*, *per definitionem*.
4. *g-hards* are not *k-elastic*. So, they must be *k-soft*. *Tertium non datur*.

His premise (3) is analytic, entailed by the very way that he defined body-species above. But (2) requires evidence to be true, and it is a mystery what his evidence is. What could count as evidence for the behavior of non-existent objects? MacLaurin's solution remained glib, and the difficulties of rigid impact eluded him.[41]

In contrast, MacLaurin's closest competitor, Johann Bernoulli, claimed "hard" bodies rebound perfectly; that is, they behave as *k-elastic* solids. To establish his result, however, he took a peculiar route: he replaced the Academy's imposed setup (of two rigids colliding head-on) with one of his own: "I will apply artificial recoils, from outside." Specifically, Bernoulli proved that, if we place a spring between two arbitrary bodies about to

---

[40] MacLaurin 1724, 16.
[41] For a detailed exposition of MacLaurin's argument, emphasizing different elements than those we highlight here, see Scott 1970, 24–9.

collide, they will rebound with the same relative speed. Incoherently, he then claims the same effect obtains even if "we suppose the spring to be so small that the two bodies . . . would count as being in contact."[42] His maneuvering displeased the jury, which disqualified his paper.[43] Years after the fact he explained why he replaced their physical setup with his:

> I thought I was free to attach to the term "hardness" the idea that seemed to me (and it still seems) the most appropriate to the nature of things. . . . In doing that, my aim was to reconcile perfect hardness with the Laws of nature. For I had shown, in my paper, that the common opinion—which takes perfectly hard bodies to lack all flexibility, even an infinitely small one— cannot coexist with these very laws of nature. (Bernoulli 1742, 81f.)

In other words, he set out from the thought that certain fundamental laws rule out the very obtaining of rigidity in nature. This brings us to the other problem in Malebranche's theory of hard-body impact: do any hard bodies exist?

We have already seen that Malebranche denied the existence of hard bodies. He had admitted it by 1687, when Leibniz pointed out incoherent results in the Frenchman's theory:

> The cause of the Paradoxes ensuing from the rules [of impact] I gave . . . is that I had reasoned from a false assumption, which I made willingly. It is the premise that, in a vacuum, perfectly hard bodies would exist. But the premise conflicts with the fact—which I take to have demonstrated—that bodies cannot be hard except when compressed by the subtle matter surrounding them. They can never be hard in virtue of their parts being at [mutual] rest, for rest has no force to resist motion. (Robinet 1955, 264)

Leibniz agreed (though from different premises): Malebranche "blames the errors" in his collision rules "on the false Hypothesis of there being perfectly hard bodies, which I grant do not exist in nature." And so did Bernoulli: "Thus

---

[42] Bernoulli 1727, 14, 24, 26. For details of Bernoulli's argument, and the principles that he appealed to in deriving his results, see Scott 1970, 30–9. Particularly important for our purposes is the relationship between Bernoulli's discussion of collisions and his treatment of the compound pendulum (on 35ff.), which we discuss in Chapter 8.

[43] For a description and critique of Bernoulli's account of how hard bodies arise, and why they rebound perfectly, see Scott 1970, 24. D'Alembert takes up the use of a spring for analyzing elastic collisions (see Chapter 9).

we conclude that hardness taken in the vulgar sense [of rigidity] is absolutely impossible." MacLaurin too: "There are no perfectly hard Bodies . . . but that does not prevent us from considering them in Physics."[44] So: no rigidity in nature, just high stiffness.

This rejection of hard bodies should give us pause for the following reason: hard bodies came to play a foundational role in the mechanics of constrained systems begun by Clairaut and d'Alembert in the 1740s. That alone would have been sufficient to keep Malebranche's theory relevant for decades, independently of the issue of hard-body collisions. Indeed, we cannot overstate how important the appeal to hard bodies turned out to be for 18th century science. In short, the rejection of hard bodies by Leibniz, Malebranche, MacLaurin, Johann Bernoulli, and others leaves us in the 1730s not only with unsolved puzzles in collision theory at the foundations of natural philosophy, but also with the seeds of difficulties for the foundations of mechanics.

2.6 After Malebranche: elastic rebound and the Paris Academy prize competition of 1726

In 1704, Louis Carré, one-time secretary to Malebranche, read a paper at the French Academy "on the laws of motion" seeking to "demonstrate a general Rule" for all outcomes of direct impact. This paper did little to go beyond Malebranche's own account. Carré's proof takes the same constructive approach that his mentor had pursued above, reasoning from matter theory, viz. premises about the causal behavior of bodies during impact. One notable difference is the absence of any hard-body collisions: only "soft" and "rebounding" bodies are considered. To "demonstrate the [general] Proposition," he proposes that colliding bodies "flatten each other a little" at the point of contact. This mutual compression requires the "stronger" body to spend some "force." Then "whatever force or motion is left over" in the system distributes itself across the bodies in proportion to their masses. This yields post-impact velocities. In explaining the causal mechanism behind elastic rebound, Malebranche had given just a qualitative picture with no exact details, let alone a quantitative treatment. Carré did no better: he

---

[44] See, respectively, Leibniz, "Lettre de M. L[eibniz] sur un Principe," in Robinet 1955, 257; Bernoulli 1742, 10; and MacLaurin 1724, 16.

restated his mentor's vague claim that, midway through impact, when mutual compression has ended, the "reaction of subtle matter" (which the impact had ejected from the pores) "restores to their natural state the bodies' parts compressed and flattened, hence it pushes them in the direction contrary to that from which they collided."[45]

Following Malebranche's death in 1715, the vagueness in the account of elastic collisions was felt as a painful lacuna. The community of research perceived the need for a *causal* account of the mechanics of elastic impact. "We have not yet given an account of the cause of Rebound," judged in 1721 the Academy's then *secrétaire perpétuel*, Fontenelle. That situation motivated two papers on the topic by Nicolas Saulmon, who proposed an alternative explanation to that of Malebranche. His speculative explanation has two parts. One is a phenomenological picture of a fluid pressing equally on a cylindrical flexible pipe (*un tuyau mou et sans ressort*) as it flows through it. At any location, pushing the pipe in allegedly causes the fluid to "act against" the pressing agency, by pushing back until the pipe is restored to its initial, undeformed configuration.[46] The other part is a picture of axis-symmetric "arcs" that, if bent, "make an effort against" the bending cause until they restore themselves to the initial shape. Saulmon combines the two parts as follows. He posits that atmospheric air is made of "small treelike parts" whose stems and twigs "recoil and can bend," just like his arcs above. And he proposes that rebounding solids are all permeated by pipe-like pores filled with "ordinary air." When such bodies collide, their mutual compression eventually bends the twig-like tiny arcs of air trapped in their pores; and as they "reprise their first figure," the arcs cause the two bodies to press against each other. Elastic rebound ensues.[47]

Notice, however, that this account is circular: Saulmon explains elasticity in mesoscopic solids by assuming it to obtain at sub-visible scales, in his arcs and "balloons," or solids of revolution, a fact not lost on some.[48] And it was not the only account on offer: Crousaz, an unreconstructed Cartesian from Lausanne, had yet another view of the "physical reason" of rebound, which

[45] Carré 1706, 446.
[46] Saulmon 1723, 105–7. This was Saulmon's second work on impact; see also Saulmon 1721.
[47] See Saulmon 1723, 130–2, and Nakata 1994.
[48] Johann Bernoulli caught it: "I wonder if those who posit air to consist in an infinite heap (of tiny branched, pliable particles in perpetual agitation . . . tending naturally to resume shape when an external cause compresses them) realize that they commit the fallacy of *petitio principii*" (Bernoulli 1742, 89).

won him plaudits from a provincial academy.[49] It relied on an analogy with a phenomenological scenario. Picture a flexible cylindrical pipe with an inviscid fluid in laminar flow parallel to the pipe walls. Crousaz assumes: if the pipe's cross section gets smaller, the fluid there presses on it "with a greater force of motion . . . on the walls as it moves through," and if the cross section increases, the fluid "force" on the wall (in the section's plane) decreases.[50] This loose relation is his key dynamical premise. The rest of his causal explanation he pictures thus: suppose that cylindrical pores permeate bodies throughout, and ether flows through them rapidly. Now imagine bending a flat straight blade from both ends, so that, of two opposite sides, one curves in concavely and the other becomes convex. Let there be pores crossing the blade, with orifices opening on either side mentioned above. Bending it causes pore orifices to get smaller on the now-concave side, and to get wider on the convex one. Per his assumption above, ether pressure increases on the pore walls at the narrow orifice, and decreases at the wide one. The higher pressure "forces" the orifices back to their initial, unbent configuration; and the lower pressure favors those walls coming closer together, thus back to their unbent configuration. When the entire pore has been restored to a cylindrical shape, the ether pressure inside it equalizes, so the pore—and so its walls, which are part of the body itself—remains in equilibrium. Crousaz submits that impact always produces a deformation (at the collision point) just like the bending of the blade above. Mutatis mutandis then, the same mechanism causes elastic impact. In turn, shape restoration causes the two bodies to rebound from each other. That concludes his explanation of perfect elasticity.

We suggest that this situation—the presence by 1725 of at least three accounts, most of them avowedly conjectural—was the impetus behind the Paris Academy's 1726 decision to end the collective uncertainty about the "physical reasons" of elastic rebound in collision. Indeed, the topic had appropriate international reach: accounts of elastic collision were on offer by leading natural philosophers outside France, from England to the German-speaking world to St. Petersburg, and nowhere was a consensus to be found (see Chapters 3 and 4 for more details). And so, the advertised topic was to

---

[49] In "Dissertation sur les causes du ressort," a paper that won him the prize for 1721 from the Bordeaux Academy of Sciences.

[50] See Crousaz 1721, 9f. To justify his key assumption, Crousaz fumbles dimly at proto-versions of the Continuity Equation and Bernoulli's Law.

give "the laws of impact between bodies with recoil, perfect or imperfect, deduced from a probable explanation of the physical cause of recoil."[51]

To its call for papers, the Academy received some ten submissions, of which two are memorable. One, by Father Mazière, an Oratorian cleric, took the first prize; another, by Johann Bernoulli, was a resubmission from 1724, and now it received special mention. Remarkably, both proposed causal explanations that are structurally the same, and relied on the same dynamical assumption: "I adopt as my principle the *centrifugal force*, but taken in an intelligible sense."[52] In essence, they speculated that rebounding bodies have pores whose cross section (in the equilibrium configuration) is circular. Crucially, the pores are closed: they are spherical, and the ethers inside are unable to escape; no pore has open orifices that communicate with the outside, cosmic ether (as they had for Saulmon and Crousaz). Inside each pore a tiny ether vortex is trapped, turning at constant speed. Now all spinning matter, even the "subtle" one, exerts a "centrifugal force," an outward pressure normal to the container surface, or body whose pore it is. By compressing the body in one direction, the impact makes the pores shrink, and thereby the vortices' radii get shorter, and that—both authors assert, separately—entails that the trapped ether's "centrifugal force" increases. Ergo, the ether pressure on the walls goes up, causing the pores—and thus the body itself—to revert to its pre-collision shape. The rest is familiar: shape restoration makes the bodies (now mid-stage through the collision) press on each other, which drives them to rebound. *Quod erat ostendendum.*[53] Later, Bernoulli explained to the reader the progress his paper meant to have made over his predecessors, Malebranche included:

> Generally speaking, these Gentlemen are right to admit a subtle matter that, through its motion, is the ultimate cause of bodies' rebound. But, it is not enough to just suppose a perpetually-agitated fluid; we must in addition explain the circumstances of that fluid, and exhibit the nature of an agitation capable of producing the rebound, for not all motion is appropriate to that end. (Bernoulli 1727, 84)

[51] Mazière 1727, 1–2.
[52] Bernoulli 1727, 89; his emphasis.
[53] See the details in Mazière 1727 and Bernoulli 1727, 90–9. Other contributions to these French debates on collision are also discussed in Schmit 2015.

That he had achieved such an explanation was a prematurely optimistic assessment.

The accounts of elastic rebound published before 1726 saw themselves as speculative and conjectural. Their dominant mood is the subjunctive, and markers of epistemic hesitancy pervade them: "could be regarded as," "could be the cause," "we may conceive it to be," "the hypothesis I am about to propose," "I posit that," "the conjecture I have just alleged." Bernoulli thought his account fared much better than the speculative "physical reasons" on offer before his 1726 *Discours*. And yet, his own paper rests on three premises labeled "Hypotheses."[54] Why did he think it evidentially superior?

Bernoulli seems to have defended his model of elastic recoil by an apagogic argument in which the options are: any account based on an action-at-a-distance repulsive force (d_i); a contact action account along the lines of Crousaz and Saulmon (c_i); or Bernoulli's own contact action model (c_B). His argument then has this structure:

1. Either d_i or c_i or c_B.
2. Not d_i.
3. Not c_i.
4. Therefore, c_B.

First, Bernoulli refutes approaches along the lines of Crousaz and Saulmon (c_i) by showing that they violate accepted principles of hydrostatics.[55] Then, he rejects all species of d_i accounts as false in principle because they rest on action-at-a-distance, allegedly an unintelligible, insightless notion:

> I doubt that those who, in elastic bodies, admit some elementary corpuscles endowed by nature with an expansive virtue—without explaining whence this property—deserve to be refuted. These Philosophers clearly assume the very thing at issue. And if this virtue (which they think innate and primitive) is independent of the arrangement of particles composing the elastic body, then it is just as easy to attribute it, in one stroke, to the largest body as easily as to their corpuscles. Everyone sees, I trust, how that would be to give renewed shelter to ignorance and new life to the occult qualities that we decry with so much reason. (Bernoulli 1727, 84)

[54] One is the Law of Inertia, another is Galilean relativity for collisions, and the last is a version of Newton's Third Law.
[55] Bernoulli 1727, 85–9.

Bernoulli thus seems to think that his recoil model was the only one left standing, consilient with his commitments in mechanics and ontology. Still, he was quite mistaken. His model requires two physical impossibilities in order to avoid self-contradiction. For one, Bernoulli claims that, as the initially spherical pore gets shrunk in impact, the speed of ether-bits in the outer layers remains constant through shrinking: this violates the conservation of angular momentum. For another, he claims that spheres get shrunk into spheres (pores, that is), which requires a radially symmetric inbound pressure. But collision results in a "sideways," asymmetric pressure wave— from the struck side, horizontally across the body—which turns a spherical pore into an ellipsoid, and thereby the centrifugal force of the outer layers gets smaller, not greater, as Bernoulli needs and claims.

The problem of elastic collisions remained unsolved, then, notwithstanding the 1726 prize competition.

2.7 Open questions, hidden problems

We have seen that Leibniz's challenges forced Malebranche to revisit his account of impact. This raised the visibility of the problem of collisions in France, and led to the Paris Academy prize competitions discussed above. Nevertheless, notwithstanding our discussion, the extent to which Malebranche's revisions, or those of his followers, successfully engaged with and addressed the challenges posed by Leibniz remains something of an open question requiring further work.

Moreover, in addition to the problems and loose ends that Malebranche and his immediate successors explicitly recognized and addressed, there were other problems in the foundations of his theory of collisions that remained unrecognized, but which pose a significant threat. Two are the most serious. The first is Malebranche's appeal to mass, despite not being entitled to any such concept. The second is Malebranche's dependence on "true speed," and the resulting conflict between his qualitative account of the collision process and (what the 20th century would come to call) the Galilean relativity already present in the quantitative treatment of collisions. Both "elusive mass" and "Galilean relativity" persist as difficulties for natural philosophy well beyond Malebranche's own theorizing, and they resurface in later chapters of this book.

(i) Elusive mass

Malebranche's account of collisions appeals to the mass of a body (section 2.4). The quantity of motion is the product of a body's speed and its mass. Likewise, this product expresses the quantity of moving force actually applied so as to produce the motion, because effects are in proportion to the forces that produce them.[56] And yet, he has no definition of mass itself; not even a nominal one, let alone a real definition. On this count no charitable interpretation can save him. To see this, consider three promising ways to credit him with a mass concept (call it Malebranchean mass, or M-mass) and how none really succeeds.

An operational definition, (i): let M-mass denote the other factor that makes a difference to collision outcomes (next to speed, his preferred quantity). Specifically, let a test body S, moving at some constant speed s, collide consecutively with some set of bodies M, N, P, Q, etc. initially at rest in each case. And, let Δq_M, Δq_N, etc. denote the Malebranchean "quantity of motion" that each body acquires after collision. Then we may define M-mass as follows: it is the quantity μ_k such that, for any k, μ_k equals $\Delta q_k / \Delta q_B$, and k ranges over bodies. Regrettably, this proposal is not feasible. Like most in his time, Malebranche restricts the scope of his collision theory—and so the range of empirical applicability—to direct impact, viz. that in which the bodies' contact point lies on the line between their M-mass centers. Absent that initial condition, his theory is silent—its empirical consequences are not well defined. Now the fatal problem is in plain view: to set up Malebranchean test collisions (needed to define M-mass operationally) we must already have a contentful, empirically applicable concept of M-mass, so as to locate the bodies' M-mass centers and ensure the collisions are treatable from his theory. This proposal is logically defective, then.

Another route is (ii) to let his M-mass be the quantity that tracks body size by co-varying: it increases and decreases in proportion as size does. To make this work, however, Malebranche would need two ingredients. First, a distinction between true- and merely apparent size; plus the claim that M-mass correlates with true size, not with apparent volume. Without the distinction, the proposed solution is trivial to defeat. Second, he would need an argument that, at true sizes, all matter has the same density (i.e., that two equal true volumes of matter contain the same amount of M-mass). He needs this

[56] OM xvii.1, 59.

clause sine qua non, or else the proposal is doomed, because true size does not track the right notion of M-mass unless we guarantee that its density is everywhere the same. But, it is far from clear how to extract the two premises above from Malebranche's system.

Lastly, he could have chosen (iii) Descartes' proto-concept of mass qua measure of corporeal rest's power to resist motion and of motion to resist rest. Supposing Cartesian momentum to be conserved—as his Rule IV indicates—this notion is close enough to Newton's understanding of inertia, which mass measures. Frustratingly, Malebranche foreclosed this option for himself early and without a second thought. As we have seen, his first and most visible departure from Descartes was to give up on his predecessor's "force to remain at rest," and he never looked back. In Search F, he sounds as resolute as ever on this count: "rest has no force to resist motion, for it is just pure privation."[57] He thus unwittingly forbade himself his best chance to let into his system the dynamical quantity needed for any sound theory of collision.

Newton's introduction of inertial mass into his theorizing was hard won and highly fruitful. Any successful theory of collisions will have to contend with the problems that led to its introduction, and either adopt Newtonian mass or offer an alternative that solves these difficulties (or at least those that are relevant to collisions). But Newtonian inertial mass is problematic from a Cartesian perspective, so there is work to be done either way. No one in France of the 1720s had seen the problem, let alone taken it on.

(ii) Galilean relativity

Malebranche's explanation of impact suffers from a weakness that subverts it from two sides. Seen from outside, the problem is that his account privileges individual speeds: it is a body's "true" speed, as it were, that determines whether that body is the stronger or weaker, in impact; whether it acts or resists; and how much force, if any, it has available as it sets out to collide with the other. The appeal to "true" speed immediately entangles Malebranche in the debates over absolute and relative motion that came to the fore in the Leibniz–Clarke Correspondence and persisted throughout the 18th century and beyond. There is ample literature on this, and we will not foreground it in

[57] OM xvii.1, 75.

this book, with the exception of one element. Impact is Galilean-relative: no fact about the observable mechanical behavior of colliding bodies can distinguish between true and merely apparent quantity of speed (relative to the observer's frame) or between true net speed and true rest. Thus, whenever two bodies C and D collide, no observer is able to determine non-arbitrarily if C or D was the "stronger," agent body. This fact causes a problem of evidence for Malebranche's theory—and for any account of impact that appeals to "true" speed in its explanations of collision outcomes.

The same problem arises from within Malebranche's theory. As he admits, the sole kinematic factor that determines the outcome (ceteris paribus) is the bodies' speed *relative* to each other. It alone (next to their mass ratio) makes a difference to the "quantity of impact" exerted by the acting "stronger" onto the acted-on "weaker." And yet, two bodies C and D could have the same relative speed—ergo, their collision will amount to the same interaction—no matter whether C happens to be "stronger" and D "weaker" or vice versa or both are equally "strong." Then Malebranche's causal terms above, and the different-in-kind agencies they convey, are explanatorily inert. The real predictive work is done by the two-body relation "relative speed." Not by single-body speeds; hence, not by one body being "stronger" either.

Malebranche does not seem aware of this serious problem, but he had ample chances to see it.[58] Inexplicably, these worries left him unfazed. Nor was it just his problem then; many natural philosophers embraced some variant of explaining collision as a contest between unequal agents exerting distinct species of powers. Behind their confident expounding—about agents overcoming patients—lies the same inability to connect univocally their metaphysics of asymmetric powers with the rules of collision. Because they think of these powers as *velocity* functions, the Galilean relativity of collision underdetermines their causal account. If there is to be evidence for one body being "stronger" or "weaker" than another, its source will not be the rules of collision or any of the empirical support those enjoy.

---

[58] A version of the Law of Continuity (as Leibniz had urged on him) states that rest and motion are "continuous" states. Leibniz explained it as follows. Suppose one and the same body K to be in two successive collisions, at rest in the first and in motion in the second, but in both cases *at the same speed relative to the other* mass. The Law of Continuity requires that K's speeds post-impact (again relative to the other mass) cannot differ by a finite amount. Leibniz saw, correctly, that Descartes' original collision rules, in *Principles of Philosophy*, violate this demand, and that Malebranche's picture of hard-body impact does too.

In sum, these two problems—of elusive mass and Galilean relativity—together subvert Malebranche's best efforts to have his metaphysics of body reach explanatorily into the natural world. To be precise, his chosen explanatory-evidential bridge is collision theory, in which he singles out two quantities as privileged: mass and true speed. And yet, his system lacks the concept of mass needed for a successful theory of impact, and true speed is inert with respect to his rules of collision.

2.8   Conclusions

The problem of collisions, as bequeathed to 18th century French philosophers by Malebranche, was this: to provide a causal-explanatory account of collisions within a single theory that unifies the matter-theoretic account of body with rules of impact. Between 1700, when Malebranche published his mature collision theory, and the 1730s, this problem resisted solution. The situation was so troubling to those involved that the Paris Academy held two prize competitions on the topic in the 1720s. These failed to yield a solution. Malebranche's taxonomy of bodies divided the kinds of bodies into three (hard, soft, and elastic) using geometric considerations (the behavior of the body's shape during collision), while his kinematic considerations (the speeds of bodies after collision) divided them into only two (rebounding or not rebounding). Hard bodies were problematic—Malebranche was not alone in denying their existence—and this problematic status threatened to infect the rules of collision for all three kinds of body, as well as the foundations of the mechanics of constrained systems. Moreover, an account of the process by which an elastic body rebounds eluded the best minds of the time. These are the most obvious problems, but there were others. Two are especially difficult, and they escaped most philosophers of the time, disciples and opponents of Malebranche alike. These are the need for a mass concept, and the Galilean relativity of collisions.

In Chapter 1 we highlighted the foundational role of collisions in Cartesian natural philosophy. In this chapter, we have argued that the problems facing philosophers in early 18th century France, as they attempted to develop a satisfactory theory of collisions, ran deep, and that by the 1730s no such account had yet emerged. In the next chapter, we argue that difficulties with handling collisions were not confined to French natural philosophers of a broadly

Cartesian bent. Rather, they were endemic in natural philosophy throughout Europe and the British Isles. As a result, the failure to solve the problem of collisions is a powerful lens through which to view developments in natural philosophy and in the relationships between matter theory, mechanics, metaphysics, and physics throughout the period.

3

Beyond Newton and Leibniz: Bodies in collision

3.1 Introduction

In Chapter 2 we saw that providing a causal-explanatory account of collisions was a foundational problem for French Cartesians. Here, we investigate the place of collisions in early 18th century natural philosophy more generally. This enables us to demonstrate three things. First, collisions were seen to be an important problem across Europe and the British Isles at this time, not just by "Cartesian" philosophers but also by "Leibnizians" and "Newtonians."[1] Second, there is a common approach to the treatment of collisions throughout natural philosophy that requires integrating rules of impact with an account of the material constitution of bodies. We call this a "philosophical mechanics of collisions." Third, as of the 1730s, a successful philosophical mechanics of collisions had yet to be provided—all the accounts on offer had serious shortcomings. The significance of this last conclusion will be made clear in the chapters that follow.

The early 18th century inherited not only Cartesian natural philosophy (developed most notably for our purposes by Huygens and Malebranche) but also the alternatives offered by its most influential critics: Newton and Leibniz. The turn of the century is spanned by the first and second editions of Newton's *Principia* (1687 and 1713, respectively) and by multiple works of Leibniz, from *A Brief Demonstration* (1686) to *The Leibniz–Clarke Correspondence* of 1715–6 (brought to a close by Leibniz's death). We show that collisions played a central role in natural philosophy for Cartesians, Newtonians, and Leibnizians alike.

---

[1] As we note below, there is no straightforward division of philosophers of the period into "Cartesians," "Newtonians," and "Leibnizians" in terms of their philosophical commitments. Nevertheless, here (as in Chapter 2) we divide the figures we treat into these broad camps, for the specific purposes of the issues we are considering. We do so by following the figures themselves, either via their own self-labeling or via specifically identifiable doctrines that they endorsed, and we note the origin of our choice of label explicitly in each case.

Philosophical Mechanics in the Age of Reason. Katherine Brading and Marius Stan, Oxford University Press.
© Oxford University Press 2023. DOI: 10.1093/oso/9780197678954.003.0003

In one sense, this is not surprising: there is no neat division of philosophers from this period into "Cartesians," "Newtonians," and "Leibnizians"; different figures incorporate, reject, and transform different elements of each, making new additions of their own as well. Nevertheless, it is striking the extent to which the treatment of collisions is *the same* throughout Europe and the British Isles, from the framing of the problem, to the particulars of the accounts, to the conception of what would count as a satisfactory resolution. We demonstrate these deep similarities, and argue that there is a common project: a philosophical mechanics of collisions.

We begin with John Keill as our starting point for a broadly Newtonian approach to collisions (section 3.2). We show the similarities to and differences from Malebranche, and argue that he (and self-professed Newtonians more generally) faced a number of difficulties in attempting to provide an account of collisions. We then turn our attention to Leibniz (section 3.3). We have already seen in Chapter 2 that Leibniz criticized Malebranche's Cartesian account. In this chapter, we outline the resources Leibniz himself sought to use in theorizing the motions of bodies. We then use *The Leibniz–Clarke Correspondence* to demonstrate the importance of collisions for both Leibniz and Samuel Clarke, through the role that the theory of bodily impact plays in their wide-ranging disagreements. Finally, we turn to Jakob Hermann's *Phoronomia* of 1716 and Christian Wolff's 1720s discussions to examine two explicit accounts of collision within the Leibnizian framework (section 3.4). We show the similarities to and differences from Malebranche and the problems that Hermann's and Wolff's Leibnizian accounts face. We end by concluding that, despite these widespread and concerted efforts, as of the late 1730s no successful philosophical mechanics of collisions had been developed. The problem of collisions (or "PCOL" as we shall call it at the end, see section 3.5) turns out to be deep and lasting, with widespread ramifications for philosophy.

3.2 Newtonian collisions

Like the French Cartesians, the English, Scottish, and Dutch Newtonians sought a causal account of impact that provided an explanatory basis for the rules of collision; also like the Cartesians, they found themselves in difficulties. To show this, we begin with John Keill's treatment of collisions in his 1700 Oxford lectures, published as *An Introduction to Natural Philosophy* in 1702 (in Latin). Keill's Oxford lectures were highly successful, going through three further editions as well as an English translation. Remembered today for his polemical pro-Newton intervention over the invention of the calculus, Keill was a strong and leading voice in early 18th century English natural philosophy, and his lectures can be taken as representative of the forefront of natural philosophy in England at the time.

As we will see, Keill's account of collisions looks remarkably familiar to anyone who has read the discussions surrounding Malebranche's attempts to revise Descartes' rules of collision (see Chapter 2). We demonstrate the similarities, and thereby establish the problematic shared amongst Cartesians and Newtonians. We also argue that, despite these commonalities, differences in epistemology lead to some distinct problems.

(i) Keill on collisions

Keill classifies bodies into three kinds according to how the *shape* of the body changes during the process of collision:

> VII. I call that a perfectly hard Body, which does not yield in the least to a Stroke; that is, which does not lose its Figure for a moment.
> VIII. That is a soft Body, which so yields to a Stroke as to lose its first Figure, and never to endeavour to recover it again.
> IX. That is an elastick Body, which yields indeed for a little while to a Stroke, yet of its own accord does recover its first Figure. (Keill 1726, 163)

Moreover:

> X. An elastick Force is that Force, whereby a Body obliged to quit its Figure, recovers it again. (Keill 1726, 163)

And:

> XI. A perfectly elastick Body is such a one, as recovers its first Figure with the same Forces whereby it was obliged to quit it. (Keill 1726, 163)

Just as with Malebranche, this is a geometric classification of bodies into hard, soft, and elastic (see Chapter 2).

With this geometric taxonomy in place, Keill turns next to developing rules of collision. He offers only two sets of rules, the first covering both soft and hard bodies and the second covering elastic bodies. For the former, Keill asserts that after impact the bodies will move together with the same speed. Considering first the case where the impacted body B is either at rest or moving with less speed than body A (in either direction), his argument is as follows:

> it cannot move slower, by reason the Body A follows immediately behind it; and it cannot move faster, because by Hypothesis there is given no other impelling Cause of its Motion besides Body A; since we suppose all others, as an elastick Force, and an ambient Fluid, to here have no Influence. (Keill 1726, 173)

In other words, in the absence of either an internal (elastic force) or external (ambient fluid) cause for Body B to move faster than Body A, impenetrability is the only quality in play, and Body B will simply move at the same speed as Body A. Notice that Keill appeals to a *causal* account of the collision process to argue for a particular outcome: the bodies will move together. The upshot is a kinematic classification, in which both hard and soft bodies fall into the same group: non-rebounding.

This is not yet sufficient to determine at *what* speed the two bodies will move following collision, and the remainder of Keill's chapter is taken up with addressing this issue. However, at this point Keill switches tack, and turns to the *laws or rules* as the primary resource for developing an adequate theory of collisions. He makes *no* further appeal to arguments concerning the causal processes involved in shape deformation or rebound, and instead uses Newton's laws of motion to determine the quantitative outcome of collisions, via the notions of relative velocity and center of gravity.

In Lecture XIV, Keill turns his attention to elastic collisions. He says that were there no such thing as elasticity, then all bodies would obey the rules given in the previous chapter, but "since there are indeed very few Bodies, wherein there is not some degree of elasticity," bodies do not move together but "fly from one another." He feels he needs to offer some kind of causal explanation for how this comes about. However, unlike Malebranche, Keill proceeds not by a reductive account of elasticity in terms of an underlying matter theory, but instead by beginning from a simple case of elastic behavior (see below), and building from there to an account of the elastic rebound of bodies. For Keill, observation of rebound compels us to introduce elasticity as a quality of bodies, and the task is to explain this corporeal behavior accordingly. He does so as follows.

His starting point is the behavior of a string fixed at two ends, stretched out to one side, and then released. He takes the shape changes that the string undergoes as an empirical given. Next, he considers a case where the initial stretching of the string out to one side is caused by an impinging body. Due to the behavior of the string previously described, this body is repelled back in the other direction, Keill claims. Notice that in this step, Keill transitions from geometric considerations (the change in shape of the string) to kinematic considerations (the motion of the body following impact with the string), and his claim is that the body is repelled away from the string. The connection between the two behaviors—the geometric shape change of the string and the kinematic rebound—is *stipulated* to occur as a consequence of the string being "elastic."[2] With this account of reflection in place, Keill replaces the string with an elastic body supposed fixed at the ends (just as was our elastic string), and then finally he allows that the elastic body is not fixed, so that it too is free to move, and thus the bodies rebound from one another. He concludes:

> And so we have demonstrated after what manner it happens, that Bodies after Impulse do not either rest, or move conjointly together, but rebounding from one another, they move with different Velocities, sometimes towards the contrary Parts, and sometimes towards the same. (Keill 1726, 184–5)

And yet, as with Malebranche and his followers, the association of the geometric conception of elasticity with the kinematic remains stipulative— just a posit.[3] In the course of discussing elastic bodies, Keill criticized the Cartesians, who he says were "ignorant of the Force of Elasticity to reflect Bodies."[4] As Keill knew, Descartes had argued that perfectly hard bodies would rebound by saying that there is no cause present by which they could lose their speed, and impenetrability prevents them from continuing in the

---

[2] That is, the *g-elastic* behavior is stipulated to yield a *k-elastic* outcome, without further justification. See Chapter 2, Table 2.1.

[3] With his account of the elasticity of bodies in place, Keill turns his attention to the rules of elastic collision (1726, 188–95).

[4] Keill 1726, 185.

same direction, so they must rebound.[5] Keill rejected this association of hard bodies with rebound, criticizing the Cartesian rules of collision for hard bodies, and more specifically conservation of motion. He pointed out that non-elastic collisions are problematic in relation to the claim that quantity of motion in the universe is conserved, writing:

> Hence the Law of the Cartesians is demonstrated to be false, whereby they contend that there is always preserved the same Quantity of Motion in the Universe: for Bodies which are not elastick, meeting each other from contrary Parts with the same Motions, mutually destroy each other's Motions. (Keill 1726, 176; theorem XXIV, corollary 2)

The issue of whether or not the total quantity of motion in the universe is conserved, and of how to define the quantities of motion and force associated with a body, became a central topic of dispute: originating with Leibniz's *Brief Demonstration* (directed at Malebranche's collision theory, see Chapter 2), this dispute is known today as the *vis viva* controversy (about which more in section 3.3).

Where does this leave us? Keill was a staunch, self-identified Newtonian. We see in his work the same taxonomy of body as we found in Malebranche (Chapter 2, Table 2.1): bodies are labeled as either hard, soft, or elastic. And he classified them as such based on the *geometric* criterion of their shape behavior during collision: no deformation, irreversible deformation, and reversible deformation. This is placed alongside a second, *kinematic* classification into just two kinds: non-rebounding (encompassing both hard and soft bodies) and rebounding (elastic bodies). Keill sought to use the first classification to give a causal account of how the second classification arises. He then moved on to state the quantitative rules of collision, and these are not derivable from the qualitative causal account of the collision process previously given. We see in Keill an example of the constructive approach (see Chapter 1), in which the properties of bodies relevant to the collision process are identified *independently* of, and *prior* to, specification of the rules of collision. These properties of bodies are used to give a causal explanation of the collision process, including the outcomes of collisions among different kinds

---

[5] Descartes 1991, II.46. This reasoning is repeated in Rohault's *System of Natural Philosophy* (1723) in his chapter on collisions (chapter XV, §2). This is the English translation of Rohault (1671), made by Samuel Clarke, and to which Clarke added extensive notes. Keill makes no mention of the Malebranchean approach in his discussion of elasticity.

of bodies, and the rules of collision are then introduced to complete the account of collisions.

However, Keill is left in the same uncomfortable position as Malebranche. He has a three-way geometric taxonomy of hard, soft, and elastic bodies, based on how their shape changes during impact, and using this classification he sought to give a causal account of the post-collision behavior. Yet this post-collision behavior comprises a two-way kinematic taxonomy of bodies based on whether they rebound or not, and the relationship of bodies in one taxonomy to those in the other is set by stipulation: elastic bodies *just are* those that undergo reversible shape deformation, but *why* this should result in rebound is left unexplained. The kinematic dual taxonomy is reflected in the dual collision rules for rebounding and non-rebounding bodies, respectively, but the relation between the qualitative kinematic taxonomy and the quantitative dynamical rules remains stipulative: Keill *says* that the former grounds the latter, but he is unable to derive the quantitative outcomes from the resources of either the geometric or the kinematic taxonomies of bodies. In all these respects, the approach to collisions offered by Keill is the same as that of Malebranche.[6]

Nevertheless, there is an important difference between Keill's and Malebranche's approaches. Despite offering the same three-way taxonomy of kinds of material body, Keill—unlike Malebranche—did not seek to explain the differences between soft, hard, and elastic bodies by appeal to a common underlying theory of matter: he said nothing about what makes a soft or hard body soft or hard, respectively, and his account of elastic rebound is non-reductive, relying on a simpler case of an elastic body: an elastic string, endowed with an elastic force. Instead, he presupposes these properties of material bodies, and starts from there. This is a weak version of the constructive approach, in contrast to Malebranche's strong one (see Chapter 1).[7] Moreover, Keill's "quantity of matter" is Newtonian mass,[8] and his impact theory is built on relative velocities, so unlike the Malebrancheans he avoids

---

[6] This situation, found not just in Keill but also in other self-professed "Newtonians" (see below, and see the discussion of MacLaurin in Chapter 2), is all the more surprising in light of Newton's own treatment of collisions. For Newton there is only one kind of body, and only one rule for collisions: hardness, softness, and elasticity are treated simply in terms of rebound characteristics by means of what today we term the "coefficient of restitution." See also the discussion of Pemberton, below.

[7] The difference in whether or not to accept elasticity as primitive reflects differences in methodology and epistemology between Keill and Malebranche; this is an important issue, much broader than our current concerns, and for now we keep our focus on the topic of collisions.

[8] For that fact, see especially Keill 1726, 114.

the issues of "elusive mass" and "Galilean relativity" highlighted at the end of Chapter 2.

We saw in Chapter 2 that Malebranche and his disciples faced serious problems arising from the epistemic status of hard bodies. The Newtonians encountered analogous problems, as we will now see. In setting out his rules, Keill himself provided no evidence for there being *any* non-elastic bodies, whether hard or soft, either by appeal to principles (such as those invoked by the Cartesians, see Chapter 2) or by appeal to empirical evidence. In order to pursue the epistemic status of the kinds of material body further, we need to turn our attention to other "Newtonians" of the period.

(ii) Epistemic access to hard bodies

The Dutch philosopher 's Gravesande discusses collisions in his *Mathematical elements of physics prov'd by experiments: being an introduction to Sir Isaac Newton's philosophy* (1720).[9] Here, he divides bodies according to the familiar three way taxonomy of soft, hard, and elastic. Following this, he offers experimental evidence for two kinds of collision process, inelastic and elastic, in line with two sets of collision rules. He writes that for making experiments on "Bodies which are not elastick, we use Globes or Balls of soft Clay," and that "Experiments concerning elastick Bodies, are made with Ivory Globes or Balls."[10] The former are clearly soft bodies, the latter he labels as elastic, and there is no example of hard bodies given. This highlights the problem facing empirically inclined philosophers wishing to include hard bodies in their taxonomy of material kinds of body: what empirical access do we have to them?

's Gravesande's compatriot Musschenbroek, in his *Elements of Natural Philosophy*,[11] is explicit about the empirical status of hard bodies and the epistemic consequences of this. In his chapter on collisions (chapter XVI) he divides bodies into the usual three cases (soft, hard, and elastic) and then says:

---

[9] 's Gravesande was professor of mathematics and astronomy in Leiden and a Fellow of the Royal Society of London. His experimental approach to "Newtonian philosophy" was highly influential in Europe.

[10] 's Gravesande 1720, 97 and 98.

[11] Published in 1726, translated into English in 1744. Pieter van Musschenbroek was a leading Dutch natural philosopher and professor of mathematics, philosophy, medicine, and astronomy. He was a fellow of the Royal Society of London, and his work helped to spread "Newtonian" ideas in Europe. For 's Gravesande's broader foundations of mechanics, see Ducheyne & van Besouw 2022; and for Musschenbroek, see Ducheyne & Present 2020.

Because there are no perfectly hard bodies, which are large enough for making experiments upon, (for these bodies perfectly hard are merely elementary, and by their subtilty do not fall under the cognizance of our senses) I shall examine the percussion of soft and elastick bodies. (Musschenbroek 1744, 176)

Perfectly hard bodies are unobservable, and so we cannot do experiments on them; therefore, we cannot find out by these means what the rules of collision are for them, and consequently Musschenbroek does not offer any rules for hard-body impact. The epistemic challenge is that the "elementary" or "ultimate" particles of matter are assumed to be hard, and Musschenbroek's position means that, despite their assumed existence, their rules of motion are epistemically inaccessible.[12]

Musschenbroek and the weak constructive approach

Musschenbroek's three-way taxonomy of bodies into hard, soft, and elastic is an example of the weak constructive approach.[13] Unlike the strong version of Malebranche, Musschenbroek does not construct his account of bodies and their properties from an underlying theory of matter. Rather, he begins with bodies, and his taxonomy is made in terms of a rich set of primitive qualities of bodies: hardness, softness, fragility, flexibility, and elasticity.[14] A perfectly hard body is one whose figure is immutable under all and any conditions. A soft body is one whose parts may be separated by a small force. While the world contains many large, soft bodies, there are no *perfectly* soft bodies: where there is a body, there is a force of cohesion uniting its parts into a body. A flexible body, Musschenbroek says, is one whose figure may be changed "without dissolving the union and coherence of the parts," and

[12] Musschenbroek 1744, 171; see also his chapter XV, §420. The status of the mentioned elementary hard bodies is also uncertain, as Musschenbroek discussed earlier in his text, in chapter II. There, he offered empirical evidence concerning the divisibility of bodies, noting that "divisibility is not yet received as an attribute of body, since it cannot be demonstrated [by experiment], that the smallest sensible particles can be further divided by any natural force," and suggested that there may be atoms or "elements" that are indivisible, without pores, and perfectly solid, hard, firm, etc. However, he maintained that the case for or against atoms was not yet decisively empirically determined.

[13] In Chapter 1, we distinguished between the constructive and principle approaches to collisions, and between weaker and stronger versions of the former. Stronger versions appeal to an underlying matter theory in constructing their account of bodies, whereas weaker versions begin directly with bodies and their physical qualities.

[14] His account of primitive qualities is in chapter XV, and in XVI he gives his taxonomy of bodies.

when a flexible body restores its own figure by its own force, it is said to be elastic. Thus, perfect elasticity is "when the force of the compressed or distended body, by which it restores itself, is equal to the force by which its figure was changed, so that the body returns exactly to the same figure it had before the change."[15]

According to Musschenbroek, the shape of a soft body changes during collisions, with its parts being squeezed inward. In this process, the force by which the parts of the soft body cohere must be overcome by the force of the striking body, this latter consequently being used up in changing the figure of the soft body. This is what accounts for the changes in speeds of the bodies as a result of the collision. However, this constructive approach yields only a *geometric* account of the shape behavior of soft bodies, along with a *qualitative* story about its dynamical underpinnings. Just as for the Malebrancheans, the challenge is to connect the geometric collision behavior with the kinematics: what will the motions of the bodies be after impact?

The assumption Musschenbroek makes is that the absence of elasticity associated with lack of shape recovery is the same as that associated with failure to rebound. With this assumption in place, two conclusions follow. First, soft bodies can be assumed to undergo no rebound and therefore move together after collision. Second, the force associated with the shape change can be equated to that associated with changes in motion, so that from the post-collision speed we will be able to determine how much force was used up in changing the shape of the bodies, given that the initial relative velocity (and mass ratio, though it seems Musschenbroek deals only with "equal" bodies) is known. Alternatively, if we know how much force is lost in a particular collision process, we will be able to calculate the post-collision speeds, provided we add a second condition. For Musschenbroek, this latter is a relationship between the force of a moving body and its speed: the force is related to the square of the speed.[16]

In the case of two elastic bodies striking one another, the parallel assumption needs to be made: geometric elasticity (shape recovery) is equated to

[15] Musschenbroek 1744, 172–3.

[16] See Musschenbroek 1744, chapter VI, "Of the Force of Bodies in Motion," §190: "The forces in a moved body are in a duplicate ratio of the velocity with which it moves." He uses this in his example in chapter XVI, §440: body A strikes body B with 10 degrees of velocity and with a force of 100. After collision, A and B move together with a velocity of 5. Therefore, the force for each mass after collision is 25, and the sum of both is 50. Here we see the "Newtonian" Musschenbroek embracing the "Leibnizian" measure of force. This is just one example of how these labels, though sometimes helpful, do not do justice to the independence of the figures to whom we apply these labels.

kinematic elasticity (rebound) such that, after the change of shape, the elastic force of the body will restore its figure and "the parts will press themselves outwards with the same violence as they were pressed inwards." Hence, "the elasticity reproduces the same force, which was extinguished by the intropression, so that no force is lost in the percussion of elastick bodies."[17] In both elastic and soft bodies, the impact force is consumed in changing the shape of the bodies, but in elastic bodies there is a further elastic force that is responsible for shape recovery and rebound. For Musschenbroek, this explains *why* bodies behave according to the rules of collision.

Like the Malebrancheans, Musschenbroek attempted to unify the geometric criteria for the shape behavior of soft and elastic bodies with the kinematic criteria associated with their behavior in collisions. In both cases, the desired outcome seems to be this: an account of the collision process which uses the properties and qualities of material bodies to explain how the outcome of impact is in accordance with the rules of collision.

Musschenbroek did not, though, go so far as Malebranche, in two important ways. First, he did not attempt to reduce the behaviors of soft and elastic bodies to an underlying account in terms of hard bodies. Since hard bodies are empirically inaccessible, Musschenbroek refused to offer any account of their collisions. Second, and relatedly, he did not attempt to explain the cause of elasticity. Rather, he took the force of elasticity as a primitive, provisionally. His chapter XV ends with arguments against those philosophers who have sought to find the cause of elasticity, and in particular those who have sought to do so using pores and subtle matter. The target here was clearly the Malebrancheans discussed in Chapter 2. Instead, Musschenbroek aligned himself with the Newtonian approach of suspending judgment as to the *cause* of elasticity until more experiments had been done.[18] In our terminology, Musschenbroek rejected the strong constructive approach to collisions in favor of the weak.

Interestingly, Henry Pemberton, who was the editor of the third edition of Newton's *Principia*, abandoned the three-way division of bodies into hard, soft, and elastic found in Malebranche, Keill, 's Gravesande, and Musschenbroek. In 1728, two years after the third edition of the *Principia*, Pemberton published his notes on Newton's philosophy under the title *A*

---

[17] Musschenbroek 1744, 181.
[18] Musschenbroek 1744, §§432–3; cf. also Newton's approach to gravity in the General Scholium (1999, 943).

View of Sir Isaac Newton's Philosophy. There he divided bodies simply into inelastic and elastic, identifying geometrically soft bodies with kinematically inelastic ones, and geometrically elastic with kinematically elastic, thereby implicitly rejecting geometrically hard bodies. As with Musschenbroek, this identification goes via a force of elasticity, by which a body both recovers its shape and causes the other body to rebound. However, in the absence of a detailed dynamical theory of the elasticity of shape recovery, it is a *stipulation* that shape recovery leads to rebound (and that shape non-recovery does not). So, despite having apparently done away with the problem of a three-way geometric taxonomy of bodies and a two-way kinematic taxonomy (via the implicit rejection of hard bodies), the underlying problem of unifying shape behavior with collision behavior remains. In Pemberton, as in Musschenbroek and others, the unification is by stipulation.

Together, Keill, 's Gravesande, Musschenbroek, and Pemberton represent the leading "Newtonians" of the early 18th century. Their work made clear that there was interest in providing a causal-explanatory account of collisions, one that integrated the rules of impact into a material account of bodies. We call any such attempted integration a "philosophical mechanics" of collisions. Despite their differences, the similarities among the French Cartesians (see Chapter 2) and Newtonians in their treatments of impact show that such a philosophical mechanics was a task they held in common. Moreover, we see that they are in the same predicament: from matter theory and natural philosophy they inherited a tripartite division of bodies into hard, soft, and elastic; from mechanics they inherited a bipartite division of bodies based on the rules of collision; and any unification was via stipulation. No-one had succeeded in using qualitative accounts of the material constitution of bodies to provide a causal account of impact that explained the specifics of the rules of collision. As a result, a philosophical mechanics of collisions remained elusive.

3.3 Leibniz on collisions

In the same year that Keill gave his lectures in Oxford (1700), Malebranche published his mature collision theory. As we saw in Chapter 2, it was Leibniz's criticisms of the earlier versions that prompted Malebranche to revisit and revise his account of collisions. The opening salvo from Leibniz was his 1686 *Brief Demonstration*, in which he argued against Descartes' claim that the conserved quantity in collisions is the quantity of motion, or size times speed. According to Leibniz, the conserved quantity is *vis viva*, or quantity of living force, measured by the product of size and speed squared. This marks the beginning of a long-running dispute, now known as the *vis viva* controversy. The entanglement of *vis viva* with a wide range of philosophical issues—from the nature of matter to God's presence in the world to human freedom—is evident in the correspondence between Leibniz and Clarke which took place in 1715–16 (see below). For our purposes here, Leibniz's account is important for two reasons. First, it shows that impact was of pressing importance for Leibniz, and that a motivating goal was the search for a causal explanation of the collision process. In making this case, we complete our argument for the foundational status of the problem of collisions within all three major strands of natural philosophy in the early 18th century: Cartesian, Newtonian, and Leibnizian. Second, the resources offered in Leibniz's approach for tackling the problem of collisions differ from those of Descartes and Newton, and this will be important as the century progresses (see especially Chapter 5).

(i) Leibniz and the forces of bodies in collision

Leibniz's arguments for *vis viva* take place in the context of his account of body and force. *A Brief Demonstration* provoked responses both in correspondence and in publications, and not just from Malebranche. During the course of these exchanges, and in the years that followed right through to the end of his life, Leibniz developed his theory of force in a series of publications.[19]

In 1695 he published the supplement to his *Brief Demonstration* as well as *Specimen Dynamicum*; in these he set out in detail his position on the force of bodies. We begin with the opening claim of *Specimen*, and build from there to the main elements of Leibniz's account that are important for our purposes.

The paper opens with the anti-Cartesian claim that there is something besides and prior to extension in corporeal things: natural force. Natural force is present everywhere in matter, even where it does not appear to the senses,

---

[19] Including 1688, when Leibniz read and took extensive notes on Newton's *Principia* (see Bertoloni Meli, 1993). For the reaction to Leibniz's *Brief demonstration*, see Iltis (1967, chapter IV). Around 1691, Leibniz followed up his controversy with Denis Papin (on the true measure of force) with *Essay de dynamique sur les loix du mouvement*, an unpublished piece in which he first made use of the terms *forces mortes* and *forces vives*. The essay remained unknown to the public until Gerhardt printed it as piece no. XII in volume VI of Leibniz's *Mathematische Schriften* (1860, 215–31).

and it consists in a striving. This striving constitutes "the innermost nature of a body, since it is the character of a substance to act, and extension means only the continuation or the diffusion of a striving and counter-striving already presupposed by it."[20] Therefore, by its very nature, a body has associated with it a force that is a continuous striving (it constitutes the nature of the body, so it is always there, always acting).

Later in the paper he provides support for this claim, as part of a broader explanation for why a notion of force must be added to the geometrica conception of body (i.e., to matter theory) in order for a theory of collisions to get off the ground. With a geometric conception alone:

> the largest body at rest will be carried away by a colliding body, no matter how small, without any retardation of its motion, since such a notion of matter involves no resistance to motion but rather indifference to it. Thus it would be no more difficult to move a large body than a small one, and hence there would be action without reaction, and no estimation of power would be possible, since anything could be accomplished by anything. (Leibniz 1956, 720)

Moreover, a conservation law must be added in order for the resulting account to satisfy the principle of sufficient reason (PSR). Lacking a conservation law:

> it was obvious that no reason could be given why the colliding body should not attain the effect to which it strives, or why the opposing body should not receive the full conatus of the colliding one, so that the motion of the opposing body would be compounded of its own original motion and of that newly received external conatus. (Leibniz 1956, 720)

The upshot is that in order to have an account of corporeal bodies we need more than geometry. We need further principles of:

> cause and effect, action and passion, in order to give a reasonable account of the order of things. Whether we call this principle form, entelechy, or force does not matter provided that we remember that it can be explained intelligibly only through the concept of forces. (Leibniz 1956, 721)

---

[20] Leibniz 1956, 712.

However, the forces by which bodies act on one another are not always manifesting, according to Leibniz.[21] Therefore, we need a distinction between the force that is always acting (primitive force) and the force that is manifest when bodies act upon one another (derivative force). The latter is the means by which bodies act upon one another, and it is this that is the subject of the laws of motion. Derivative force is "that force which is connected with motion (local motion, that is)."[22]

We have yet to arrive at *vis viva* (i.e., *force vive* or "living force"). Prior to that, we need one more distinction, between active and passive force. Primitive active force is "in all corporeal substance as such," and corresponds to the soul or substantial form. Derivative active force is exercised when primitive active force is limited by the conflict of bodies with each other, and is the means by which one body acts on another. Primitive passive force corresponds to prime matter, and is the basis of why bodies are impenetrable and possess inertia. Derivative passive force is the manifestation of impenetrability and inertia by a body when another body acts on it; it is the means by which a body resists the action of another body.[23]

And so, we come at last to the distinction between dead and living force. It seems to us that the problem that Leibniz is trying to solve by means of this is (a) how a body goes from not moving to moving without a discontinuity, and (b) why that motion is in the particular direction that it is. Though he does not say so explicitly in *Specimen Dynamicum*, part (a) of this problem arises due to the requirement that our theory satisfy the law of continuity, and part (b) arises from the requirement that our theory satisfy PSR. Leibniz solves both parts of this problem in one go by introducing the notion of solicitation, which is a momentaneous tendency toward motion in a particular direction. When all that exists is this momentaneous tendency (without any resulting motion), what we have is dead force; once motion results, we have living force. Leibniz writes:

[21] In the *Principia*, Newton had written that the inherent force of matter is exerted by a body "only during a change of its state, caused by another force impressed upon it" (1999, 404, definition 3). Leibniz does not refer to Newton in his *Specimen Dynamicum*, but we take Newton's discussion of force to be a key source for the puzzle that Leibniz is seeking to resolve in this portion of his argument.

[22] Leibniz 1956, 715.

[23] It is not clear from *Specimen* itself what problem Leibniz was trying to solve with this distinction. Given his other writings, we take it that the driving concern is the problem of unity. For bodies, this is the problem of how an extended body is possible at all, given the divisibility of matter. With admirable conceptual economy, though not with obvious success, Leibniz sought to solve this problem, and the problem of how it is that bodies are able to act on and be acted upon by one another, with one tool: the notion of force playing a role akin to substantial form.

Hence force is also of two kinds: the one elementary, which I also call dead force, because motion does not yet exist in it but only a solicitation to motion . . . the other is ordinary force combined with actual motion, which I call living force [*vis viva*]. An example of dead force is centrifugal force, and likewise the force of gravity or centripetal force; also the force with which a stretched elastic body begins to restore itself. But in impact . . . the force is living and arises from an infinite number of continuous impressions of dead force. . . . But even though impetus is always associated with living force [*cum vi viva semper sit conjunctus*], the two are nonetheless different, as we shall see below. (Leibniz 1695, 148f.; 1956, 717; translation slightly amended)

Distinguishing between dead and living force allows Leibniz to solve the above problem, via the notion of solicitation. The upshot of this solution is that when the motion is tiny (i.e., in problems of statics) only first-order velocity effects show up (dead force problems), but once motion is fully underway the second-order effects become apparent (such as in Galileo's treatment of free fall). This is why, once we move beyond statics to consider the force of bodies in motion, we can no longer use the quantity "bulk times velocity" as the measure of force, but must use the quantity "bulk times velocity-squared." This is the place of *vis viva* in Leibniz's new theory, which he calls "dynamics."

Why go to all this trouble? Leibniz makes explicit what he takes his contribution in *Specimen* to be. We have the correct rules of collision, thanks to Huygens, Wren, and others, he says, but there is a remaining problem as yet unsolved, concerning the causes of collisions: "But there is no unity of opinion about the causes. . . . It would seem, indeed, that the true foundations of this science have not yet been revealed." Specifically, he says, no-one "before me" has "explained the concept of force itself."[24]

In Chapter 1 we noted that the question of causes was eschewed, explicitly by Huygens, in the 1660s papers on the rules of collision, and that this was raised as an issue at the time, most forcefully by William Neile. Leibniz too complained at the time that the rules of collision failed to make clear why motion changes in the way that it does in impact. Henry Oldenburg relayed that

---

[24] Leibniz 1956, 719. Then he continues: "a matter which has always disturbed the Cartesians and others who could not understand that the sum of motion or of impetuses, which they take for the quantity of forces, can be different after collision than it was before [i.e., not conserved], because they believed that such a change would change the quantity of forces as well."

dissatisfaction to Huygens: "[Leibniz] seems to think that neither you nor Mr Wren have assigned the causes of the Phenomena that you considered while establishing your rules of motion."[25]

In sum, the problem of collisions lies behind Leibniz's complex theory of forces that we have outlined above, and in developing this theory Leibniz sought a causal account of the actions and passions of bodies such that collision processes are rendered intelligible.

It is well known that the developments in Leibniz's philosophy yield a picture in which collisions between material bodies are not foundational in the way that they are in Cartesian philosophy. Nevertheless, they remain critical to his project in two important respects. First, collisions remain the only means by which change in the material world is intelligible. While bodies themselves may not lie at the basis of Leibniz's philosophy, insofar as we are to have a *physics* then collisions are central. Second, the *vis viva* controversy was so controversial precisely because it was not simply about collisions: the issue of conservation of motion and of force is tied to much bigger questions such as God's action in the world and human agency. That this is so becomes vividly evident in *The Leibniz–Clarke Correspondence*.

(ii) *The Leibniz–Clarke Correspondence*

In their letters, Leibniz and Clarke began from such problems as the nature and extent of God's presence and action in the world, but in discussing these issues they soon found themselves moving into the issues surrounding *vis viva*, and drawing on empirical work in mechanics. The correspondence makes vivid the connections between *vis viva* as a measure of force and a wide range of philosophical issues. As we will see, *The Leibniz–Clarke Correspondence* shows explicitly that the preferred solution to the problem of collisions—or, worse, a failure to find any solution—would have ramifications for Leibniz's philosophy far beyond mechanics. This conclusion serves to emphasize the importance of collisions for Leibnizian philosophy.

In his very first letter, Leibniz contrasted his view that the total quantity of force in the world is conserved with the Newtonian position according to which God intervenes from time to time to add new motion into the world,

[25] Oldenburg to Huygens, April 7, 1671, in Huygens 1889–1950, vol. VII, 56. This concern—to "assign the causes" for the rules of impact—persists into Kant; see Chapter 6.

stating that the latter is "a very odd opinion concerning the work of God." Clarke responded that God "is himself the author and continual preserver of their [i.e., material things'] original forces or moving powers," and the view that the world continues without God's intervention in this way leads to atheism.[26] Leibniz responded in turn by distinguishing between God's power (shown by the existence and continuation of his creation) and his wisdom (shown by the perfection of his creation), and by grounding the global conservation of force in his wisdom: if God needed to inject more force into the world at some point, this would show want of foresight on his part. He rejected Clarke's charge of atheism afflicting his own view by appealing to God's power: "God preserves every thing continually, and nothing can subsist without him." In his third letter, Leibniz summarized what he took to be the issues at stake in the dispute with Clarke, including the question of whether or not God acts in the most perfect manner, or whether "his machine is liable to disorders."[27]

This issue of God's action in the world is explicitly connected to *vis viva* in Leibniz's third letter: there, he introduced the term "active force" without saying what he meant by it, and argued that if active force should naturally diminish in the universe such that God was required to replenish it, this would be an imperfection in God. Clarke responded with his own interpretation of the term "active force," and by quoting Newton's *Opticks*, query 31, in support of his claim that active force diminishes in the world. For Clarke, "*active force* signifies nothing but motion, and the impetus or relative impulsive force of bodies, arising from and being proportional to their motion."[28] In other words, Clarke claimed the term "active force" for a quantity proportional to velocity rather than velocity-squared. As with other topics in *The Leibniz–Clarke Correspondence*, Leibniz and Clarke seem to misunderstand each other, whether deliberately or not.

Leibniz's response is unhelpful in clarifying the underlying issue concerning the appropriate measure of force. He wrote: "They who fancy that active force lessens of itself in the world, do not well understand the principal laws of nature, and the beauty of the world of God." Clarke was unimpressed: "This is a bare assertion, without proof."[29] And then collisions are explicitly introduced: "Two bodies, void of elasticity, meeting each other

---

[26] See, respectively, letters *L* 1 §4 and *C* 1 §4 in Alexander 1970.
[27] See, consecutively, *L* 2 §7, *L* 2 §11, and *L* 3 §16.
[28] See *C* 3 §§13–14, footnote b.
[29] See *L* 4 §38 and *C* 4 §38.

with equal contrary forces, both lose their motion." Notice the steps: Leibniz and Clarke began from the issue of God's action in the world and, more specifically, whether there is a quantity of force or motion that is conserved in the world; they move from this to the appropriate measure of that force; and from there to the problem of collisions. In this way, a problem that might seem to modern-day readers to belong squarely within mechanics is shown to be directly relevant to the issue of God's action in the world.

Leibniz responded by engaging with Clarke on the problem of collisions, and by appealing to his old distinction (from *Brief Demonstration*) between quantity of motion and quantity of force:

> I don't here undertake to establish my dynamics, or my doctrine of forces; this would not be a proper place for it. However, I can very well answer the objection here brought against me. I have affirmed that active forces are preserved in the world. The author objects that two soft or un-elastic bodies meeting together, lose some of their force. I answer no. 'Tis true, their wholes lose it with respect to their total motion, but their parts receive it, being shaken by the force of the concourse. And therefore that loss of force, is only in appearance. The forces are not destroyed, but scattered among the small parts.[30] . . . However, I agree that the quantity of motion does not remain the same; and herein I approve what Sir Isaac Newton says, page 341 of his Opticks, which the author quotes. But I have shown elsewhere that there is a difference between the quantity of motion, and the quantity of force. (Alexander 1970, *L* 5, §99)

Clarke was once again unimpressed. He responded in two ways.[31] First, he issued Leibniz with a challenge: what about a case "when two perfectly HARD un-elastic bodies lose their whole motion by meeting together"? The motion cannot be dispersed among the parts of the bodies in this case, so where does it go? Second, he rejected without argument Leibniz's distinction between quantity of motion and quantity of force, simply asserting that force is "always proportional to the quantity of relative motion: as is constantly evident in experience." He then went on to assert that, thus understood, active force

---

[30] Interestingly, the appeal to motion being dispersed among insensible bodies was used earlier *against* Leibniz by Papin, and Leibniz responded by accusing Papin of making an ad hoc maneuver (see Freudenthal 2002, 574, and Rey 2010, 83). Nevertheless, others such as Wolff and Du Châtelet followed Leibniz in making this move.

[31] *C* 5, §99.

"does naturally diminish continually in the material universe," and that this is as it should be for inert matter. And he re-asserted that this is no imperfection in God's creation. Here the discussion ended, due to Leibniz's death.

Our purpose is not to adjudicate between Leibniz and Clarke. Rather, the point is this: in Leibniz's philosophy, the argument over whether *vis viva* is conserved during collisions has consequences far beyond the problem of collisions itself, making impact a topic of central importance for his system. As *The Leibniz–Clarke Correspondence* shows, the problem of collisions remained a crucial and pressing issue for Leibniz right until the end of his life.

Yet despite its importance, and for all the resources that Leibniz introduced—and for all his stated aims in *Specimen Dynamicum*—he failed to provide a causal account of the collision process. Moreover, his followers failed too.

3.4 Leibnizian collisions in Hermann and Wolff

In pursuing the fate of a broadly Leibnizian foundational program, two figures stand out; each owed their position to his patronage, and each adopted some of his central ideas in natural philosophy. Specifically, they shared his commitment to PSR and the Law of Continuity as constraints on admissible dynamical foundations and to his claim that the nature of body includes force—active and passive, living and dead. They also agreed that the basic laws of mechanics must be laws of these corporeal powers, and that all action is by contact. Hence the central importance of impact.

(i) Hermann's *Phoronomia*

In the Germanic community of research, the first collision theory influenced by Leibniz is in chapter VI of *Phoronomia*, a 1716 treatise by Jacob Hermann. At Leibniz's recommendation, Hermann had been hired at Padua into Galileo's former chair; *Phoronomia* was intended as a textbook in mechanics for advanced students. Two things are particularly relevant for our argument: the book offers a Leibnizian approach to collisions, and the treatment is broadly familiar from the accounts we have already seen in the Cartesians, the Malebrancheans, and the Newtonians—complete with corresponding difficulties. Moreover, while we do not yet have a critical and comprehensive study of this very influential work, a synoptic survey reveals that for the most part it is a collation of extant results in mechanics ca. 1715, when it went to press. As such, it represents the state of the art at the time.

Hermann sets out by defining his basic notions for that theory. First is a division of bodies into "inert" and "actant" [*inertia et actuosa*]. The latter species are bodies that "have the actant force of repelling [other bodies] with the same force that they had been themselves impelled," he explains. Hermann adds that this dichotomy between "inert" and "actant" corresponds to a then-current division of bodies into "non-elastic and elastic," but the distinction remains unclear.[32] This is because, as we saw in Chapter 2, the terms "elastic" and "inelastic" were ambiguous at the time, between geometric shape behavior (during collision) and kinematic rebound behavior (after impact), and Hermann does not tell us which he means. Rather, he seems agnostic about the actual existence of the two kinds; to justify his reliance on them he is content to simply mention the pragmatic consensus. So, while he goes beyond his predecessors in choosing a *dynamical*, force-based criterion to divide body into two basic kinds (not three), corresponding to two sets of collision rules, his criterion fails to specify those kinds unambiguously. Still, he must connect force terms with space and time parameters, and yet he fails to do so unambiguously. That is, he fails to specify whether his "actant" and "inert" dichotomy just overlaps with "elastic" and "inelastic" qua descriptors of shape behavior (during collision) or of kinematic-rebound behavior (post-impact).

His next ingredients are "proper velocity," a signed quantity of individual bodies, and "relative velocity," which is the vector sum of the proper ones. Another notion is "absolute force of bodies," which is a key effective parameter, it will turn out. He defines it, much like Huygens and Leibniz, in terms of Galilean heights. Namely, the heights from which the center of gravity has descended (before impact) or would ascend (after impact), supposing all vertical motions to be "naturally accelerated," viz. in accordance with Galilean gravity, constant and parallel.[33] Lastly, there is an undefined notion, namely mass [*massa*], which emerges unannounced halfway through Hermann's account of collision. This tacit notion shows in retrospect that, by absolute force of bodies, Hermann means the *vis viva* of the collision system's mass center.

[32] Hermann 1716, 112.
[33] Hermann 1716, 111.

Having presented his conceptual tools, Hermann went on to treat impact, by giving a causal explanation of the collision process, and a derivation of the rules for exit velocities.

Inelastic collision

Hermann treats inelastic collision first, as he thinks he can dispatch it easily with "just one brief" theorem: "When inert bodies collide frontally, their [proper] velocities after impact equal the velocity that their common gravity center had before impact."[34] Then he expresses that (common) exit velocity as a function of their masses and relative speed as defined above. Because he calls it a theorem, Hermann then adds a proof of sorts—as opaque as it is interesting, for he evidently aims there to rely on the dynamical notions (of body types) that he began his entire account with.

> *Demonstration.* In collision, the motion of bodies does not change except on account of, and in proportion to, the inertia of matter. Hence, when [inert] bodies collide, the outcome is that they move together—as if they were glued to each other—with the same quantity of motion. This quantity equals the amount whereby the greater body's [quantity of motion] exceeds the lesser's, if they collide head-on. In catch-up collision, it equals the sum of their individual quantities of motion. (Hermann 1716, 114)

This seems disappointingly brief, and raises the not unreasonable suspicion that Hermann had no clear and compelling way to connect his matter-theoretical notions with the kinematics of collision. And yet, we can develop his verbal proof into quantitative reasoning. The clue is Hermann's use above of a telling phrase, namely, "on account of, and in proportion to, the inertia of matter" (*pro ratione inertiae materiae*). Elsewhere in *Phoronomia*, he sheds some more light on this confounding phrase and its role in collision theory. Specifically, early in the book Hermann gives a picture of force, or rather of forces plural, that all bodies exert in interactions; for undeclared reasons, his paradigm interaction is inelastic collision. In it, he explains, one body counts as "passive," and so, he infers hastily, that body must exert a passive force:

> In addition to active force, there is in bodies also a certain passive Force, from which neither motion nor tendency to move results, but consists in

[34] Hermann 1716, 114.

that *Resistance* whereby it opposes any external force striving to change
the bodies' state of rest or motion. . . . This force of inertia is plain enough
in resting bodies. Indeed, when a [moving] body A strikes another one B
at rest, A loses something of its force, whereas B, drawing out some of A's
force and motion, acquires it. Whence it is clear that the resting body really
has some passive force, which the incoming's force must break and overcome. . . . This force of resistance the great Astronomer Kepler called *Force
of inertia.*

In this force of inertia of matter is grounded the law of Nature whereby *to
every action there is an equal and opposite reaction.* For in every action there
is a struggle [*luctatio*] between an agent body and a patient one, and without
such struggle no action, properly so called, of the agent upon the patient
can be conceived; for otherwise there would be no stable foundations for
Mechanics, but any effect would follow from any cause. (Hermann 1716, 3)

And then later:

Hence in all corporeal action there is a clash between an agent force and the
resistance of the patient body, an application of the agent's force onto the
body receiving the action; that is, action itself is equal and contrary to the
resistance of the patient, which is its reaction, because this resistance—or
this force of inertia—by the patient body must be removed, so that the patient might be set in motion by the agent. . . . Therefore, when we say that
any action is equal and contrary to the reaction of the patient body, all we
mean is: in all corporeal action, as much of the agent's forces is lost as it is
gained by the body receiving the action. (Hermann 1716, 378–9)

Here is how we construe his words. In almost any direct collision, the two
bodies play different dynamical roles, viz. they exert distinct causal powers.[35]
One body counts as the agent, and the other as the patient. The agent acts
on the latter by exerting active force; and the patient resists by exerting *vis
inertiae*, a passive force of resistance. Moreover, the agent acts exactly to the
extent that the patient is able to resist it—neither more, nor less (because,
Hermann believes, there is no sufficient reason why the agent should act
more than needed to defeat the resistance encountered). Ergo, the agent's

---

[35] The one exception is collision in the mass-center frame, where the two bodies meet with equal
momentum, and neither counts as passive.

action equals the reaction of the resisting patient body. And, the common measure of these two parameters is the net linear momentum—*not vis viva*—lost by the agent and gained by the patient, respectively. Once the patient's resistance has been overcome, the agent then pushes it along, unresisted—which for Hermann explains the kinematic appearance of the two bodies moving together after the inelastic collision.

Finally, a word of caution: Hermann's distinction, active versus passive, appears unstable: it relies on empirically inaccessible quantities; and it appeals to an equivocal criterion.[36] Still, that did not keep it from becoming very popular and influential in post-Leibnizian Germany, thanks to Christian Wolff and his school.

Hermann then quickly dispatches oblique impact, in which the bodies' lines of motion are at an angle. For two bodies A and B, he resolves their velocities into "motions they contain virtually," viz. into a central component (aD and bD, along the line between the two bodies' mass centers) and a normal component Aa and Bb, respectively. Now, he reasons, "because the (normal-component) motions Aa and Bb are parallel, the two bodies cannot act on each other by means of them." So, he infers, they interact "only through their (central-component) motions."[37] Thereby, oblique impact reduces to head-on collision, whose outcome Hermann has just taught above how to compute. Finally, these central motions must be added to the (unchanged) normal components Aa, Bb. Thus, we get individual exit velocities.

This algorithm yields for Hermann a complete solution to the problem of oblique inelastic collision—but with a severe limitation: it predicts the post-impact motion *for just one point* within either body, namely, the velocity of its individual mass-center. The motion of the *body as an extended whole* is beyond the power of his algorithm.

---

[36] When both bodies move (before impact), the one with less linear momentum counts as the patient—because it has less active force, presumably. When one body is at rest, it counts as the patient, but now because it is endowed with nothing but passive force (of inertia). That makes the criterion of passivity equivocal. Moreover, the quantity of active force—which determines the respective ascription of activity and passivity—is linear momentum, hence it is a function of velocity. And so, Hermann's picture of collision (as a clash between agent and patient) in the end relies on the premise that a colliding body has a distinguished, or *true*, velocity. But impact is Galilean-relative, as Huygens and Newton had made clear; thus, true velocity is not detectable by any observation or experiment that appeals to impact. So, Hermann's picture of collision lacks *determinate empirical* content. For the post-Leibnizian fortunes of his account, see Stan 2017.

[37] Hermann 1716, 115.

Elastic collision

To handle elastic impact, Hermann sets out with a "Hypothesis: *when actant bodies collide, their absolute force post impact remains the same as it was before the collision.*"[38] From this hypothesis alone, Hermann seeks to derive several important results. Namely, that in collision with rebound, the system's net quantity $m_i v_i^2$ remains constant; that its center of gravity's velocity is not changed by the collision; and that the *relative* speed (of the colliding bodies) is the same, before and after impact. All these results serve as his premises from which he infers without effort an algorithm for computing the exit velocities of "actant" bodies.

We shall not try to assess here the validity of his reasoning; in particular, his above corollary on relative-speed conservation seems paralogistic. More important for our theme in this book is Hermann's heuristic. It really is astonishingly similar to Huygens' earlier treatment of elastic collision. Specifically, they both set out with a hypothesis—in effect, a disguised energy principle—about a privileged parameter which they construed purely kinematically, despite the overt dynamical connotations of Hermann's term "absolute force."[39] Beyond that, both Huygens and Hermann rely on diagrammatic reasoning alone to infer their corollaries; see also Fig. 3.1.

At the same time, however, Hermann's heavy reliance on kinematics and geometry qua driving engines for his elastic-collision theory raises some questions that deserve mentioning.

First, Hermann's account is disconcertingly sinuous. It has all the early signs of a constructive approach—in that it sets out with definitions of body types and their forces—and then it turns abruptly into a minimalistic, very austere *principle* approach: in effect, a conservation law and diagram-based reasoning is enough to secure his sought outcome. But then his notions of inert and actant body play no role in his collision mechanics—whether explanatory or evidential—and so they sit in the book idly, as mere decorations.

Second, this being so raises a puzzling question: why does he break from tradition so as to define body types in *dynamical*, causal-property terms, instead of hewing to precedent, viz. defining their specific difference in terms of shape change or kinematic behavior in collision?

---

[38] Hermann 1716, 112; his emphasis.

[39] In truth, his assumption is stronger than Huygens' (a statement now called the "Torricelli-Huygens Principle"), which claims merely that a system's gravity center does *not rise* throughout the collision—though it can descend, clearly. Hermann's hypothesis entails that it rises to the same height as that from which it must have descended. See Fig. 3.1.

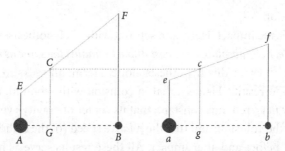

Fig. 3.1. Hermann's geometric solution to the direct-impact problem. *EA, eA, FB,* and *fB* are heights of free fall. The heights are proxy measures for v^2, based on Galileo's kinematics of free fall. They represent the two bodies' speeds, before and post impact, respectively. *CG* and *cg* are the respective heights of ascent and descent for the mass center of the two-body system. To prove his collision rules, Hermann assumes that the mass center falls from, and rises to, the same height through the collision.

Third, his force concepts are explanatorily unmoored from his account of the collision's outcomes: they do not help *understand why*, post-impact, the bodies move together or rebound. By all appearances, Hermann's picture of body-force is just initial *evidence* for the truth of the dynamical laws from which he *infers* values for exit speeds. However, channeling truth (from premises to conclusion, which inferences do) is not the same as causing us to *understand*—and so Hermann's appeal to force notions, in his impact mechanics, is explanatorily inert.

Fourth, *Phoronomia* is home to two notions of force, not one, and they have different measures. "Inert" bodies spend "active force" in non-elastic collision; its measure is the scalar momentum, Δmv, lost by each in the process. In contrast, Hermann's "absolute force" is the same as *vis viva*, and so its measure is different, viz. mv^2. This split foundation forces on us the conclusion that, in regard to collision theory, Hermann did not make any progress beyond Huygens and Leibniz. To sum up, then, in the terms of Chapter 1, Hermann's approach is weakly material. He starts in medias res with body species as given, with no further grounding needed. From there, he seeks to give a causal-explanatory account of the collision process and to thereby arrive at the rules of collision. Had he been successful, the result would have been what we call a "philosophical mechanics" of collisions. But his account has problems, as we have seen.

(ii) Wolff on collisions

Another figure whom Leibniz greatly influenced took the strongly material path to collision theory, by design. That was Christian Wolff, an academic don whose philosophical mechanics was part of a comprehensive system of rationalism that runs the gamut from substance metaphysics to civil law. A substantive part of that system was his philosophical mechanics of collisions. As to why he needs one, he would give three rationales. For one, he subscribes to the mechanists' reducibility program: "There can be no change in bodies except by means of motion."[40] That is, all corporeal properties and changes of state must reduce by explanation to kinematic and dynamical terms. For another, he denies any action at a distance. All momentum transfer—indeed, all physical action—is by contact only: "A body does not act on another unless it presses against it."[41] That makes collision the fundamental process in his physics. Lastly, the rules of collision are explanatorily unsaturated:

> We call rules of motion those whereby motive force is changed by the collision of bodies. . . . In the rules of motion there are general principles, from which these rules can be derived. These principles once established, the rules of motion, i.e. of *impact*, were proved from them in several ways. Mathematicians assume these laws without proof; but it behooves the Metaphysician to demonstrate them. (Wolff 1731, §§302–3)

Those principles are his two laws of motion below, which he derived from an ontology of body. And so, his project in effect requires him to reconcile rules of impact with a theory of matter: a philosophical mechanics of collisions.

The basis of his matter theory is a list of privileged concepts that pick out the "internal principle of [a body's] actions and passions."[42] This principle is threefold: the nature of bodies is to be extended, inert, and endowed with active force:

> All bodies are extended. Every determinate body is endowed with a shape [*figura*] and a definite size [*magnitudo*]. Each body fills a determinate space.

---

[40] See Wolff 1731, §129.
[41] Wolff 1731, §320.
[42] Wolff 1731, §145. To legitimize his notion, Wolff appeals to Aristotle and latter-day followers (such as Daniel Sennert), whose term "nature" allegedly matches his.

Every body resists motion. The corporeal principle of resistance to motion is called *Force of inertia*, or also *passive Force*.

Any body in motion is endowed with a force of acting [*vis agendi*]. . . . Hence, active force is the principle of changes [in bodies]. Mathematicians try to measure this active force of bodies; in so doing, they assume something [to exist] that it behooves the philosopher to prove.

The active force of bodies is called *motive force*, as it is associated with local motion. And so, all the changes in bodies can be explained through [*per*] extension, force of inertia, and active force. (Wolff 1731, §§122–30, 135–8)

In sum, Wolff argues that to be a body is to be extended (or fill a definite volume), resist by a (passive) force of inertia, and act by exerting motive force. These three body attributes are distinct in the sense that "neither can be determined from the other."[43] Two of them count as powers, while extension is causally inert.

The two powers (inertia and active force) ground—explanatorily, via the PSR—two dynamical principles: a law of inertia, and a law of action-reaction. As he declared above, to discover and confirm these laws is a task for metaphysics of matter, not rational mechanics. His two laws are one explanatory bridge from matter theory to the kinematics of collision. The other bridge is a taxonomy of body types; it too straddles the physics of body and the mechanics of impact. We say that it straddles these two domains because Wolff uses two distinct sets of criteria to classify bodies, and hence to explain collision. Let's call the resulting classes Group A and B, respectively. In his doctrine, "hard" and "soft" bodies share a commonality about power: when soft bodies collide, the sole cause of their motion-changes is their "mutual action and reaction." And so do hard bodies; hence they belong together in one group (say, A). In contrast, elastics are dynamically different: when they collide, an additional power gets exerted, viz. "elastic force." So, the philosophical mechanics of their collision would require a different account, hence they belong in a group by themselves. Besides this dynamics-first grouping of body types, Wolff's collision theory makes clear that he also relies on a kinematic criterion to classify bodies, viz. those that move together post-impact versus those that rebound.

---

[43] See Wolff 1731, §§129 and 130, in the notes.

Between these two criteria Wolff wedged yet another one, geometric (i.e., based on shape behavior during impact). This particular criterion yields for him three classes of objects: hard bodies are non-deformable, soft bodies change shape irreversibly, and elastics "revert to their state" (i.e., self-restore the shape they had before impact). He uses this geometric classification to defend a claim about impact dynamics, namely, that in the collision of "non-elastic bodies" (viz. hard and soft) the "only changes of motion" come from these bodies' action and reaction.[44] When elastics collide, a third agency (elastic force) is at work, in addition to the action and reaction they exert simply qua bodies.

The above claim is absolutely crucial in his philosophical mechanics, because it aimed to connect explanatorily his theory of matter with the rules of impact, via his laws of motion:

You ought to take into account this [force of] resistance if you wish to understand distinctly the *communication of motion*. Now as all change occurs by means of motion, we must classify this resisting force among the principles of change. Accordingly, we must not leave it out of the concept of body [*notio corporis*], especially as we have just shown that this force is not at all entailed by bodies being extended. And, we allow this resistance into the concept of body merely because *the communication of motion must have a sufficient reason*, and thereby every change be explicable from a completely distinct reason. (Wolff 1731, §129, note; cf. also §134; emphasis added)

Examined in more detail, his two accounts work as follows.

Inelastic collision

Like many predecessors, Wolff partitions the class of possible initial conditions—head-on, catch-up, and resting collision (with a stationary body)—and accounts for them *separately*. For head-on impact, his proffered causal-explanatory mechanism is "destruction of equal forces." Namely, when inelastics collide, equal and opposite amounts of "force" get destroyed in the "agent" and the "patient" body, respectively. At that point, with the leftover motive force, the agent then pushes along the patient, whose ability

---

[44] Wolff 1731, §145. His warrant is: "if a body is not elastic, then in collision its shape either does not change or if it changes it is not restored. Hence, there is no reason for the motion that results from bodies' action and reaction to change any further. Because nothing can be without a sufficient reason, the resulting motion is not further changed."

to resist (by further slowing down the agent) has been neutralized. Call the agent's leftover force Q, as Wolff does. After contact,

> it is entirely as if body A is urged to move by the force Q. Now *because there is nothing that impedes force Q from producing motion*, then motion will be produced. . . . As nothing is without a reason why it is rather than not, so there must be a reason why a body acts on another. And, since a body does not act on another unless it touches it, the only reason is if the one resists the other's actual motion or endeavor to move. Therefore, when there is no longer a resistance, the [agent body's] reason to act has been removed, and so its action must cease too. (Wolff 1731, §§342–3)

That sums up his destruction-of-force mechanism, which he might have borrowed from Jakob Hermann's earlier opus, *Phoronomia*.

For catch-up and resting collisions, Wolff gives a different explanation, based in a common mechanism; we call it "mental gluing." The gist is this: when (the moving body) A catches up with (the slower or resting body) B, they "seem to cohere, as it were," hence *can be regarded* as a single body, AB. Wolff then claims that if something *is* a single body, all its parts must move with the same velocity, *celeritas*. From these two premises, he concludes that, post-collision, the two "parts" A and B (of the as-if single body AB) must "move as one body, in the same direction. Therefore, the speed must be distributed throughout [A and B], such that either acquires the same speed."[45]

Elastic collision
Again, like many predecessors, Wolff too regards it as a two-stage process. Stage I is mutual compression: upon contact, bodies press on each other (at their common surface) and thereby use motive force to squeeze together the other body's parts. The result of *that* is that "each body acquires," or builds up, "the same elastic force."[46] Stage II is acceleration away from each other, which causes rebound. Wolff is quite evasive about the mechanism behind this action. In passing, he suggests that, as *vis elastica* restores one body's shape, it pushes against the adjacent body, thereby accelerating it away: "Now as body A's compressed parts get restored, body B (which resists that restitution) is

---

[45] Wolff, 1731, 381. See below for analysis.
[46] Wolff 1731, 415.

urged in the direction in which it was already moving. Hence, the motion of B is accelerated."[47] And vice versa.

Assessment

In turning bodies into substance-likes, Wolff aimed to spell out a notion of body fit for dynamics. It is doubtful that he succeeds—which serves again to emphasize just how difficult was the task of articulating an internally coherent philosophical mechanics centered on collisions.

In particular, his theory of soft-body impact falters at several important junctures. The critical weakness is that it falls short explanatorily. Recall, the generic project is to integrate successfully matter theory with rules of collision. Wolff's particular strategy is to argue that the nature of body comprises two corporeal powers (motive force and the passive force of resistance) that yield sufficient reasons for the rules of impact: "you ought to take account of this [force of] resistance if you wish to understand distinctly the communication of motion."[48] But he does not live up to that promise. His two powers above—their action and reaction—account just for the colliding bodies' loss or gain of momentum. They do *not* explain why those particular, determinate momentum exchanges obtain, or why the bodies have the particular (common) exit velocity post-collision. To account for that, Wolff introduces, out of the blue, a very different picture of how bodies act on one another, namely, "mental gluing" as we explained it above. In plastic collisions, once the moving body A runs into the slower (or resting) body B, they "detain each other in the same place," and thus "appear to cohere" into a "single body," he says.[49] They can be *regarded* as just one compound body AB that we mentally glue out of them.[50] From that act being conceivable, he concludes, fallaciously, that there is a "diffusion of speed." Namely, that the stronger body's leftover speed diffuses throughout the entire compound, AB.[51] It follows that A and B regarded *individually* will move at the same speed in the same direction after contact. Note, however, that mental gluing makes no appeal to the bodies' natural powers of action and resistance. It is completely unconnected

---

[47] Wolff 1731, 418
[48] Wolff 1731, §129.
[49] Wolff 1731, §347.
[50] This conclusion is false by his own criteria. To count as one body, two matter bits must "press against each other" (Wolff 1731, §285). But he has not established that they *do* press so. In catch-up collision, all they can exert (based on his ontology) are "motive forces." But those point in the *same* direction—not in contrary ones, as they should do (if they are to press mutually). So, A and B do not count as pressing on each other, hence we may *not* regard them as one body, AB.
[51] Wolff 1731, §366.

to the nature Wolff ascribes to bodies. That makes his philosophical mechanics inadequate.

Incidentally, Malebranche had the same problem. He too furtively reached for mental gluing at the critical juncture: "we can reduce all the laws of the communication of motion to just one, namely, that—because we may *consider* the colliding bodies to make up just one body at the instant of their impact—the moving force divides itself among them, in proportion to their respective sizes."[52]

Serious problems lurk in Wolff's theory of elastic collision as well. For one, he never explains how shape-restoration relates to the nature of body—whether it is an essential property of the species of matter he counts as elastic or whether it is an effect (at phenomenological scales) induced by some micro-mechanism of plastic matter. For another, it is very hard to see how *vis elastica* accounts for perfect rebound in collision—for the conservation of relative speed. That is because Wolff gives at least three different clues about the measure of this force. He thinks of it as a Hooke's Law type of force, proportional to distance, viz. the depth of compression at the end of Stage I of impact. But he also thinks that this force generates *vis viva* in the colliding bodies, hence it must be proportional to the quantity Δv^2 gained by each from rebounding. Finally, he wishes to argue as well that the *vis elastica* generated in compression equals the "impetus," or scalar momentum, lost by the "stronger" body as it strikes the "weaker."[53] There is no clarity on how these various measures connect inferentially, if at all, with the exit velocities of bodies in elastic impact. Which makes it doubtful that he lives up to the main goal of philosophical mechanics, namely, to integrate matter theory with the collision rules *coherently*.

Our remarks here give a flavor of Wolff's approach to collisions, showing that he, too, sought a causal-explanatory account of the collision process, based in matter theory, and which would yield the rules of collision. He took the strong constructive approach (see Chapter 1). A similar, but much clearer and—for our purposes—much more informative attempt is found in the work of Émilie Du Châtelet (see Chapter 5). So we leave it as an exercise for the interested reader to pursue Wolff's account of collisions in more detail, and move on.

---

[52] Malebranche, in *OM* iii, 215; emphasis added.
[53] These various measures of the elastic force are scattered throughout his discussion in Wolff 1731, §§409–19.

3.5 The problem of collisions

We saw in Chapter 2 that in France the problem of collisions persisted into the 1730s unsolved. For the Cartesians, it is clear why collisions are foundational: all change comes about through impact, and so collisions lie at the foundation of Cartesian natural philosophy, both ontologically and explanatorily. In this chapter, we have seen that collisions also posed a serious problem for figures working within a broadly Newtonian or Leibnizian framework. Throughout natural philosophy, impact remained a basic mode of material activity and change. This was so even for those who, following Newton, added action-at-a-distance forces to their explanatory resources, and even (though not always) to their ontologically basic categories. And it was so for those who, following Leibniz, made a fundamental realm of non-extended simples metaphysically prior to the physical realm of bodies, for change in the physical realm was rendered intelligible via collisions.

This is not the only way in which collision theory was a foundational concern in natural philosophy. As we have seen throughout Chapters 2 and 3, the requirements of the theory played a role in determining the basic properties of bodies. For example, simple extension (measured by volume) turns out to be inadequate for a successful theory of collisions, and Newtonian mass must be attributed to bodies instead. Similarly, Leibniz argued that "quantity of motion" is inadequate and that quantity of force (his *vis viva*) must be attributed to bodies as the quantity that is conserved during collisions. In short, collision theory had a vital role to play in investigating the nature and properties of bodies, including their activity and causal powers. Moreover, acceptable laws of nature or of motion, whether in natural philosophy or mechanics, needed to incorporate a successful impact theory, and so collisions provide privileged evidence for the truth of postulated laws.

We pursue these lines of argument in later chapters, and it will prove helpful to highlight the problem of collisions with a label, PCOL.

PCOL: What is the nature of bodies such that they can undergo collisions?

For all parties, a successful solution to PCOL required the integration of a material account of bodies with the rules of collision into a single treatment of impact. Almost everyone divided bodies into three kinds (hard, soft, elastic) in terms of their *geometric* behavior during impact (e.g., irreversible change of shape), whereas the rules of collision offered a distinct dual taxonomy in terms of the *kinematic* behavior of bodies after impact (rebound and non-rebound). The challenge was to unify these systematically—by argument and causal explanation, rather than by stipulation. In Chapter 2, we saw the efforts of French Cartesians to bridge the gap. In this chapter, we have seen Newtonians and Leibnizians attempting the same. Why, after all, should irreversible shape change result in non-rebound after impact? A successful account of collisions was expected to use the material nature and properties of bodies to provide a causal explanation of the impact outcomes expressed in the rules of collision. We call such an account a "philosophical mechanics" of collisions, and we have argued that by the 1730s, almost 100 years after Descartes' *Principles*, no such account was to be had: no-one had been able to integrate matter theory and collision rules coherently.

The remaining question is, so what? To understand the ramifications of this predicament, we need to widen our purview to the broader philosophical context in which PCOL finds its home. We do that next.

4

The Problem of Bodies

4.1 Introduction

What is a body? And how can we know? This is the Problem of Bodies (*BODY*), and this entire book is ultimately about it. Why is it so important? Because in the early 18th century physics was a sub-discipline of philosophy, and *BODY* was its central concern and engine of innovation. Entangled therein were two quintessentially problematic issues of early modern philosophy: substance and causation.

In the two previous chapters, we uncovered protracted efforts to address the problem of collisions (PCOL).[1] Now, we broaden our scope: we argue that the philosophical significance of PCOL is best understood within the context of *BODY*. This, in turn, has consequences for *BODY* itself for, as we saw in Chapters 2 and 3, PCOL requires a philosophical mechanics for its solution. We indicate the significance of this at the end of the chapter; the remainder of our book wrestles with its implications.

Our approach to PCOL in the preceding chapters was diachronic and genetic. In widening our scope to *BODY*, we change tack. We analyze *BODY* into its conceptual parts, and claim that its solution required philosophers to discharge four distinct tasks: we call these Nature, Action, Evidence, and Principle (NAEP); we spell out their content, and indicate how these tasks had to cooperate toward a single answer to *BODY*.

Our claim is that NAEP is a powerful tool of analysis for widespread debates falling under the umbrella of *BODY* in the 18th century. We give some indications in support of this with our remarks on gravitation, and in our analysis of Newtonian debates over method at that time. The main source of evidence is the remainder of this book, where we deploy NAEP to analyze developments in *BODY* from Du Châtelet and Euler to Kant and Lagrange. Articulating it is one of the major achievements of this book.

---

[1] PCOL: What is the nature of bodies such that they can undergo collisions? See Chapter 1.

Philosophical Mechanics in the Age of Reason. Katherine Brading and Marius Stan, Oxford University Press.
© Oxford University Press 2023. DOI: 10.1093/oso/9780197678954.003.0004

One more piece of the puzzle is needed before we can begin. At the beginning of the 18th century, BODY was a problem in physics, and physics was a sub-discipline of philosophy: it was importantly different from physics as we conceive it today. So, a pre-emptive warning: wherever we use the term "physics," it denotes the 18th century sub-discipline of philosophy (unless otherwise noted). And, because the term "physics" carries significant risk for anachronistic misunderstanding, we will sometimes use the label "philosophical physics" as a reminder. Philosophical physics is the context in which to understand BODY.

We proceed as follows. We begin this chapter by documenting the scope and remit of philosophical physics (section 4.2). Then, we move to break down BODY analytically, starting with its ontological elements, Nature and Action (section 4.3), before continuing to the epistemic, Evidence and Principle (section 4.4).

Had BODY been easy to solve, there would have been no reason for us to write this book. But the expected cooperation among Nature, Action, Evidence, and Principle proved elusive. In particular, the epistemological aspects of BODY—its logic of evidence and principles of adequacy—did not fit smoothly with its metaphysical dimensions: the material pictures offered exceeded the epistemic allowances. We illustrate this by turning our attention to discussions of method for philosophical physics. In particular, to the struggles of self-proclaimed Newtonians as they sought to address Nature and Action while meeting the demands of Evidence and Principle (section 4.5).

Next, we re-frame our results within the context of debates over substance and causation (section 4.6). Our goal is simply this: to make vivid the importance of BODY for philosophy in the 18th century by explicitly demonstrating its place in this familiar landscape. We make clear which questions about substance and causation are at stake in addressing BODY, and which are not.

Finally (section 4.7), we return to PCOL, to show why it deserves the attention we are giving it in this book. The reason is that PCOL is a necessary condition for solving BODY. The consequences of this are threefold. First, every philosopher who sought to offer a solution to BODY needed to address PCOL. Second, PCOL was a powerful tool for investigating BODY. And third, since PCOL requires not just physics for its solution, but a philosophical mechanics of collision, BODY also becomes a problem to be solved within philosophical mechanics.[2] And that is the topic of Chapter 5.

---

[2] For the term "philosophical mechanics," see Chapter 1.

4.2 The scope and remit of physics

In the early 18th century, "physics" was a sub-discipline of philosophy. The terms "natural philosophy" and "physics" were used broadly interchangeably. To see what was meant by this term, we look at several leading figures from the period, beginning with Rohault.

Rohault's *System of Natural Philosophy* (1671) was the standard French Cartesian textbook into the early 18th century, and it became popular also in England following Samuel Clarke's Latin translation (to which he added extensive notes) and the later English translation of this version by John Clarke.[3] Chapter 1 is entitled "The Meaning of the Word Physicks, and the Manner of treating such a Subject," and here Rohault says that in his book he will use the word "Physicks" to signify the "Knowledge of natural Things, that is, that Knowledge which leads us to the Reasons and Causes of every Effect which Nature produces."[4] Three years after the Latin edition of Rohault, the self-proclaimed Newtonian John Keill likewise stated that the goal of natural philosophy is to investigate "the Causes of Natural Things." Similarly, 's Gravesande opens his introduction to Newton's philosophy by writing:

> Whoever shall compare the Works of several Philosophers who have treated of *Physicks*, must readily own quite different Sciences are meant by that Title, tho' they all pretend to give us the true Cause of the *Phenomena* of Nature. ('s Gravesande 1720, in the Preface[5])

We see here a key feature of 18th century physics: the search for *causal* knowledge of the natural world.

In his *Elementa Physicae*, Musschenbroek felt the need for a careful articulation of the remit of physics as a branch of philosophy. The opening chapter is highly important for its clarity in articulating the place of physics within philosophy. He begins with a characterization of philosophy in general:[6]

---

[3] Rohault 1697 and 1723, respectively.

[4] Rohault 1723, 1.

[5] Quotations here are from the Keill translation; there was another one, by Desaguliers. We find Keill's version much clearer than the latter. For 's Gravesande's broader natural philosophy, see van Besouw 2020.

[6] Quotations are from the English translation of 1744, with indications of substantial differences from the Latin version of 1726. For a deeper look into Musschenbroek's Newtonian commitments in regard to method, see Ducheyne 2015.

Philosophy is the knowledge of all things both divine and human, and of their properties, operations, causes, and effects; which may be known by the understanding, the senses, reason, or by any other way whatever. Its end is to promote the real happiness of mankind, as far as may be attained in this life. (Musschenbroek 1744, 1)

He then divides all of philosophy into six different areas:

The first is *Pneumaticks*, which comprehends whatever belongs to spiritual existences, their attributes and operations. The second is *Physicks*, which considers the space of the whole universe, and all bodies contained in it; enquires into their nature, attributes, properties, actions, passions, situation, order, powers, causes, effects, modes, magnitudes, origins; proving these mathematically as far as may be done. The third is *Teleology*, which investigates the ends, for the sake of which all things in the universe have their existence, and all their actions, changes and motions are performed . . . . The fourth is *Metaphysics*, which explains such general things as are in common to all created beings. As what is being, substance, mode, relation, possible, impossible, necessary, contingent, etc. . . . Fifthly, *Moral Philosophy* . . . Sixthly and lastly, *Logick*. (Musschenbroek 1744, 2–3)

Two things are most important for our purposes. First, as in the above texts, physics includes *causes*: "actions, passions, . . . powers, causes, effects."[7] Second, physics is the study of *bodies*: their nature, attributes, properties, actions, passions, situation, order, powers, causes, effects, modes, magnitudes, and origins.

Physics as the study of bodies is a widespread commitment—from "Cartesians" (such as Rohault and Crousaz), to "Newtonians" (such as Keill and Musschenbroek), to "Leibnizians" (such as Wolff).[8] Here is Wolff's description of physics as that sub-discipline of philosophy concerned with bodies, from his "Preliminary discourse on philosophy in general":

[7] Ducheyne and van Besouw (2017, 10) state that Musschenbroek rejected exploration of causes in his physics, but we disagree with this assessment. For more on this point, see below.
[8] There is no simple and clean division of philosophers of the period into "Cartesians," "Newtonians," and "Leibnizians" in terms of their philosophical commitments. Nevertheless, it is sometimes helpful to divide figures into these broad camps, for the specific purposes of the issues at hand. When we do so, we follow the figures themselves, primarily via their own self-labeling.

The part of philosophy that studies Bodies is hailed by the name of Physics. For that reason, I define Physics as the science of those things that are possible through bodies [*per corpora*]. I do so based on the same argument as above. For philosophy is the science of possibles insofar as they can be; and physics studies bodies; therefore, physics shall be the science of those things that are possible through bodies. (Wolff 1740, 30)

For some, such as Rohault, physics begins with matter ("matter" being a mass noun) and from there constructs bodies ("bodies" being a count noun). For others, such as Musschenbroek and Wolff, physics begins with bodies as its primary object of study. All agreed that physics studies bodies. This is the second key feature of 18th century physics.

Musschenbroek's discussion of physics has two further characteristics important for our purposes. First, notice that the placement of the division between physics and metaphysics is such that physics includes a great deal of what today we might think of as metaphysics. While physics studies bodies, including their natures, properties, causes, and effects, metaphysics studies only what bodies and created spiritual beings have in common. For example, if both bodies and angels are substances, then metaphysics studies substances as such, but it is physics that studies the nature of bodies in every respect in which they differ from angels. Similarly, if bodies and angels are both causally efficacious, then it is metaphysics that studies this causal efficacy, but it is physics that studies causation among bodies, insofar as this differs in any way from causation among angels. We must not assume that the resulting relationship between metaphysics and physics is that the former supplies a foundation for the latter. In particular, we must not presuppose a Leibnizian (let alone a Kantian) view of this relationship.

Second, Musschenbroek explicitly places teleology outside of both physics and metaphysics. This move, familiar from the 17th century rejection of teleology by mechanical philosophers,[9] will be disputed as the 18th century progresses.[10]

Returning to the two key features of 18th century physics noted above, we arrive at the following general claim: *physics is a sub-discipline of philosophy that aims to give a causal account of the nature, properties, and behaviors of bodies in general.* This is what was meant by the term "physics" in the 1700s,

---

[9] For an introductory summary of this issue, see Shapin 1998, chapter 2.
[10] See the end of section 4.5.

not only in the early decades but throughout the century. It is what we may call "philosophical physics."

Examples abound. Here are just a few, to illustrate. In Germany, Fr. Baumeister describes physics thus:

> Chapter II: On the Parts and Division of Philosophy.
> In theoretical philosophy, one part is natural philosophy, or *Physics*, which deals in explaining natural bodies and their forces. . . . The other part is *Metaphysics*, which deals in common notions and the most general truths. (Baumeister 1747, 10)

This is philosophical physics. Moving on through the century, we find M. Chr. Hanov, a compendious disciple of Wolff:

> Physics is the science of bodies (viz. of their essence and nature). We may also define it as the science of those that are or can be by means of bodies. From Latin we have another name for it, viz. natural Philosophy, for it is the science that studies the things and forces of nature. (Hanov 1762, 13)

Examples continue throughout the century. Across the country, at Göttingen, Erxleben explained to his students, in a bestselling textbook that ran through seven editions:

> Bodies around us have manifold effects on us. . . . A proper knowledge of bodies must indisputably have an effect on our wellbeing, and so the doctrine of nature, or Physics (*philosophia naturalis, physica*), qua science of the properties and forces of bodies, is one of the most useful sciences. (Erxleben 1772, 2)

For one final example, we turn to Fr. Jacquier. He is particularly relevant for this chapter because he, along with Thomas Le Seur, published an edition of Newton's *Principia* with extensive commentary in 1739–42, and it was he who encouraged Du Châtelet to embark upon her own French translation, rather than translating a commentary by John Keill.[11]

---

[11] See Du Châtelet 2009, 9.

We have arrived now at the part of Philosophy that we call *Physics*, or the *Science of nature* . . . . In *general* Physics I explain the universal properties of bodies, and then I move on to *particular* Physics, in which we observe and study individually the various species of body. . . .

   We call *Physics* the part of philosophy that studies the properties of natural bodies. . . . Hence physics may also be called *natural philosophy*, or the *science of nature*. (Jacquier 1785, 3, 6, 13)

This is philosophical physics, and it is the home of BODY.

4.3 The problem of bodies: nature and action

The preceding description of physics places BODY at the heart of physics, for the central goal of such a physics will be to address the question "What is a body?"

   This question divides immediately into two, concerning first the nature of bodies, and second their actions. Crousaz, in a prize essay of 1720 that won over the Royal Academy of Paris, expresses the pair eloquently:

> I take a physicist [*un Physicien*] to be a man who tries his best to see if he can come to understand how the bodies around him are constituted; and to form the right idea of the manner in which these bodies act on him, and the several ways they act on each other. (Crousaz 1752, 3)

Hence the first two of our four criteria for any satisfactory solution to BODY.

Nature: Determine the nature of bodies. Ascertain their essential properties, causal powers, and generic behaviors.

Action: Explain how bodies act on one another. Give an explanation of how, if at all, one body changes another's state (where specifying the "state" of a body is addressed by Nature).

We maintain that these criteria are shared among natural philosophers at the time, as evidenced explicitly in such writers as Musschenbroek and Crousaz, as well as implicitly in the attempts to solve BODY (as we will see going forward).[12]

---

[12] Further examples are easy to find, throughout the 18th century. Here is an example from later in the century. "To the nature of body belong two aspects. First, its essence, i.e. the way and manner

For many philosophers in the early 1700s, addressing Nature and Action required giving an account of the matter out of which bodies are made: a matter theory. But the theory of matter was the locus of pressing metaphysical and conceptual difficulties:

A complex of interrelated problems plagued the theory of matter during the seventeenth and eighteenth centuries: problems concerning matter's divisibility, composition, and internal architecture. Is any material body divisible ad infinitum? Must we posit atoms or elemental *minima* from which bodies are ultimately constructed? Are the parts of material bodies themselves *concreta*? Or are they merely potentialities or possible existents? Questions such as these—and the press of subtler questions hidden in their ambiguities—deeply unsettled natural philosophers of the early modern period. They seemed to expose serious conflicts in the most basic metaphysical commitments of the new science. (Holden 2004, 1)

Many Newtonians, for example, took atomism as their underlying matter theory, with the view that these ultimate parts of matter are indivisible and perfectly hard. Yet atomism was subject to metaphysical and conceptual challenges, such as the apparent conflict between the concept of extension as infinitely divisible and the concept of an extended atom as indivisible. A central question falling under the umbrella of Nature was simply this: How can bodies be extended?

A second question is: What other essential properties do bodies have? While Descartes had rested with extension as the sole essential attribute of matter, others found that additional properties, such as impenetrability, hardness, mass, and so forth, were necessary in order to arrive at an adequate account of bodies. Which of these, if any, should be admitted as essential? What is their metaphysical status? All of these questions fall under the scope of Nature.

A successful solution to Body will address not just Nature, but also Action. For Leibniz, active and passive force are included in Nature precisely because they are required for meeting the demands of Action. In Chapter 5, we see an example of how this plays out in detail. For now, the general point is this: at

---

of its composition. The other is corporeal powers, forces, and capacities. Now all bodies have a force of inertia and a moving force; hence both the body's inertia and its moving force belong to its nature. . . . Sometimes, by the term 'nature' we mean just the nature of body. Hence, the science of nature, or Physics, denotes simply the doctrine that studies the nature of body" (Meier 1765, 201–2).

the heart of philosophical physics lies the quest for causal knowledge of the natural world, a world populated by bodies; and so we seek an account of how it is that one body is capable of acting on another, if indeed such bodily action is possible at all. We return to this issue below. First, we introduce two more criteria for successfully addressing BODY.

4.4 The problem of bodies: evidence and principle

Cartesians, Newtonians, and Leibnizians alike divided bodies into hard, soft, and elastic. Hardness, as we saw in Chapter 3, posed an especial difficulty for those favoring an empiricist epistemology, as hard bodies are empirically inaccessible. Moreover, if the ultimate parts of matter are perfectly hard, then the observed elasticity and softness of composite bodies requires explanation. Some appealed to specific arrangements of hard bodies; others postulated an elastic force in bodies (see Chapter 3). But if the latter, why not postulate elasticity all the way down? For some, following Leibniz, the ultimate parts of matter were indeed perfectly elastic, but then the problem is reversed: why isn't *every* collision perfectly elastic—why is there impact without rebound? Furthermore, for the Cartesian-inclined, this leaves an unanswered question at the very basis of matter theory: how can a part of extension be elastic? Voluntarists might allow that God simply made the parts of matter that way (e.g., he endowed them with a force of elasticity, just as he endowed them with a force of inertia, and with the force of gravity, etc.), but those with a more rationalist epistemology owe us an explanation of how it is that elasticity can be an inherent, perhaps even essential, property of matter. These brief remarks are intended to highlight one thing: any purported solution to BODY will be judged within the context of epistemological (and allied methodological) commitments; and these varied a great deal from one philosopher to another. We therefore introduce two further criteria on successful solutions to BODY, both of which are epistemological:

> **Evidence:** Elucidate the evidential reasoning behind Nature and Action. Spell out what counts as evidence for these claims, and what patterns of inference take us to them as conclusions.

> **Principle:** Elucidate the constraining principles appealed to in attempting a solution to BODY, and check that proposed solutions conform.

Such principles include the Principle of Sufficient Reason (PSR), the Law of Continuity, the restriction to contact action, the criterion of clear and distinct ideas, and so forth. Different protagonists understood such principles disparately as a priori philosophical requirements, defeasible heuristics, and so forth, but such principles are always in play, whether implicitly or explicitly.

Together, these criteria require that we make explicit the justificatory reasoning behind Nature and Action.

4.5 The methods of Newtonian physics

How should we go about solving BODY? How can we address NAEP? The most famous 18th century locus for this question is the debate over Newtonian gravitation. The issues at stake included: whether gravity is an essential property of bodies (Nature); whether bodies can act on one another at a distance (Action); whether Newton's methods of gathering evidence from the phenomena are sufficient to decide these issues (Evidence); and what, if any, constraining principles—such as mechanical philosophy's commitment to contact action and Leibniz's Principles of Sufficient Reason (PSR)—should be brought to bear (Principle).[13] The latter two issues concern epistemology and, more specifically, the appropriate methods for doing physics. This question animates natural philosophy from Descartes' *Discourse on Method* through *The Leibniz–Clarke Correspondence* to Du Châtelet's chapter on hypotheses in her *Foundations of Physics*, and beyond.

Examination of the debates over the methods of physics reveals just how difficult it is to address Nature and Action while adhering to the demands of Evidence and Principle. We can see this by analyzing the response of self-identified Newtonians to the weaknesses they perceived in Cartesian methods of pursuing physics, and their proposals for arriving at knowledge of bodies and causes. As we will see, these figures encountered serious difficulties in their attempts to reconcile self-imposed epistemic strictures with ambitions for a causal account of bodies. Moreover, the obstacles to a satisfactory solution of BODY, via a coherent package addressing NAEP, pertain not only to the debates over gravitation, but to collisions, and to physics quite generally.

---

[13] See Janiak 2008, Smith 2002 and 2014, Harper 2012, Biener 2018, and references therein.

(i) Evidence and principle

A famous commonality among self-proclaimed "Newtonian" philosophers such as Keill, 's Gravesande, Musschenbroek, and Pemberton is their rejection of the Cartesian method of theorizing in natural philosophy. Pemberton describes the Cartesian method thus:

> The custom was to frame conjectures, and if upon comparing them with things, there appeared some kind of agreement, though very imperfect, it was held sufficient. Yet at the same time nothing less was undertaken than entire systems. (Pemberton 1728, 3)

In this brief description, Pemberton highlights two issues. First, the method requires only very weak agreement between theory and observation. Second, it licenses theory construction far beyond observational evidence. The upshot, according to the Newtonians, is a high risk of error. The method invites us to make extravagant use of our imaginations without sufficient constraints on those imaginings.[14] 's Gravesande admonishes:

> We must be careful not to admit Fiction for Truth: By such a false Step we shut up the Door against all future Examination. No true Explication of the Appearances of Nature can be deduced from a false Principle. How vast a Difference is there between learning the Fiction of a humane Wit, and considering the Work of the most wise Mind . . . it is needless to offer further Arguments, that we ought not to reason from imagined Hypotheses. ('s Gravesande 1720, iii)

There are two kinds of constraints that we might appeal to. The first is evidence, and this may be either empirical or theoretical. Empirical evidence might include loose qualitative agreement with the phenomena, and theoretical evidence might include fertility for solving further problems; whether either of these proposals is adequate as an evidential constraint on physics

---

[14] This problem was already highlighted by Keill in his 1700 Oxford lectures (Keill 1726, 10). In the English edition of his *Elements*, Musschenbroek extends his earlier (1726) remarks prohibiting hypotheses as follows: "Therefore hypotheses are to be utterly banished from Physicks; for whatever is deduced from them must be uncertain, nor can be esteemed as demonstrated. And besides science will rather be oppressed than advanced by feigning hypotheses. Useless controversies will be raised, and the phenomena will be distorted, nay perhaps feigned, that the hypotheses may be defended and confirmed" (Musschenbroek 1744, 8).

is precisely what is in dispute. The second kind of constraint is the principles appealed to in doing physics, such as the restriction to clear and distinct ideas in Cartesian physics, Leibniz's use of PSR and conservation of *vis viva*, Newton's first rule of reasoning, and so forth.

In addition to the room for error, there is a problem of underdetermination: too many conjectures are compatible with the Cartesian constraints on theorizing. In the late 17th century, Huygens had sought to remove, or at least reduce, the underdetermination problem by means of a method he explicitly set out in his *Treatise on Light*. He required a much tighter quantitative fit between theory and observation, emphasized the need for a large number of observations, and placed special importance on novel predictions. With all these conditions met, he thought "it must be ill" were the proposed causal explanation of the phenomena not, in fact, true.[15] Huygens' articulation of these additional requirements on theorizing in physics was an attempt to reduce the risk of error in the Cartesian method of conjecture and, more specifically, to address the problem of underdetermination arising from lack of adequate constraints.

Some responded to these problems by advocating that hypotheses be banished from philosophy entirely.[16] More generally, early 18th century Newtonians argued that a new method was needed. At stake was the evidential basis (Evidence) and constraining principles (Principle) appropriate for pursuing physics.

Newton's own explicit remarks in the *Principia* on method are exceedingly sparse. The work of elucidating the method and associated epistemology responsible for the achievements of the *Principia* fell to later philosophers (and continues to the present day).[17] Though discussed already by Cotes in his

---

[15] Huygens 1912, vi–vii.

[16] Pemberton describes the risks in terms of Bacon's idols, writing: "for in this spacious field of nature, if once we forsake the true path, we shall immediately lose our selves, and must for ever wander with uncertainty." He goes on: "all men are in some degree prone to a fondness for any notions, which they have once imbibed, whereby they often wrest things to reconcile them to those notions, and neglect the consideration of whatever will not be brought into an agreement with them" (Pemberton 1728, 5–6). Both of these motifs are later picked up by Du Châtelet 1740.

[17] For Newton's method, we have to look at his *practice*, rather than solely his explicit statements. We have had to wait until recent years, and especially the work of Harper 2012, and Smith 2002 and 2014 for a philosophical analysis of the depth, sophistication, and sheer epistemic power of the method developed and employed by Newton in his *Principia*. Even those working closely with this book have struggled to see what's going on, and have then been bowled over once the realization dawns (e.g., Maupertuis; see Beeson 1992). The relationship between Newton's Rules of Reasoning and the method deployed in the *Principia* is a matter of dispute (see Biener 2018 and Ducheyne 2012). An alternative view is that Keill had it right: the rules do not so much offer a method as provide constraints on theorizing; Keill was right to place them among his Philosophical Axioms, whereas 's Gravesande, Musschenbroek, and Pemberton were wrong to offer them as providing a method in and of themselves.

preface to the second edition of the *Principia*, it was John Keill who placed method center stage. In his lectures of 1700, Keill opens with "Of the Method of Philosophizing," and there reviews the previously available methods from different schools of philosophy, along with their successes and especially their failures. He sets out an alternative method, drawing elements from each but adhering to none. The method he proposes is explicitly tied to an epistemological goal: "But that we may proceed in this Affair with the greater safety, and, as much as possible, avoid all Errors: we shall endeavour to observe the following Rules."[18]

The lesson from the failures of the past cataloged by Keill is that we need a method that, as far as possible, allows us to proceed in a "safe" manner, keeping us from making errors.[19]

Keill's first rule states that we are to begin with definitions that are based on experience, and are sufficient for us to recognize (without ambiguity among them) the things that are our subject-matter, via their properties. That is, we are to begin with nominal definitions, rather than with real definitions. The rule combines methodology—asserting that definitions provide a means for us to distinguish one thing from another—with epistemology—asserting that insofar as we know things, we know them through their sensory properties. The second rule suggests using abstraction to simplify the case (e.g., by first treating bodies in motion as points rather than as extended bodies). And the third rule stipulates that we begin with the most simple cases and advance from there to compound cases (e.g., by treating bodies moving first in a vacuum and then in resisting media).

Keill's explicit target with these rules is Cartesian philosophers who have rushed ahead to give accounts of complex systems—"a World, a Planet, or an Animal"—before mastering simple systems first.[20] This leads to error:

---

[18] Keill 1726, 7.

[19] Keill's first major publication (*Examination of Dr. Burnet's* Theory of the Earth, 1698) begins with a fiery and provocative attack on philosophers ancient and modern, with especial vitriol reserved for Malebranche and Descartes. Keill asks how we are to proceed in light of these appalling and insufferable failures of both the ancients and the moderns. His response: by means of appropriate method. At this point in time, Keill had little positive to offer by way of method, other than that we need a philosophy "which is founded upon observations and calculations, both which are undoubtedly the most certain principles that a Philosopher can build upon. It is in vain to think that a system of Natural Philosophy can be framed without the assistance of both, for without observations we can never know the appearances and force of nature, and without Geometry and Arithmetick we can never discover, whether the causes we assign are proportional to the effects we pretend to explain" (Keill 1698, 18). These initial thoughts are given some elaboration in his lectures of 1700, as we will see next.

[20] Keill 1726, 10.

[O]ur Theorists, not having laid a good Foundation, raise up but a weak Superstructure; whence it is not to be wondered, if their great Building immediately sinks, not without the disgrace of the Architects. (Keill 1726, 10)

These problems are to be avoided by adopting the proper method:

But those who philosophize aright, ought to take another course, and proceed in quite a different Method; and tho' they do not pretend to form a World, an Earth, or a Planet, yet they may be able to lay down a sure Foundation. (Keill 1726, 10–11)

His rules concern the methods for gathering evidence in doing philosophy. He is clear that the epistemological goal is to proceed with physics while avoiding error, which requires a new method distinct from that of the Cartesians, with their tendency to rush toward system-building. What is less clear is whether his rules are enough methodological guidance for his goal to be met.

In Lecture VIII, Keill adds more advice: he offers some philosophical axioms with accompanying remarks. He asserts that the objects of physics—bodies and their actions—are to be distinguished from the objects of geometry in being not so "easily and distinctly conceived." This distinguishes his position from that of the Cartesians, with respect to the epistemological status of the axioms or principles of physics, and the demonstrations that are drawn from them:

I would not have any one in physical Matters, insist so much on a rigid Method of Demonstration, that is, Axioms so clear and evident in themselves, as those that are delivered in the Elements of Geometry: for the Nature of the thing will not admit of such. But we think it sufficient, if we deliver such as we apprehend are congruous to Reason and Experience, whose Truth shines out, as it were, at first view, which procure the Belief of such as are not obstinate, and to which nobody can deny his Assent, unless he professes himself to be altogether a Sceptick. (Keill 1726, 88)

In other words, the requirement on the principles is that reasonable people will find themselves persuaded to accept them, on pain of falling into skepticism otherwise. This is clearly a rather under-developed account of the epistemic status of the principles, but it is all that he has.

Next, Keill deals with the epistemic criteria for satisfactory demonstrations, noting that "also in demonstrating, it is necessary to make use of a more lax sort of Reasoning, and to exhibit Propositions that are not absolutely true, but nearly approaching to Truth."[21] There are two issues at stake here concerning evidence. First, how rigorous does a demonstration in physics need to be for it to count as supporting the conclusion? Second, given a demonstrated conclusion, what is its truth status? These issues arise also in Newton's fourth rule of reasoning: "In experimental philosophy, propositions gathered from the phenomena by induction should be considered either exactly or very nearly true, notwithstanding any contrary hypotheses."[22] But Keill does not offer anything new. He dismisses any remaining concerns about the epistemic status of such axioms and demonstrations:

> But if there are any who harden their Minds against such Principles and Demonstrations, and will not suffer themselves to be convinced by Propositions sufficiently manifest; we leave such to enjoy their supine Ignorance, nor do we think them worthy to be admitted to the Knowledge of the true Philosophy. (Keill 1726, 89)

Fighting words indeed, but rather inadequate as a guide to the epistemic and evidential standards appropriate for the principles and conclusions of physics.

Keill's philosophical axioms are themselves also of interest. He follows Rohault's lead.[23] Rohault's first axiom is that "*Nothing, or that which has no Existence, has no Properties*," and Keill's states that "There are no Properties or Affections of a Non-entity or Nothing." For his second axiom, Keill skips to Rohault's third, with an additional restriction to the case of bodies ("No Body can be naturally annihilated"). Then, like Rohault, Keill turns his attention to causes and effects (Axioms IV–VII; see discussion of *Action*). Axioms VIII–XII concern the composition of motion, momenta, and forces. And then the three final axioms (XIV–XVI) state that all matter is everywhere of the same kind; that different forms of bodies arise from the different arrangements of the particles composing bodies, and from the properties of those particles; and that the "Qualities, or Actions or Powers" by which bodies act on one

[21] Keill 1726, 88.

[22] Newton 1999, 796. Newton did not state this rule until the 1726 edition of the *Principia*.

[23] Chapter V of Rohault is entitled "The principal Axioms of Natural Philosophy." Quotations are from Clarke's 1723 translation of Rohault 1671.

another "arise only from the former Affections and Motions conjointly."[24] Together, these axioms form a set of principles that are to be adhered to when practicing physics.

Later Newtonian textbooks followed Keill's lead, beginning their discussions with a treatment of method, and of associated epistemology, working through the issues in more detail. Moreover, the 1713 edition of Newton's *Principia* saw the highlighting of his rules of reasoning (through their re-labeling as such), and the replacement of the original hypothesis 3 with the new third rule of reasoning.[25] In their textbooks, 's Gravesande (1720), Musschenbroek (1726), and Pemberton (1728) all use method as a means to distinguish themselves from the Cartesians, and each offers Newton's rules of reasoning in an opening chapter on method. All three philosophers treat several of the themes touched on by Keill and, despite the important differences among them, jointly they offer a further articulation of the methodological and epistemological commitments associated with this self-professed alternative to Cartesianism.

Our point is twofold. First, at stake was the articulation of the principles that are to guide theorizing in physics, and the kinds of evidence by which its conclusions are to be supported. Second, this articulation was in its infancy.

(ii) Nature

When it comes to our knowledge of the nature and properties of bodies, a central commonality among Newtonian philosophers is their rejection of a priori reasoning in favor of observation.[26] Musschenbroek argues specifically against a Cartesian a priori approach to the nature of body as extension, using four claims. First, he states that mental abstraction does not uniquely

[24] Keill 1726, 92.
[25] The fourth rule was added in the third (1726) edition. For the development and interpretation of Newton's rules of reasoning, see Biener 2018, and the references therein.
[26] For example, see 's Gravesande: "The Properties of Body cannot be discovered a priori. Body itself is therefore to be examined, and its Properties exactly consider'd, that we may be able to determine what must follow from these Properties in the Phaenomena of Things. Upon a further Examination of Body we find there are some general Laws, according to which Bodies are moved ... which can in no wise be deduced from the Properties, which are said to constitute Body itself" (1720, Preface). To be sure, this is not to deny all a priori knowledge, nor to rule out any role for reasoning in physics. See, for example: "As to what regards Spiritual Beings, this can be known only by reason and the understanding. What is corporeal can only be known by the means of our senses. The knowledge of each of those, being first acquired by our understanding and senses, may be much promoted and extended by reasoning and reflection" (Musschenbroek 1744, 2).

yield extension, since if we started with the sensation of the weight of a body, we could compress its extension to a point and we would not thereby remove its weight. Second, we cannot know how many attributes, and which ones together, constitute a body—and so having knowledge of some does not mean we have ascertained them all. Third, we cannot derive solidity, inactivity, mobility, gravity, and attraction from extension alone, so extension alone cannot exhaust the nature of body.[27] And finally, he claims he will prove in later chapters that space is distinct from body. According to Musschenbroek's critique, the Cartesian a priori method fails to deliver either the essential properties of bodies by showing them necessary, or the further properties of bodies by showing how they follow from extension. In short, he disputes their approach to Evidence.

More generally, the reason we cannot begin with a priori knowledge of the properties of bodies lies in a voluntarist conception of God, widely shared among Newtonian philosophers (and by Newton himself), and according to which he can associate any properties he so chooses with matter:

> Who dares assert whether or how many other Properties may not be given to Body, of which we have no Idea? Or who has been able to discover, whether besides those Properties of Body, which flow from the Essence of Matter, others may not be given it, depending on the free Power of GOD? And that Substance extended, and solid, (for this we call Body) may be dignified with some Properties not necessary to its meer Existence. ('s Gravesande 1720, viii)

This position amounts to a denial of Leibniz's PSR, and therefore takes a stand on Principle. The epistemic consequence of this voluntarism is that reason alone gives no means for us to ascertain the properties of bodies, let alone the essence of matter. It therefore has consequences for Evidence. We can have epistemic access to some of the properties of matter through our senses, but we have no access to the nature of the substance in which they inhere. Therefore, we must begin with observations, and from *phenomena* (i.e., from that which we are able to observe using our senses).[28] The observations are to

---

[27] Musschenbroek 1726, §18.

[28] Musschenbroek writes: "All the situations, motions, mutations, and actions of bodies, which we may observe by our senses, whether by one or more of them, are called *Phenomena* or *Appearances*." He gives examples of each kind of phenomenon (i.e., of situation, of motion, of mutation, and— interestingly—of action). For this last he offers "bodies impinging upon one another, or by powers mutually drawing each other" (1726, 4; original emphasis).

be detailed and to proceed slowly, and by this means we may arrive at knowledge of some properties of bodies.[29] We should not expect to attain complete knowledge of the nature of bodies: our epistemic situation in the world does not allow this.[30]

Such a view does not require a voluntarist justification. Maupertuis, in defending Newtonian attraction, adopted the same empiricist epistemology with respect to the nature and properties of bodies as Musschenbroek, writing:

> If we had perfect Ideas of Bodies; if we well knew what they are in themselves, and what their Properties; how, and in what number resident in them; we should be able to pronounce whether or no Attraction is a Property in Matter: But we, so far from being sufficiently informed; know Bodies but by some Properties only, without the least knowledge of the Subject in which these Properties are reunited. (Maupertuis 1734, 12)

He suggests that the only *rational* grounds we have for ruling out a candidate property as a property of bodies is contradiction with other properties already known to pertain to bodies. Beyond this, our knowledge of the properties of bodies must come from experience.[31] At stake was what counts as evidence when addressing Nature.

---

[29] 's Gravesande, Musschenbroek, and Pemberton all claim epistemic caution and modesty for their proposed method. For example: "Whereas the only method, that can afford us any prospect of success in this difficult work, is to make our enquiries with the utmost caution, and by very slow degrees. And after our most diligent labour, the greatest part of nature will, no doubt, for ever remain beyond our reach" (Pemberton 1728, 4).

[30] Pemberton distinguishes between essential and non-essential properties of matter in terms of those that always belong to bodies, and those that depend on the particular composition of the body. In seeking the properties of bodies we should be looking for those that pertain to all the bodies that we examine. However, he cautions against concluding that such properties are essential to matter, noting Newton's distinction between universal and essential properties (1728, 17–8). Musschenbroek makes no attempt to provide epistemic access to essential qualities, distinguishing only between those that we find in all bodies (his attributes) and those that we find only in some (his properties or qualities). Maupertuis 1732 follows suit, urging that, even though we do not find any bodies without extension and impenetrability, from this we cannot infer a necessary connection between them, or what other properties a body may or may not have.

[31] Maupertuis 1734, 14. For more on his defense of Newtonian attraction, see Beeson 1992, Terrall 2002, and Downing 2012.

(iii) Action

Developing a philosophical physics requires more than making observations and cataloging observable properties of bodies. We need a means of reasoning from these observations to causes. What kinds of causal claims about bodies are within reach for natural philosophy, and how are they justified?

How we proceed will depend on how we approach Evidence and Principle. One consequence of beginning from observations is, according to Pemberton, that we need to proceed step by step, starting with the most immediate cause of each appearance before pursuing more remote causes. Pemberton contrasts this way of proceeding with that of making conjectures about causes:

> Causes assumed upon conjecture, must be so loose and undefined, that nothing particular can be collected from them. But those causes, which are brought to light by a strict examination of things, will be more distinct. (Pemberton 1728, 12)

The first edition of the *Principia* had been immediately criticized for failing to provide a physics—that is, for failing to provide a causal account of gravitational phenomena.[32] Newtonians such as Pemberton responded that the Newtonian method allows for the discovery of *true* causes of phenomena, even without the discovery of their *ultimate* causes.

Pemberton interprets Newton's rules as guides to inductive reasoning, through which we arrive at knowledge of causes. He views the first rule as an elaboration of the claim that "nature does nothing in vain," as applied to causes and effects. He justifies it by saying that, if we deny it, natural philosophy would be impossible: it would "reduce all philosophy to mere uncertainty, for the only proof, which we can have, of the existence of a cause, is the necessity of it for producing known Effects."[33] Since we are seeking a method that neither requires (*per impossibile*) the same standards of proof as in mathematics, nor leaves us with skepticism, we are well advised to accept this rule.[34]

---

[32] See the review in Anonymous 1688: 238.

[33] Pemberton 1728, 28, 20.

[34] Pemberton writes that reasoning in mathematics differs from that in natural philosophy because the objects differ. In mathematics, the objects are free creations of our minds and can be completely known by us. In natural philosophy, by contrast, the objects are outside our minds, known to us through observations, and therefore not so well known. As a result, we cannot demand the level of certainty in physics that we expect in mathematics. "But in natural knowledge the subject of our

Musschenbroek also offers Newton's rules of philosophizing as a guide to inductive reasoning from effects to causes. His comments on Newton's first rule, in the 1744 English edition of his *Elements*, are especially interesting. As Newton stated the rule, the causes are required to be true, but the epistemic problem of identifying true causes is not addressed, making the rule problematic as a methodological requirement. Musschenbroek addresses this epistemic issue explicitly: causes are to be derived from the phenomena, and will be true if three conditions are met: (1) "it appears that they really obtain"; (2) "the phenomena are manifestly proved to flow from them; and (3) "the bodies under examination being tried in various ways, always exhibit the same causes of the same phenomena."[35] Condition (1) concerns observation, condition (2) concerns embedding of cause and effect in theory such that the latter can be *deduced* from the former, and (3) concerns a requirement of varying the observations (to test for robustness under counterfactuals, as we might say today). Negatively, the first rule also requires restraint in assigning causes:

> But if the causes cannot be assigned, nor certainly proved to exist, it will be better ingenuously to acknowledge it, than to invent imaginary causes, from whence we may endeavour to explain the phenomena. (Musschenbroek 1744, 8)

This is a prohibition on hypotheses, on which Musschenbroek elaborates vigorously, as did other Newtonians. Following his presentation of rules 2 and 3, Pemberton writes:

> In these precepts is founded that method of arguing by induction, without which no progress could be made in natural philosophy. For as the qualities of bodies become known to us by experiments only, we have no other way of finding the properties of such bodies, as are out of reach to experiment

contemplations is without us, and not so compleatly to be known, therefore our method of arguing must fall a little short of absolute perfection." He adds: "It is only here required to steer a just course between the conjectural [Cartesian] method of proceeding, against which I have so largely spoke, and demanding so rigorous a proof [adopting standards of demonstration appropriate in mathematics], as will reduce all philosophy to meer scepticism, and exclude all prospect of making any progress in the knowledge of nature" (Pemberton 1728, 19–20).

[35] In the first edition (1726) he offers a short prohibition on hypothesizing, but in the later English edition there is an extended discussion of the first rule as a guide for reasoning about causes. See Musschenbroek 1744, 8.

upon, but by drawing conclusions from those which fall under our examination. The only caution here required by us, that the observations and
experiments, we argue upon, be numerous enough, and that due regard be
paid to all objections that occur. (Pemberton 1728, 21)

Notice that he is not trying to justify induction as itself securing knowledge. Rather, he offers inductive reasoning, cautiously pursued and guided
by Newton's rules, as a means of escaping the dilemma between requiring a
higher standard of certainty than the objects of physics allow, and giving up
on the possibility of physics altogether.

For 's Gravesande and Musschenbroek, why we must reason from
observations to the laws of nature inductively, and why these laws of nature are not to be discovered by reason, is that they depend on the arbitrary will of God and knowledge of his will is beyond our epistemic reach.
Musschenbroek offers an explicit account of our epistemic access to laws. He
writes:

> All bodies are observed to move according to stated laws or rules, whatever
> may be the cause of their motions. By the name of *Laws* we call those con-
> stant appearances, which are always the same, whenever bodies are placed
> in like circumstances. (Musschenbroek 1744, 5)

The epistemic status of these laws is that they are discovered by the senses
and "constant appearances," such that "we may expect" the same effect as we
have previously observed to follow on future occasions "if the circumstances
are the same." The justification for the inductive practice based on "constant
appearances" is the wisdom of God:

> §8. These laws are discoverable only by the use of our senses; for the wisest
> of mortals could not have discovered any of them by reason and medita-
> tion, nor can pretend to have any innate ideas of them in his mind. For they
> all result from the arbitrary appointment of the Creator, by which he has
> ordered, that the same constant motions shall always obtain on the same
> occasions. (Musschenbroek 1744, 5)

He offers some examples of constant appearances, and then goes on:

All these things might have been otherwise constituted, if God had so pleased. And why he thought fit to constitute them in this manner, we can by no means apprehend. It is sufficient for us to know, that they are thus constituted, and to adore the infinite wisdom of the Creator, in this most admirable order and constitution of the universe. Therefore the cause and reason of these laws are entirely unknown to us, but we know they will perpetually be observed, because the divine will act always in the most constant and uniform manner. (Musschenbroek 1744, 6)

The mode of reasoning whereby we must arrive at the laws of nature is inductive. We have no independent means of checking whether the regularities we identify are in fact laws, and the only basis we have for believing that the laws will continue to hold lies in our belief in the constancy of God's actions.

Moreover, both Musschenbroek and 's Gravesande hold that the laws of nature are the epistemic terminus of our enquiries in physics. For the latter, this is because we cannot know whether the laws we infer follow from (i) the essence of matter; (ii) properties that God happened to give the bodies in this world but which are "by no means essential to Body"; or (iii) some other cause that we cannot even comprehend. Once we have reached the laws of nature, depending as they do on the will of God, we "cannot proceed any farther into our Inquiries into Causes."[36] Musschenbroek explains why not:

In the investigation of corporeal causes we meet with insuperable difficulties, because when we arrive at the ultimate cause, which depends only on the will and power of God, we shall not perceive any clear connection between the cause and the effect. For it will never be conceived by us, how God, as a spirit, operates on matter. Again, we have no rule or criterion certainly to know, when we have arrived at the ultimate corporeal or natural cause of things. (Musschenbroek 1744, 8)

We face a dilemma: our inquiries in physics will ultimately lead to causes for which we can give no other explanation than appealing to the will of God, yet we can never know whether or not we have arrived at that point of our enquiries where such an appeal is appropriate.[37]

---

[36] 's Gravesande 1720, xii, xv.

[37] Ducheyne & van Besouw suggest that both 's Gravesande and Musschenbroek exclude the search for causes from the remit of physics, writing that they "omitted causal talk entirely in their physics" (2017, 10f). We think it more accurate to say that they include the search for causes, but are attentive to the limitations on enquiry that arise from their epistemic commitments. This is perhaps what they

In short, we are to begin with careful and detailed observations of the phenomena—that is, the observable behavior of bodies—and move from there, by means of slow and cautious inductive reasoning (including detailed quantitative investigations of the fit between theory and observations), to laws of nature. The laws of nature are the ultimate causes to which we have epistemic access; beyond that, we are not to enquire. The laws are the way they are because God willed them to be that way. To justify our use of inductive reasoning, we are offered two choices: accept it because without it we cannot proceed at all in physics (Pemberton), or ground it in the nature of God (Musschenbroek).

The empirical, inductive practices of the Newtonians discussed here are advocated as an antidote to the speculative excesses of the rationalist, deductive method, whose risk of error had epistemic consequences the Newtonians would not tolerate.[38] However, this alternative inductive method has epistemic problems of its own. We have the epistemic risks associated first with inductive reasoning, and second with the unknowable will of God. This seems to be our predicament in attempting to articulate what is required in Evidence and Principle such that we can successfully pursue Action.

There is a further problem with Action highlighted by Musschenbroek. To see this, we first return to Keill, who in 1700 used observable constant conjunction as a means of reasoning from effects to causes. Keill offered four axioms (IV–VII) that concern causation. They are worth going through in full.[39]

Axiom IV. Effects are proportional to their adequate Causes.

Statements of this kind were widespread among natural philosophers at the time, whether "Cartesian," "Newtonian," or "Leibnizian" in inclination.

have in mind when they indicate that the exclusion applies to "metaphysical" causes and "underlying causes." In our view, 's Gravesande's and Musschenbroek's positions reflect Newton's dividing physics into a two-step enquiry into causes (see Brading, forthcoming). So, when Ducheyne & van Besouw write that "the aim of physics was to uncover the rules according to which effects occur," we should recognize that these rules themselves provide us with causal knowledge.

[38] The central division here is between a rationalist epistemology, in which truth is deduced from a priori principles, and an empiricist epistemology, in which truth is induced from observational evidence. The rationalist may use observation and experiment as a tool to assist in developing theory, and the empiricist may use reasoning as a tool to develop theorizing; but the epistemological basis of the two approaches to science differ, and their methods must also differ correspondingly.

[39] Keill 1726, 89–92.

Indeed, some such principle is needed in order to reason from effects to causes. Axioms V–VII provide additional constraints on such reasoning.

> Axiom V. The Causes of Natural Things are such, as are the most simple, and are sufficient to explain the Phenomena: for Nature always proceeds in the simplest and most expeditious Method; because by this Manner of operating the Divine Wisdom displays itself the more.

The source of most of this rule is clearly Newton's first rule of reasoning.[40] However, the final clause is not in Newton, and we don't know what the source is. Interestingly, the middle and final clauses, without the first clause about causes, is the reasoning that Maupertuis uses when arguing for his least action principle. Reasoning about efficient causes concerns the power of God; reasoning about teleology (not part of physics for Musschenbroek, recall) concerns the wisdom of God.

> Axiom VI. Natural Effects of the same kind have the same causes.

This is very close to Newton's statement of his second rule of reasoning, and the examples that Keill gives are those that Newton gives in illustrating his rule. This rule plays a crucial role in the argument for universal gravitation, identifying the cause of the planetary orbits (a celestial phenomenon) with the cause of trajectories of falling bodies (a terrestrial phenomenon). While Axioms IV–VI are (at least) parsimony constraints on the attribution of causes, Axiom VII concerns the epistemic warrant for the inference from effect to cause, analyzed in terms of causal constant conjunction and covariation:

> Axiom VII. If two things are so connected together, that they perpetually accompany each other, that is, if one of them is changed or removed, the other likewise will be in the same manner changed or removed; either one of these is the Cause of the other, or they both proceed from the same common Cause. (Keill 1726, 90)

---

[40] First published as "Hypothesis 1" in the 1687 *Principia*. See also the second edition of 1713. The wording is unchanged in the later editions. It reads: "No more causes of natural things should be admitted than are both true and sufficient to explain their phenomena. As the philosophers say: Nature does nothing in vain, and more causes are in vain when fewer suffice, for nature is simple and does not indulge in the luxury of superfluous causes" (Newton 1999, 794).

Keill offers examples, including the covariation of movements of a lodestone and a needle, as well as the tides and the moon. The evidence for a causal relation in terms of constant conjunction therefore includes cases where the covarying phenomena are spatially separated. This is crucial to Keill's goal of maintaining that Newton's theory of universal gravitation provides a causal account of gravitational phenomena, and is therefore a *physics* (and not merely beautiful mathematics awaiting an accompanying physics).

There are two distinct problems facing Keill's strategy of using constant conjunction and covariation as an epistemic warrant for causal knowledge. The first is in the inference from constant conjunction to the presence of causation—something with which Hume famously took issue. The second concerns the causal process. Musschenbroek makes explicit this latter epistemic difficulty, and his example is collisions:[41]

> we do not understand the manner of operating of any one thing; and all we can do is to observe the effects that constantly flow thence. When two bodies impinge against one another, how do they operate upon each other? What is the force, and how is it transferred from one to the other? . . . All these mysteries are concealed from us mortals. Those that think they very well understand the impulse of bodies, because manifest effects of it continually present themselves before their eyes, seem not at all to consider what produces those effects, nor to distinguish between the cause and the effect. (Musschenbroek 1726, ix)

Constant conjunction may allow us to infer the existence of a causal relation, but this does not tell us *how* the effect comes about: we do not arrive at a causal-explanatory theory of collisions.

For all the fuss about Newtonian gravitation and action-at-a-distance, the real problem was both broader and deeper: how, if at all, does one body act upon another? Despite their pronouncements that physics is the investigation of causes, there are limitations to how far Newtonian philosophers are able to proceed by means of their method. It seems there may be a mismatch between Action, on the one hand, and the epistemic resources available in their specifications of Evidence and Principle, on the other.[42]

---

[41] Maupertuis (1732) makes the same point in France in his defense of Newtonian attraction. See Chapter 5.

[42] Newton himself had split physics into two steps: in the first he inferred the existence of gravity as the cause of a wide range of observed motions of bodies; but for the second—ascertaining the cause of gravity—he never reached a conclusion. We see here the beginnings of this same issue unfolding

(iv) Alternatives

Early champions of Newtonian natural philosophy in France were Émilie Du Châtelet and Pierre Maupertuis. Du Châtelet is most remembered for her translation of Newton's *Principia* (1756) with its extensive commentary, and Maupertuis for his contributions to least-action approaches in mechanics. Du Châtelet's early advocacy of Newtonianism came through her work with Voltaire, but by the time of her *Foundations of Physics* (1740), Du Châtelet had come to the conclusion that neither the Cartesian nor the Newtonian method was adequate for the demands of physics. In other words, she sought a new approach to Evidence and Principle in order to achieve Nature and Action.

Revising her text just as it was going to press, she replaced a discussion of Newton's rules of reasoning with a new approach to method in physics.[43] The rules of reasoning, viewed as guides for theorizing about *causes*, had turned out to be inadequate. In her *Foundations*, she appeals instead to Leibniz's PSR, offering a two-pronged methodology that also uses empirical resources to test hypotheses.[44] By these means, Du Châtelet attempts to deliver on Nature and Action while meeting the epistemological challenge facing Newtonian philosophical physics in the early 18th century (see Chapter 5).

Du Châtelet's contemporary Maupertuis, with whom she was in frequent correspondence, also sought an additional principle by which to constrain theorizing in physics, and his proposal also has a Leibnizian flavor. For Maupertuis, however, the objective is not to strengthen our resources for reasoning to efficient causes. Rather, in his assessment, efficient causation is insufficient for developing a full causal-explanatory account of the phenomena, for it takes into account only the power of God, and not his wisdom.[45] Rejecting Musschenbroek's placement of teleological reasoning outside of physics, Maupertuis took "nature does nothing in vain," which Pemberton had interpreted as the basis of Newton's first rule of reasoning

---

in the case of contact action, with the consequent pressure to revisit what kinds of causal knowledge, if any, are epistemically accessible. See section 4.6 for further discussion of causation, as well as Chapter 5 for continuation of this theme.

[43] See Brading 2019, chapter 2.
[44] See Du Châtelet 1740, chapter 4. See also the essays in Hagengruber 2012, as well as Detlefsen 2019, Brading 2019, and references therein.
[45] See Terrall (2002, 179) for this interpretation of Maupertuis 1744.

concerning efficient causes, and used it instead to re-introduce final causation into physics with his principle of least action (PLA).[46]

Importantly for our purposes, among the processes Maupertuis treats via his PLA is collision.[47] It is only familiarity that makes impact seem intelligible, he says; once we reflect, we realize it is utterly mysterious how motion could pass from one body to another.[48] He elaborates on his point with a vivid example, of someone who had never touched a body or seen them collide, but who had experience of the mixing of colors. Presented with a blue body moving toward a yellow one, and asked about the outcome, such a person might say that the blue body would turn green when the two bodies meet, Maupertuis suggests.[49] But could such a person guess that the two bodies would join together and move with a common speed, or that they would reflect off one another in opposite directions? Maupertuis thinks it would not be possible for the person to guess these things in the absence of experience. Moreover, even once we know from experience that bodies are impenetrable and that they reflect and displace one another, we still do not know how these changes come about: we lack a causal explanation of the collision process in terms of the nature of bodies.

And so, Maupertuis seeks to shift the debate away from causal explanations in terms of the nature of bodies, and toward a search for the principles and laws by which motion is distributed, conserved, and destroyed. Maupertuis explicitly rejects the Leibnizian "law of continuity" as a suitable principle because, he says, we do not have good reasons for excluding hard bodies. Similarly, conservation of *vis viva* does not hold for all bodies and so cannot be accepted. What we need, according to Maupertuis, is a single, universal principle. Such a principle would reveal what all physical processes have in common, and in this way render them intelligible. This is what his PLA provides, he claims.[50]

---

[46] This is not to say that the principle of least action must, or even should, be interpreted in terms of final causes, but this is how Maupertuis understood it. See McDonough 2020 and Veldman (n.d.).

[47] Maupertuis 1746.

[48] See also Maupertuis 1740.

[49] Maupertuis 1746, 280.

[50] Maupertuis claims to show that PLA encompasses (i) equilibrium in statics (1740); (ii) the propagation, reflection, and refraction of light (1744); and (iii) both hard and elastic body collisions (1746). Note that the 1740 paper contains no mention of final causation. The search for unification through a single principle (nomic unity) emerges as an important theme in rational mechanics in the mid to late 18th century, see Chapters 9–12. For Maupertuis, the epistemic status of such principles is inductive (see especially Maupertuis 1740). For details of Maupertuis' PLA, and its subsequent development by Euler, see Veldman (n.d.).

For our purposes here, the important point is this: both Du Châtelet and Maupertuis, despite their expertise in Newtonian physics, came to believe that the methods of the Newtonians were insufficient for us to arrive at knowledge of causes. This led them to revisit the demands of Nature, Action, Evidence and Principle.

To sum up: In the Newtonian response to Cartesian physics, we see the importance of Evidence and Principle for articulating conditions on acceptable accounts of Nature and Action. And we see the preliminary nature of these discussions: a great deal more work was needed to provide a satisfactory articulation of Evidence and Principle, and to evaluate the alignment of these criteria with the proposed accounts of Nature and Action.

4.6 Substance and causation

BODY can be situated within the early modern debates over the metaphysics of substance and causation. All created substances were presumed (by almost everyone at the time) to depend on God, the primary substance and primary cause of all things. BODY is exclusively concerned with secondary substances (bodies) and secondary causation (agency among bodies). For our purposes, this is really all that matters.

That said, such a brief statement risks underplaying the significance of BODY for the metaphysics of substance and causation, because standard narratives would have us believe we know where this leads. By the mid-18th century Hume had shown us that we lack epistemic warrant for endorsing any metaphysics of secondary substance or causation for bodies, and meanwhile science progresses just fine, striding boldly on without either. And so we know what to expect: any metaphysical elements of NAEP will, in the end, fade away as Hume's epistemic lessons take hold and science establishes itself as autonomous from metaphysics. Were this right, then when it comes to the debates over substance and causation our book would fill in details but it would not change the narrative. Our investigations suggest a different philosophical trajectory within which to locate the significance and relevance of BODY for the debates over substance and causation.[51]

---

[51] The following discussion draws from and engages with Clatterbaugh's 1999 study of causation in early modern philosophy.

Suppose we begin from a familiar theme of early modern philosophy: the increasing attention to epistemology—in particular, the use of epistemology in constraining acceptable metaphysics. Metaphysical claims should not outstrip our epistemic resources, as the kinds of creatures that we are. A widely shared view was that our knowledge of secondary substances is limited to knowledge of their properties. Thus bodies were presumed to be (secondary) substances, and the contribution of Nature to addressing BODY, in identifying the essential properties of bodies and so forth, is to uncover all there is that we can know about such substances. At first sight, this approach *presupposes* that bodies are substances, thereby implicitly committing us to claims about what makes bodies *substances* in the first place. Such claims may include: secondary substances have independent existence (they depend on nothing other than God for their existence); they are unities (in some sense to be specified); and/or they are active (either with respect to the unfolding of their own states, solely, or also in relation to other substances). The relationship between these metaphysical criteria for substancehood and BODY seems to be that the latter presumes the former: metaphysical commitments underpin BODY, even when they do not receive explicit attention. However, things are not so straightforward, as we will see. First, however, we need to add causation into our discussion.

It is broadly uncontroversial that protagonists in the late 17th century causation debate worked with four desiderata for a metaphysics of causation:

1. The account should be consistent with God being the primary cause of all things, and with creatures being, at most, secondary causes.
2. It should make the mind–body interaction intelligible. This was an especially pressing problem in light of Cartesian dualism.
3. It should ensure that body–body interaction is intelligible. Seventeenth century philosophers claimed this intelligibility as the primary advantage of the new "mechanical philosophy" over its rivals.
4. The preceding desiderata should all be achieved by a single, unified metaphysics of causation.

At first, the debate focused on (1), and by the early 1700s it had yielded occasionalism, concurrentism, pre-established harmony, and physical influx as contenders for secondary causation among bodies. Ironically, this makes (3)—viewed as unproblematic as of the mid-17th century—highly problematic. Given occasionalism or pre-established harmony, secondary causation among bodies fails to be "true causation" because in both cases God is the only true cause. Alternatively, it becomes unintelligible: for physical influx, this is the conclusion of Leibniz's critique; for concurrentism, it follows because the position is unstable (or at least seemed to many to be so).[52]

As the 18th century progressed, debates attempting (4), viz. a single, unified metaphysics of causation, faded. The question is why, and what philosophical lessons we should take away. One view, expressed by Clatterbaugh, is that "science" successfully redirected attention away from the search for a metaphysics of causation to the task of identifying regularities as a means of discovering secondary causes. Hume's epistemic critique showed that constant conjunction is all that we have access to epistemically; this was in line with the needs of science, and so Hume laid to rest the quest for a metaphysics of causation.[53] We think there is an alternative narrative that brings rather different philosophical lessons to light.

In our view, the fading away of (4) is not the demise of the quest for a metaphysics of causation. Rather, it is a fragmentation of the debate: discussions of God's action in the world, of mind–body interaction, and of bodily causation continue, but as increasingly independent lines of enquiry. Moreover, the emphasis on epistemology engenders a shift in method, so that an investigation of *particular* causes becomes a prerequisite for pursuing a *general* metaphysics of causation. And so, it's not that attention to the "scientific" task of discovering secondary causes *replaces* the search for a metaphysics of causation; rather, this task becomes a first step *toward* pursuing the latter.

Within this fragmented debate, the metaphysics of causation becomes a localized enquiry, one venue of which is the metaphysics of body–body (secondary) causation. There is an ongoing line of philosophical development that, without presupposing a general metaphysics of cause, seeks an account of bodily causation. In our view, these attempts to make bodily causation intelligible can be understood as exploring and developing a metaphysics of causation, albeit one that may be limited in scope (viz. to bodies).

Such a localization in turn has an unexpected outcome arising directly from BODY, as we shall see in later chapters. Shortly put, difficulties with BODY lead to a disconnect between the bodies of physics and those of ordinary life,

---

[52] See Clatterbaugh 1999. See also Watkins 2005, chapter 1, for attempts to save physical influx from Leibniz's critique.

[53] Though, as Clatterbaugh notes, there is a further irony that Hume's analysis does not serve the purposes of science, but that is an additional story.

which in turn severs the metaphysics of causation developed within the particular contexts of the former from that appropriate for the latter.

We can offer some support for this narrative of localization in the metaphysics of causation. First, notice that even those who opted for occasionalism (e.g., Malebranche) or for pre-established harmony (e.g., Leibniz himself)—according to which bodies do not "truly" act on one another—sought a physics that would make body–body action intelligible. Of course, the criteria for intelligibility were themselves up for dispute, but the commonalities across all discussions of bodily interaction are striking, as we have seen (Chapters 2 and 3). Philosophers attempted to render collisions intelligible by means of a "causal" explanation of the process whereby one body acts on another, such that outcomes are in accordance with the rules of collision. The point is that those who responded to (1) by adopting occasionalism, for example, did not then desist from pursuing (3). On the contrary, aside from the caveat about secondary causation not being "true and proper" causation, their accounts of secondary bodily causation are indistinguishable from those who take such causation to be genuine.

Second, and as a consequence, discussions of causation that focus on (1), and thereby on occasionalism, pre-established harmony, and so forth, become disconnected from discussions intended to address (3). To the extent that discussion of (1) fades, so too do debates over the metaphysics of causation incorporating (1). But notice: this does not entail that discussions concerning secondary causation among bodies also cease to be a locus of debate over the metaphysics of causation. Rather, debates over (1) come to seem irrelevant to the task associated with (3). In a revision to the "standard" narrative, we suggest that the disconnection of (1) from (3) allowed philosophers to set aside the former and focus attention on the latter; that is, on developing an adequate metaphysics for (secondary) causation among bodies. That task explicitly required the intelligibility of body–body action. And so, insofar as intelligibility was achieved, we should expect the account to bring with it an appropriate metaphysics of causation, albeit one that is localized to bodies. What fizzles out is not the search for a metaphysics of causation, but rather (1) and (4), leaving (3)—and separately (2)—to be pursued.

If this is right, then our enquiries into BODY open the way to a narrative in which bodily action (Action) becomes an important locus of enquiry for the metaphysics of causation in the mid to late 18th century, one that brings no metaphysical preconceptions to the table, not even Aristotle's taxonomy of four causes. The criterion of success is that it renders bodily action, as identified and explored in BODY, intelligible (where the criteria for intelligibility are themselves controversial). What looks like "science" from our 21st century vantage point, is simultaneously a richly loaded vein of metaphysical enquiry, as yet underexplored.[54]

Third, having first localized our inquiries into the metaphysics of causation to the particular case of bodies as secondary causes, one might hope to build outward to a general metaphysics of causation. In fact, this is not what happened. On the contrary, further localization and fragmentation followed. As we will see in later chapters, by the late 18th century the "bodies" of physics had proliferated into a variety of objects of theorizing (mass points, mass elements, ideal fluids, etc.) and these had become disconnected from the bodies of our everyday lives (human bodies, animals, rocks, trees, chairs) in the sense that there was no explicit route by which to recover the latter from the former. This is one of the lessons of this book, and of our study of BODY. And one consequence is this: for any metaphysics of causality developed for the objects of physics, we will lack the means to transfer it to the bodies of everyday experience. Where the standard narrative asserts that the search for a metaphysics of causation was rightly abandoned, we instead offer an alternative perspective: our epistemic situation is such that our enquiries into the metaphysics of causation will have to be piecemeal, at least in the first instance.

So much for causation. What about substance? BODY inquires into causation via Action, and Action in turn draws on Nature. Nature, as we know, is above all a search for the generic properties of bodies—those that pertain to all bodies universally, those that are essential to bodies and without which a thing would not be a body, and so forth. This seems to leave metaphysical questions about the nature of substance untouched, for it presumes that bodies are (secondary) substances. We suggest that there is an alternative perspective here too. In parallel with the causation debate, we suggest that the 18th century began with a unified approach to substance, both primary and secondary, and then became localized. The criteria for substancehood were a matter of dispute, and BODY provides one context in which they can be probed. Suppose we are interested in independence, unity, and/or activity as

---

[54] There is a further wrinkle. For some, such as Descartes and Musschenbroek, it is *laws* rather than bodies that are secondary causes. Sorting out the metaphysics of the relationships between laws, bodies, and the properties of bodies (or between laws and the entities they "govern") was then and remains now a topic of much discussion. See Ott & Patton 2018 and references therein, as well as Ott 2009, Brading 2012, and Biener & Schliesser 2017.

conditions. We can use our study of BODY to investigate what these amount to for bodies, and the extent to which they satisfy them. Our methods will be the familiar feedback between empirical inquiry and conceptual clarification.[55] Again, the local becomes prior to the general in the order of investigation, and detailed empirical research is the first step, prior to a general metaphysics of substance and causation. Metaphysical inquiry becomes localized, but it is no less a metaphysical inquiry for that.

Rather than seeing in the 18th century an abandonment of the metaphysics of secondary substance and causation (as epistemically inaccessible, at best), we see the emergence of an alternative route forward: (a) presume that secondary causation exists, (b) identify cases of secondary causation, and (c) render them intelligible, including by providing a sufficient metaphysics; and (a) presume that secondary substances exist, (b) identify these secondary substances, and (c) render their status as substances intelligible by, for example, showing that they can serve as the objects of laws.[56]

We do not pursue the metaphysics of substance and causation explicitly in the remainder of this book (though it will appear again when we discuss Kant and Boscovich), and our remarks here cannot be more than suggestive. Nevertheless, the route lies open for this work to be done. In our opinion, there is a strand of philosophical development that remained live and productive, though transformed into localized debates that were highly entangled with detailed empirical enquiries. This is the place where BODY engages with the metaphysical debates over substance and causation, and it does so concerning secondary substance and secondary causation.

4.7 The goal: a philosophical mechanics

A successful solution to BODY will address both Nature and Action. In Chapters 2 and 3, we saw the importance of collision theory for addressing PCOL. In accounts of the natural world where collisions are a means of

---

[55] For example, we might try out the claim that to be a secondary substance (e.g., a body) is to be something of which the laws can be predicated, and then explore what independence, unity, and activity amount to in that case. We begin from an initial conception of these three conditions, and then refine them in response to what we learn through our study of bodies. This makes metaphysics empirically responsive (as Aristotle's was, for example, but now in ways that reflect the epistemological lessons of the 17th and 18th centuries), but it is metaphysics nonetheless.

[56] Insofar as such an approach requires a rejection of Humean epistemology, alternatives were available at the time from those who rejected the widespread commitment to "ideas" as being the starting point of epistemology.

action among bodies (and especially where they are the *only* means of action, as in most "mechanical" philosophies eschewing action-at-a-distance), addressing Action will therefore require solving PCOL. In Chapters 2 and 3 we argued that PCOL was a foundational problem in natural philosophy. We now see the breadth and depth of its significance. The overarching goal of physics was to solve BODY, and solving PCOL is a necessary condition for completing that task.[57]

This result has three consequences. The first two concern the importance of PCOL. First, any philosopher who sought to offer a solution to BODY needed to address it. PCOL was important not just in its own right, therefore, but had even greater significance due to its role in relation to BODY. Second, it becomes a powerful tool for investigating BODY. As noted above, Leibniz includes active and passive force in addressing Nature precisely because they are required for addressing Action: for providing an explanation of how it is that one body acts upon another. Collisions are the exemplar of bodily action, and solving PCOL places demands upon the resources we make available in Nature. Turning this around, by investigating PCOL we have a route toward addressing BODY by means of a problem that is more clearly articulated—both conceptually and with respect to empirical evidence. It was recognized and pursued as such, as we will see in later chapters. We begin this project in Chapter 5, where we look in detail at several attempts to address BODY that use PCOL as an investigative tool and test case for adequate solutions. There, we complete the case for our claim that PCOL was the most important, far-reaching, and pressing challenge at the foundation of early 18th century philosophical physics. The consequences of this persist throughout the remainder of our book.

The third consequence concerns BODY. We have seen in Chapters 2 and 3 that PCOL requires not just philosophical physics for its solution, but a philosophical mechanics of collision. That is to say, an adequate solution will integrate resources from both philosophical physics and rational mechanics so as to provide a causal-explanatory account of the collision process (see also Chapter 1). PCOL is a necessary condition for solving BODY, and thus BODY too becomes a problem in philosophical mechanics. Any solution to BODY must provide bodies that are adequate not only for the purposes

---

[57] From a metaphysical perspective, at stake is secondary substance and causation, pursued independently of—or even prior to—a general metaphysics of substance and causation. More specifically, with our metaphysical enquiry localized to BODY, we seek accounts of Nature (bodily substance) and Action (bodily causation). The primary tool for this pursuit is PCOL.

of philosophical physics, but also as the objects of rational mechanics. We show how this plays out—and with what ramifications—in the remainder of our book.

To sum up and conclude. We began this chapter with BODY—what is a body, and how can we know?—plus four criteria for success: Nature, Action, Evidence, and Principle. We have explored these within the context of 18th century natural philosophy, and have shown the relationship of BODY both to PCOL and to debates over the metaphysics of substance and causation. Together, the results of these inquiries enable us to formulate the principal goal to be met by purported solutions to BODY.

> **Goal:** to provide a single, well-defined concept of body that is simultaneously (i) consistent with an intelligible theory of matter, (ii) adequate for a causal-explanatory account of the behaviors of bodies, and (iii) sufficient for the purposes of mechanics.

This is a project in philosophical mechanics, and it was central to natural philosophy in the early 18th century. It also persisted throughout the century, as later chapters will show.

5

Body and force in the physics
of collisions: Du Châtelet and Euler

5.1 Introduction

At the beginning of the 18th century, the problem of bodies (*BODY*) lay at
the heart of natural philosophy: What is a body? And how can we know?
Solutions to *BODY* were expected to ascertain the properties, causal powers,
and generic behaviors of bodies (Nature); explain how, if at all, bodies act
on one another (Action); and provide justification for their results (Evidence
and Principle).[1] By the middle of the century, prospects for a solution looked
bleak, and the result is an intriguing philosophical predicament, or so we
will argue.

In the preceding chapters, we have seen the central role of collision theory
in addressing *BODY*: we have made vivid the foundational role of collisions
in philosophical physics,[2] and have claimed that solving the problem of
collisions (PCOL) is a necessary condition for solving *BODY*. Any satisfactory treatment of impact was expected to give a causal-explanatory account
integrating philosophical physics with rational mechanics: a *philosophical
mechanics* of collisions. In this chapter, we show just how difficult this task
proved to be. We argue that the two main lines of approach available in the
middle of the 18th century, defended by Émilie Du Châtelet and Leonhard
Euler, respectively, arrived at insurmountable obstacles. The consequence
is that the branch of philosophy charged with the study of body in general
fails: if it cannot address PCOL, then it cannot answer *BODY*.

---

[1] See Chapter 4 for *BODY*. For explication of all the shorthand terminology that we use in this
chapter, including *BODY*, PCOL, Nature, Action, Evidence, Principle, and "philosophical mechanics,"
see Chapter 1.

[2] Today the term "physics" means something very different from what it meant then. Throughout
this book, we use the term "physics" as it was used at that time, to indicate the subdiscipline of philosophy charged with the study of body in general (for more details, see Chapter 4). To remind the
reader of this unfamiliar usage, now and then we use the label "philosophical physics." Physics in that
sense was also called "natural philosophy." The two terms were often used interchangeably (again, see
Chapter 4).

Philosophical Mechanics in the Age of Reason. Katherine Brading and Marius Stan, Oxford University Press.
© Oxford University Press 2023. DOI: 10.1093/oso/9780197678954.003.0005

We begin (section 5.2) with Nature: with the properties of body presumed in any physics of collisions. We argue that the three most important ones—extension, quantity of matter, and impenetrability—all involve long-standing conceptual difficulties that neither Du Châtelet nor Euler fully resolved.

Next, we turn our attention to Action: to the philosophical physics of bodily action, and to the specific case of how one body acts on another through collision (section 5.3). Beginning from primarily Leibnizian and Newtonian resources, respectively, Du Châtelet (section 5.4) and Euler (section 5.5) each offered an account of how it is that one body acts on another. Their proposals involve radically different conceptualizations of body and force, and we argue that both face serious difficulties.

In our view, this represents the end of the road for a philosophical mechanics of collisions based on a philosophical physics of extended, impenetrable, material bodies acting on one another by means of contact forces. Or so we conclude (section 5.6). *No such project is viable.*

This result takes some digesting, both in itself, and for its wider significance. We elaborate on some of the ramifications in later chapters, where the dramatic consequences for philosophy, and for its relationship to physics and mechanics, begin to emerge.

5.2 Nature: extension as a property of bodies

Philosophers in the early 1700s agreed that extension is a property of bodies: the fundamental objects of physics were bodies, and they were taken to be extended. Such bodies are important not just for physics, but also for rational mechanics, and so the topic of extension as a property of bodies pertains to philosophical mechanics quite generally. Moreover, for some of its purposes (e.g., constraints) rational mechanics takes extended *rigid* bodies as primitive (see later chapters). Thus, insofar as physics at that time was expected to supply mechanics with its objects (viz. the material bodies that mechanics then treats mathematically), it was expected to provide *extended bodies*, including *rigid* ones (or at least bodies that could be treated as ideally rigid in specific contexts).

From the 17th century, philosophers had inherited two basic types of matter theory from which to construct spatially extended bodies: Descartes' infinitely (or indefinitely) *divisible* matter, and the *indivisible* particles of the atomists. Both faced difficulties arising from the concept of extension.

Insofar as extension is a geometric concept, all parties agreed on the following premise: that which is extended is divisible. Cartesian matter theorists embraced geometric divisibility as a property of matter; atomists rejected it. Either way, problems ensued. By the end of the century, attempts to overcome these difficulties posed a direct threat to physics as the science of extended bodies.

For the Cartesian matter theorist, the divisibility of extension presents the following problem for constructing a viable notion of body. If bodies are the subject-matter of physics, then they should persist through time, and through interactions with one another (such as impact); but if matter is infinitely divisible, then what holds a region of such matter (a body) together over time and through collisions? Descartes claimed that mutual rest of the parts is sufficient for those parts to jointly constitute a body, but extension itself lacks the resources to keep the parts of a body at mutual rest when a collision occurs, and so cannot be what holds a body together. Cartesian bodies are doomed to be ephemeral, dividing at the slightest collision.[3] The upshot is that such "bodies," having only the property of being extended, in fact lack what is needed to make them bodies at all.

The atomist overcomes this problem by insisting that matter is not divisible all the way down; rather, there are ultimate parts of matter that are extended yet indivisible. This avoids the ephemerality of the smallest bodies, and provides stable units of matter out of which larger bodies can be constructed. However, atomism faces its own conceptual difficulty, for it must escape the contradiction present in the following set of claims: (1) that which is extended is divisible; (2) atoms are extended; (3) atoms are indivisible. Claim (1) was widely held to be true on the basis of geometry. Claims (2) and (3) are needed to avoid ephemerality and arrive at bodies. Yet (1), (2), and (3) are mutually contradictory, and to begin with a logically incoherent concept of extended body is a risky place to start when doing natural philosophy. It seems, then, that neither Cartesian matter theory nor atomism has the resources for a coherent concept of extended body. This was the situation facing natural philosophers in the early 18th century.

---

[3] This issue worried Leibniz already in his youth: if extension is the sole essential attribute of matter, then any region of matter lacks the unity required to prevent "division to dust" (see Garber 2009, 62, where he discusses Leibniz's papers of 1675–6 collectively known as *De summa rerum*). Moreover, such a region lacks the cohesion a body must have in order to survive collisions, and this worried Newton (see Brading 2018, 8–9). Both concluded: something more must be added to extension in order for us to have an extended body.

One response to the difficulty facing atomism is to distinguish mathematical from physical extension. Musschenbroek, for example, adopted this approach. While conceptual arguments show mathematical extension to be infinitely divisible, for physical extension the issue is an empirical one, he claimed. It may be the case, he suggested, that God created least parts of matter that are indivisible "by any power of nature," and whether or not he has done so is something that can be determined only by experiment.[4] This position is in line with those "Newtonians" who accepted that explanations in science come to an end in the will of God (see Chapter 4). Once we have extended atoms, made by God, we can build extended bodies. In the 1730s, Du Châtelet took a similar approach, though she later changed her mind, as we will see.[5]

Leibniz offered a different solution. He sought to arrive at extended bodies (at least phenomenologically) from non-extended simples, or monads. His proposal involved the desperate move of adding immaterial souls to the material world; "desperate" because immaterial souls lie outside the domain of physics (see, e.g., Musschenbroek's characterization of physics and metaphysics).[6] And, in making this move, Leibniz thereby conceded that physics is unable to provide its own objects. If we follow Leibniz, then we give up on the primary goal of philosophical physics, for we admit that physics lacks the resources for delivering a well-defined concept of body that is both consistent with an intelligible theory of matter and adequate for a causal-explanatory account of the behaviors of bodies.

This was the state of the debate as of the 1730s, at the time when Du Châtelet and Euler got involved.

[4] See Musschenbroek 1744, 19. In this, he follows Newton; see Brading 2018, 15, footnote 17.
[5] To Maupertuis she wrote (September 29, 1738): "Your idea that God has not made (for could not make is a big word) bodies without elasticity, gave me another one; it is that the first parts of matter could be indivisible not by the complete privation of elasticity but by the will of God, for one is often obliged to have recourse there, and I think this indivisibility of the first bodies of matter to be one of the indispensable necessities in physics" (Besterman 1959, letter 146).
[6] Musschenbroek 1744, 1.

(i) Du Châtelet on non-extended simples and extended bodies

In her *Foundations of Physics* of 1740,[7] Du Châtelet tackles the extension and divisibility of bodies head on. She rejects the Newtonian appeal to the inscrutable will of God as justification for the existence of extended atoms, believing that they cannot be postulated as primitive. She maintains that the best (and perhaps only) available alternative is the postulation of non-extended simples. Unlike Leibniz, however, she makes no appeal to minds or souls. In Du Châtelet's account, simples stand in causal (though *not* spatial) relations to one another, and extended bodies arise from our perception of multiplicities of causally related simples. Extended bodies, and the causal relations both among their parts and between distinct bodies, supervene on the simples and their causal relations.

More specifically, Du Châtelet's bodies are phenomenal: they arise from the way in which we perceive simples. That is to say, bodies arise from multiplicities of non-extended simples through how we represent those multiplicities to ourselves, in perceiving them. If we could perceive each of the simples separately, disconnected from the others, the phenomena of extended bodies would disappear. But we cannot perceive those simples, and moreover we cannot even perceive them distinctly in their relations to one another. The upshot of our confused perception is a blurring over, and the emergence of bodies in the phenomena. This brief introduction to Du Châtelet's account of bodies will have to suffice here.[8] For our purposes, the important question is how this enables Du Châtelet to tackle the infinite divisibility of extended bodies.

Du Châtelet addresses the issue by distinguishing between geometric and physical extension. She accepts the divisibility argument as applied to the former only, as a concept that we abstract from our experience of physical bodies. In our minds, we can divide geometric extension however we please: this concept contains no resources for denying infinite divisibility. However, physical extension is different. A finitely extended physical body arises from a finite and determinate number of causally interacting simples. Therefore, this body can be divided only a finite number of times. Moreover, at some point prior to division into individual simples, too few simples

---

[7] Du Châtelet 1740. Translated as Du Châtelet 2009, 2014, and 2018b. References use original chapter and paragraph number.

[8] For more details, see Brading 2019, chapter 3.

remain for the phenomenology of extension to arise. Either way, the physical extension of bodies is not infinitely divisible. In this way, Du Châtelet is able to offer an explanation for a claim that Musschenbroek had left as a brute, inexplicable (and perhaps self-contradictory) assertion: geometric extension is infinitely divisible, whereas physical extension is not. There is a price to be paid: the extended bodies that are the subject-matter of both physics and mechanics are not primitive, but are instead a derivative ontology, arising from an underlying primitive ontology of non-extended simples, or monads.

Du Châtelet was not alone in appealing to non-extended simples in her attempt to provide an account of extended bodies. She drew on Christian Wolff, and both she and Wolff were responding to, and building on, Leibniz. Notwithstanding the important differences between their accounts, all addressed the problem of the infinite divisibility of extension by means of non-extended simples, and so all were committed to some version of monads. In 1745, the Berlin Academy of Science set a prize competition on the topic of monads. The submissions were required to first either refute or support the doctrine of monads, and then deduce "an intelligible explanation of the principal phenomena of the universe, and in particular of the origin and motion of bodies."[9]

(ii) Euler's objection to non-extended simples

The 1745 monad prize competition led to a dispute in which far-reaching philosophical, social, and political issues were at stake.[10] Notwithstanding these wider issues, the heart of the dispute was the relevance of monads for natural philosophy, and the philosophical questions were of two kinds. On the one hand, we have the first-order philosophical problems themselves, such as the divisibility of extension. On the other, we have the second-order philosophical problem of the appropriate method for resolving first-order questions. This latter issue reveals itself in conflicts concerning domains of authority, such as that between the "mathematician" Euler—the central figure in the controversy—and followers of the "philosopher" Wolff.[11]

[9] Broman 2012, 2.
[10] See Broman 2012 and Calinger 1969.
[11] See Brading & Stan 2021.

The dispute over monads involved several properties of bodies, but for now we concentrate on extension. Euler explicitly rejected two features of monadic theory that we drew attention to above. First, he rejected the claim that geometric but not physical extension is infinitely divisible. Second, he rejected any "solution" to the problem of extended bodies that makes their extension phenomenal, so that it is "only an appearance of extension which is perceived in bodies, but that real extension by no means belongs to them."[12]

For Euler, extension is a property of bodies in themselves, and geometry is a legitimate means of knowledge about the properties of bodies; so bodies are infinitely divisible because they are extended.[13] The atomist approach to accounting for the extension of bodies is therefore precluded. Nevertheless, Euler also rejected the monadic approach. His positive account of bodies is discussed below. Here, we focus on his rejection of monads. In this, he remained consistent from the days of the "monad dispute" in the 1740s through to his *Letters to a German Princess* in the 1760s. Letters 125–132 offer an extended, colorful, and probing series of arguments and objections against non-extended monads.[14] Among the themes of these arguments is the question of composition, Euler offers a series of arguments against the view that monads, or non-extended simples of any kind, compose bodies.[15] One version of this argument, found in his *Natural Philosophy*, is particularly striking for our present purposes:

The proponents of simple things argue that the simple things, that constitute a body, are at a distance from each other, and because of this distance
have extent. However, if all these simple things were at a distance from each
other, with nothing in between them, there would be nothing to prevent

[12] Euler 1802, letter 124 [IX]. See also the preceding letters.
[13] More on this later.
[14] Regrettably, we must pass over Euler's *Letters* without engaging with them in detail. In the 18th century, Euler was an intellectual giant, and his *Letters* a widely read bestseller (see Breidert 1983, Calinger 1976, Hult 1985, and Grigorian & Kirsanov 2007). The subsequent division of disciplines has placed him outside philosophy, and so recovering him as an important figure for philosophy in the 18th century requires us to read his contributions across these later divisions. This is work that largely remains to be done. Calinger's 2016 biography provides philosophers with help in navigating Euler's vast corpus, and it has been invaluable for us in the work we have done here.
[15] For Du Châtelet, the phenomenal experience of bodies arises from the existence of monads and our experience of those monads, and in this way she evades the composition question. However, the cost is that, as for Leibniz, the monads themselves are out of reach of physics and bodies become phenomenal. Euler rejects this approach, takes the monads as physical, and poses the composition question.

them from being driven together so closely that no distance between them remained. (Euler 1862, chapter 2, §13)

Thus, he concludes, since the simples themselves are non-extended, no extension would remain.[16] The challenge is explicit: what prevents non-extended simples from coming together so closely that no extension results from their composition?[17] It is here that Boscovich and Kant entered the conversation, as we will see in Chapter 6.

(iii) Euler on the extension of bodies

Where Du Châtelet saw a problem in the divisibility of extension, Euler did not. He embraced the divisibility of bodies, writing "all parts of bodies, however small they might be, still are composite things as the whole body itself."[18] For Euler, every part of a body, no matter how small, is extended. That it is further divisible poses no problem, he thinks. Yet, given infinite divisibility, there are no finite-sized ultimate parts that we can glue together to form stable wholes, so how can we have extended bodies?

Euler elaborates on this issue in *Introduction to Natural Philosophy*, a manuscript written in the late 1750s or so but not published in his lifetime (see section 5.4), and in his *Letters* of 1760–2. In *Natural Philosophy*, Euler asserts that extension is a general property of bodies—nothing is a body that is not extended—so that bodies are species of the genus "extended things." Moreover, Euler accepts that all the properties of extension also pertain to bodies, since they have the property of extension. Therefore, bodies are infinitely divisible.

Nevertheless, Euler argues, "the statement that every body consists of infinitely many parts is simply wrong, and is even in contradiction with infinite divisibility," for while a body is infinitely divisible, there are no "last parts" into which it is divisible. On the one hand, due to infinite divisibility, bodies

---

[16] This passage does not seem to be in Euler's *Letters*, and since the *Natural Philosophy* was not published at the time, it is possible that this particular criticism of the monadic account of extension was not in the public domain. Nevertheless, the issue of how it is that non-extended monads could compose an extended body was a topic for Euler (see, e.g., Euler 1802, letter 126 [XI]).

[17] The analogous issue of whether points can compose a line was a familiar topic at the time. See Chapter 6 for further explanation.

[18] Euler 1746b, §74. Euler's main concern in this paper is with the force of bodies, to which we return below.

must be viewed as composite, in the sense that they can always be divided. On the other hand, any given body has only a determinate number of actual parts, notwithstanding that those parts are themselves infinitely divisible.[19] Or so he claims.

From this, one might conclude that a body has an infinite number of potential parts, and that Euler has therefore failed to evade the problem of how such parts are unified into a finite extended body. However, we interpret Euler as rejecting this way of conceiving of infinite divisibility in relation to extended bodies. For Euler, the concept of body (at which we arrive from experience) allows us to conclude that extension is a property of all bodies universally, and that bodies are extended in three dimensions and have a determinate shape. However, unlike Du Châtelet, Euler does *not* require a further account of how it is that extended bodies are possible in order to proceed with physics. In particular, we should not commit ourselves to the view that there are "ultimate" or "fundamental" wholes out of which "composite" wholes are made. Rather, it is metaphysically legitimate to treat *any* body as a unified whole, or as having actual parts, depending on the problem at hand.

Euler writes that the concept of continuity applies to extension (as well as to motion), and this allows us to *think* of the division of extension however we please; but this does not in turn allow us to make any inferences concerning the *parts* of extension.[20] Applying this idea explicitly to bodies, we can think of a body as being divided however we please, but there is no inference from here to the body having those parts. There is no inference from the infinite divisibility of an extended body to an infinite number of "ultimate parts" out of which the "derivative whole" is composed.

One implication of this view is a rejection of the distinction between "ultimate" or "fundamental" indivisible wholes and the composites out of which they are made. This is implicit in Euler's work just as it is in Newton's *Principia*.[21] With hindsight, it becomes visible as an example of a non-foundationalist approach to physics, emerging already in the 18th century. But this was missed by their contemporaries and takes us beyond our topic here.[22] It is also suggestive with respect to Euler's overall epistemology, but his philosophy awaits a comprehensive treatment beyond our scope here.

---

[19] Euler 1862, §§12, 14.

[20] Euler 1862, §19.

[21] For Newton, see Brading 2013 and 2018.

[22] Moreover, the *metaphysical* import of rejecting a distinction between ultimate parts and composite wholes is something that Newton's and Euler's contemporaries missed, and it continues to be neglected today.

And so, the situation as of the mid-1700s was that there were only three explicit alternatives available for constructing an account of the extension of bodies: continuous matter, atomism, and some form of monad theory.

(iv) Extension and quantity of matter

Extension as a property of bodies was an issue not just because of the problems arising from divisibility. In the context of mechanics, the extension of a body interests us in two further ways. First, extension was used (following Descartes) as a measure of the "quantity of matter." Newton transformed this idea with the introduction of his concept of mass, but the radical nature of this innovation escaped most at the time (see our discussion of "elusive mass" in Chapter 7). Second, the shape of a body affects its motion, and in ways that are related to the distribution of its mass throughout its volume. The latter issue we return to in later chapters.[23] With respect to the former, it was obvious to all that extension simpliciter was inadequate for the purposes of mechanics: two balls of equal volume, one lead and one wood, do not behave in the same ways, so their "quantity of matter" must differ despite their apparent equal volumes. For Du Châtelet, as a plenist, the solution is that all bodies contain two types of matter, proper and foreign, and it is only the proper matter that rests, moves, and—crucially—*is weighed and acts* with it. She claims: "The reality of the existence of two matters is easily demonstrated by experience; for experience teaches us that bodies have different densities and weights."[24] This conclusion follows only if we assume that proper matter is in itself of uniform density. Such an assumption—that fundamental matter (whether atoms or a plenum) is uniformly dense—was widespread at the time, and Du Châtelet never made it explicit.[25]

With her commitment to the plenum, her proper/foreign matter distinction, and the uniform density assumption implicitly in place, Du Châtelet then concludes:

[23] The importance of the issue for mechanics became increasingly evident as the century progressed.
[24] Du Châtelet 2014, 9.177–8.
[25] For more on this assumption, and on proper and foreign matter, see the discussion of "elusive mass" in Chapter 7.

it is necessary that these pores be filled with foreign matter that is not weighed with these bodies, and which does not enter into collisions with them if they encounter other bodies in their path. (Du Châtelet 2014, 9.178)

Her appeal to weight and to collisions is important because it connects quantity of proper matter to the dynamical roles of mass in Newtonian physics.[26] In chapter 11, Du Châtelet tells us that a body's resistance to changes in its state of motion or rest arises as a necessary consequence of its inertial force (which in turn arises from the forces associated with simples, about which more below), and that the inertial force of a body is proportional to the quantity of its "own matter"; the greater the quantity, the more it resists. Clearly, the relevant quantity here is quantity of proper matter, and this is where Du Châtelet introduces the term "mass," without any fanfare whatsoever: "That is to say, that the more mass, the more it resists."[27] Notice that while the mass of a body is its quantity of proper matter, its additional dynamical role is acquired through the *stipulation* that a body's mass is proportional to its inertial force; the discussion in the *Foundations* then moves on without pause. Deeper conceptual difficulties associated with reconciling this dynamical role for mass with extension as the measure of "quantity of matter" remain hidden from view, and we discuss this issue in Chapter 7.

(v) Extension and impenetrability

In the first half of the 1700s impenetrability, like extension, was widely adopted and uncontroversial as a property of bodies. There is some ambiguity in use among the terms "impenetrability," "solidity," "hardness," and so forth in different authors, but the idea that bodies exclude one another from spatial overlap was widespread.

Whether or not impenetrability must be postulated as an additional property beyond extension was debated, and the positions divide roughly into two camps, depending on whether the broader picture accepts or rejects a plenum of continuous matter. Those who accepted the plenum seem to implicitly infer impenetrability directly from extension (on pain of incoherence, for if the parts of extension were not impenetrable, then they could all

---

[26] See the discussion of "elusive mass" in Chapter 7.
[27] Du Châtelet 2009, 11.257.

overlap with one another, and extension would collapse upon itself in such a way that there would be no such thing). Therefore, in granting that bodies are extended we also grant that they are impenetrable. Those who accepted the vacuum postulate impenetrability as an additional property of bodies (or rather, of atoms) as a means of distinguishing space, which is penetrable, from body. Beyond this, impenetrability remained untheorized.

From the extension and impenetrability of bodies, it is a short step to consideration of collisions. Whether plenists or atomists, those who accepted impenetrability also accepted the argument that impenetrability implies contact action. Two bodies on a collision course must (given Newton's first law of motion) either remain in the same state, which would require them to pass through one another, or change state. Given impenetrability, they cannot pass through one another, so they must, on coming into contact, change state. Therefore, the postulate of bodies whose sole attributes are extension, impenetrability, and mobility implies contact action. The project of physics (as it was understood at the time) required a causal account of contact action in terms of the properties of bodies, and this was pursued via the postulation of various forces associated with bodies, as we will now see.[28]

5.3 Action

While it was agreed that bodies are extended and impenetrable, it was also widely accepted that they cannot be *merely* so. The reason for this is concern that extension and impenetrability fail to provide sufficient resources for a causal explanation of how one body is capable of acting on another. Where Descartes sought to eschew all notions of force in his physics, Leibniz and Newton found this to be not viable given the problems that they were interested in; in consequence, they each introduced notions of force (about which more below). Debates ensued. A goal widely shared was a causal-explanatory account of bodily action; the test case was collisions.

Newton and Leibniz appealed to "force" for a variety of different reasons, and their notions of force play several different roles. For our purposes, the most important features of their views are as follows.[29] Newton associated

[28] For what becomes of impenetrability, and relatedly of contact action, in Boscovich and Kant, see Chapter 6.
[29] For Newton on force see, for instance, Westfall 1971, McMullin 1978, Janiak 2008. For Leibniz on force, see Garber 2009 and references therein.

an "inherent force" with each body, and introduced a concept of "impressed force" as the source of all changes of states of motion in bodies. The two are connected via changes in the quantity of motion: the "inherent force" is proportional to the inertial mass; the inertial mass is proportional to the quantity of motion; and changes in motion of a body come about via an impressed force.[30] In the wake of the *Principia*, Newtonian philosophers used quantity of motion, or momentum, as the measure of the force of motion of a body. Leibniz, on the other hand, constructed a complex theory of forces, active and passive, living and dead. According to Leibniz, living and dead force are the means by which one body acts upon another, and he and his followers argued for living force as the correct measure of the force of motion of a body (about which more below).[31] The *vis viva* debate was over the correct measure of this motive force: does it depend on speed (the measure that Leibniz associated with "dead force") or on the square of the speed (the measure that Leibniz associated with "living force")?[32] Broadly speaking, Newtonians favored the former; Leibnizians the latter. Either way, both sides accepted the appeal to a motive force in the explanation of contact action, and maintained that by means of motive force contact action among bodies is rendered intelligible.[33]

In the 1730s, Maupertuis questioned this intelligibility, and Du Châtelet responded to the challenge with an account of collisions that uses the Principle of Sufficient Reason (PSR) as a criterion of intelligibility, as we shall see in the remainder of this section.

(i) The intelligibility of contact action

In the early decades of the 18th century, contact action was widely held to be intelligible (in accordance with "mechanical philosophy"), and it was Newtonian gravitation—as an unintelligible action-at-a-distance—that was challenged. However, as the century progressed, the former assumption came under increasing scrutiny. In 1732, Maupertuis attacked the claim that

---

[30] See Newton 1999, definitions 3 and 4, and the laws of motion.

[31] See also Chapter 3 for Leibniz, Hermann, and Wolff on active and passive forces.

[32] See Chapter 3, and below, for more on the "*vis viva* dispute." We use "force of motion of a body" as an umbrella term, recognizing that the dispute at the time involved a range of terminology designed to mark a variety of distinctions, including "quantity of motion," "quantity of force," and so forth.

[33] A good point of entry into early 18th-century discussions of this issue is *The Leibniz–Clarke Correspondence.* Criteria of intelligibility differed from one philosopher to another, of course, falling broadly into three camps: Cartesian (clarity and distinctness), Leibnizian (PSR), and Newtonian (empirically accessible lawlike behavior and the will of God).

contact action via a motive or "impulsive" force is intelligible. His argument was addressed primarily to his fellow French philosophers, for his point was that collisions are no more intelligible as a form of action between bodies than is Newtonian action-at-a-distance.[34] He writes:

> Ordinary folk are not astonished when they see a moving body communi-cate its motion to other bodies. Accustomed as they are to this occurrence, they can no longer see it as wondrous. But some Philosophers—having resolved they can decide a priori which properties bodies can have, and which must be denied to them—believe recklessly that the impulsive force is more conceivable than the attractive. But what is this impulsive force? How does it reside in bodies? Who could have guessed that it does reside in them, before he ever saw bodies collide? (Maupertuis 1732, 16f.)

Contact action *seems* intelligible only because we have seen collisions among bodies so often, and so we are accustomed to it.[35]

Du Châtelet, in her 1740 *Foundations*, offered an account of the forces of bodies explicitly intended to meet the requirements of PSR, and thereby make contact action intelligible. Kant, too, was troubled by the issue of in-telligibility, and offered his own response (see Chapter 6). Maupertuis took a different route, via what became the Principle of Least Action (see Chapter 4).

Du Châtelet's engagement with this issue begins with the *vis viva* de-bate, and with a query that she posed to Maupertuis in 1738 concerning the contrasting positions of Bernoulli (pro *vis viva*) and Mairan (against).[36] At this time, Du Châtelet had yet to develop her theory of forces (see section 5.4), but her discussion with Maupertuis reveals some of the motivations be-hind her eventual position.

---

[34] In the Cartesian picture, the essential attribute of bodies is extension, and from this it follows a priori that bodies are impenetrable and therefore that they act on one another via collisions (see above). Maupertuis accepts the view that bodies are extended, but denies that impenetrability is de-rivable from extension. Moreover, he denies that, from these resources of extension and impenetra-bility, the manner in which bodies act on one another in collisions is intelligible. Maupertuis also points out that a retreat to Malebranchean occasionalism does not help those who favor contact ac-tion and reject Newtonian gravitation. For, Maupertuis asks, if it is God who moves bodies according to the rules of collision, why should God not also move bodies according to the law of gravitation?

[35] This theme is developed most famously in Hume.

[36] Du Châtelet's *Foundations of Physics* went through two French editions (1740, 1742) and was translated into German (1743) and Italian (1743). The *Foundations* is bookended by her interest in *vis viva*: from her 1738 correspondence with Maupertuis to her clash with Mairan (consisting of her criticism of him in the 1740 edition, his response in 1741, and her further reply, all of which were published in the 1742 edition). See also Reichenberger 2012, 165.

(ii) How body acts on body: wider issues

On February 2, 1738, Du Châtelet wrote to Maupertuis as follows:

> Permit me to ask you a question. I have read many things recently on living
> force, and I want to know if you are for Mr. Mairan or for Mr. Bernoulli. I do
> not have the indiscretion to ask you on this all that I would like to know,
> but only which of the two opinions is yours. (Du Châtelet 2018a, 329–30,
> letter 134)

Maupertuis must have replied immediately, because by around February
10th she wrote to him again, responding to the opinion he had sent her. Du
Châtelet presents the two reasons on the basis of which she had previously accepted the product of mass and velocity as the measure of the force of bodies
(Leibnizian "dead force"). She then indicates her change of opinion, declaring
that in her view Bernoulli's essay (1727) proves the case for living force.[37]
Maupertuis' letter, however, has left her with a remaining puzzle. He must
have asserted that *vis viva* is conserved in the universe, and this troubled her.
In the Newtonian picture, the quantity of force (measured by the product of
mass and speed) need not be conserved.[38] Du Châtelet finds the conservation claim hard to reconcile with empirical evidence (our translation):

> I tell you that it gives me great grief when you tell me that, if we take living
> force for the force of bodies, the same quantity of it will be conserved al-
> ways in the universe. That would be more worthy of the eternal geometer,
> I admit, but how would this way of measuring the force of bodies prevent
> motion from getting lost by friction, prevent free creatures from begin-
> ning motion, prevent the motion produced by two different motions from
> being greater when these two motions conspire together than when they
> are along lines perpendicular to one another, etc. It is perhaps daring of me
> to ask you to tell me how it is that it follows that the same quantity of force
> would exist in the universe, if the force of a body in motion is the product of
> its mass and its speed squared. I suppose that we ought to distinguish per-
> haps between force and motion, but this distinction weighs on me greatly,

[37] She also endorses Maupertuis' view that "the rest" (whatever exactly that may be) is nothing but
a dispute of words, and writes that with respect to *vis viva* (but in this respect *only*) Leibniz uncovered
a secret of the creator.
[38] See Chapter 3 for discussion of this point in *The Leibniz–Clarke Correspondence*.

and because you are the one who threw this doubt in my mind I hope that you will clarify it. (Du Châtelet 1958, letter 118)

The phenomena of friction, free will, and collisions seem to Du Châtelet to conflict with the claim that the total quantity of the force of bodies is conserved. She presses Maupertuis on the issue of free will:

> But the only thing that puzzles me at present is liberty, for in the end I believe myself free and I do not know if this quantity of force, which is always the same in the universe, does not destroy liberty. Initiating motion, is that not to produce in nature a force that did not exist? Now if we have not the power to begin motion, we are not free. I beg you to enlighten me on this point. (Du Châtelet 2009, 109)

Her concern is as follows. Suppose that I am sitting in my chair, not moving, and then choose to get up and walk around. In so doing, my body goes from rest to motion and acquires a force of motion where before it had none. Importantly, it does so without any other body acting upon it to put it into motion. Free action seems to require the spontaneous injection of new quantities of force into the world, but how can this be in a universe where the quantity of force is conserved?[39]

Shortly afterward in her letters, Du Châtelet pursues the issue of collisions: what happens to the force of bodies in a head-on collision between two perfectly hard bodies in a vacuum? As we have seen in Chapters 2 and 3, those who treated hard bodies stipulated that they do not deform on collision, and most agreed that they do not rebound. In specifying that the collision takes place in a vacuum, Du Châtelet also rules out the force being dispersed into the bodies of a surrounding medium. She asks Maupertuis: "What then would become of their force?"[40]

The standard Leibnizian response is to deny that there are hard bodies (see Chapters 2 and 3). Du Châtelet does not endorse this solution. While she admits that at that time there was no observational confirmation of the existence of perfectly hard bodies, she does not accept this as proof that they are

---

[39] One response might be to assert that since the conservation law is global, so long as an equal quantity of force disappears elsewhere, the total quantity of force will remain constant. This requires an account of how it is that exactly the right quantity of force spontaneously disappears somewhere in the universe whenever I act freely. Without this, the response posits a spectacular (not to say miraculous) global conspiracy.

[40] Du Châtelet 1958, letter 124.

impossible. Moreover, she asserts that "it is even very likely that the primary bodies of matter are such."[41] And so, we should theorize about what would happen were hard bodies to collide in a vacuum.

The next move she makes is telling, since it seems to concede that there is no answer to be had from within physics. She writes approvingly of Maupertuis' suggestion that we should appeal to metaphysics. She suggests that the force produced in hard-body collision is measured by its effect (i.e., the bodies coming to rest): it is the *loss* of *vis viva* that measures the force used up in the collision. This is a clear indication that, despite accepting *vis viva* as the true measure of the force of bodies, she continued to reject its conservation. One motivation was free will, as is evident later in the same letter. She suggests that conservation may hold for inanimate bodies yet be violated by animate beings, even if the latter were to lie beyond our comprehension:

> God could have established laws of motion [governing] the collisions of inanimate bodies, according to which they conserve, or communicate, or absorb the effects of the force impressed upon them; but this does not preclude the existence of a self-moving power in animate beings, which is a gift of the creator like intelligence, life, etc. For if I am free, it is absolutely necessary that I be able to begin movement, and if my freedom were demonstrated, it would be necessary to concede that my will produces force, even if the how-really is unknown to me. Isn't the case of creation, which must be assumed if one assumes that a God exists, the same? Are there not a thousand things that will always be as impossible to deny, and to understand? (Besterman 1959, letter 124)

All of this makes vivid Du Châtelet's concerns over how bodies act on one another, the true measure of the force of bodies, and whether the total force of bodies in the universe is conserved.

In the manuscript version of the *Foundations*, Du Châtelet appealed to God, albeit reluctantly, at least for the case of perfectly hard bodies acting on one another:[42]

> In the end it seems to me that it is no easier to conceive the simple communication of movement between bodies supposed to be completely hard,

---

[41] Du Châtelet 1958, letter 124.
[42] Both the following quotations from the manuscript are found in Janik 1982, 101. Our translation, with thanks to Lauren LaMore.

than to know what their forces will be after the collision; one must, I think, leave both questions to God.

She went on:

The simplest case of them all is that of one body that hits an immovable obstacle, and this case is subject to the greatest difficulties. I am quite afraid that we must resort to God for the collisions of bodies.

In this, she was not alone.[43] Ultimately, however, Du Châtelet found the appeal to God as the solution to the problem unsatisfactory: she came to believe that God has no place in scientific theorizing.[44]

These exchanges with Maupertuis concern BODY, and the primary issue is the appropriate measure of the "force" of bodies in motion: Nature. In their discussion, the focus is on how best to resolve the issue, by means of Evidence and Principle: Du Châtelet appeals to empirical evidence associated with collisions, free will, and friction, and she asks whether this evidence can be made consistent with the principle of conservation of *vis viva*. She thereby shifts our attention from Nature to Action, for Du Châtelet wants to know how it is that the action of one body upon another can be made consistent with the principle of conservation of *vis viva*. By 1740, and the published version of her *Foundations*, she had worked out her answers to these questions.

5.4 Du Châtelet and Action

The depth and significance of the issues surrounding Action, as they stood in the 1730s, were recognized vividly by Du Châtelet. She places them at the heart of her *Foundations*. Writing for a French audience, she begins her treatment of bodies by rejecting Cartesian matter theory as leading to occasionalism, which she maintains undermines the possibility of genuine action among bodies. In this, she sympathizes with Keill, for whom the upshot of Malebranchean occasionalism is that when one man is attacked by another

---

[43] For a discussion of the epistemic limits of inquiry into the nature and actions of bodies, see Chapter 3.

[44] See Brading 2019, chapter 2. For an indication of the ways in which appeal to God permeated the debates over collisions, see Scott 1970, 29.

with a stone, it is God Almighty himself who will "dash out his brains."[45] Crousaz, as we have seen, made similar objections to occasionalism in his prize-winning 1721 article on the nature and communication of motion (see Chapter 3). If it is God who does everything, then there can be no genuine bodily action, and no human action either, or so goes this criticism of occasionalism. Du Châtelet offers her fellow philosophers an alternative account of Nature, one that enables her to address Action such that bodies genuinely act upon one another.[46]

(i) Du Châtelet on the forces of bodies

In the opening chapters of the *Foundations*, Du Châtelet argues for PSR as an important methodological principle for theorizing in physics.[47] Then, in Leibnizian vein, she argues that any account admitting only extension to the essence of matter violates PSR, via the principle of the identity of indiscernibles, since such matter would be entirely homogeneous and so all parts of matter would be similar.[48] On the basis of PSR, we are to conclude that there must be something more to the essence of matter, such that the parts are discernible. Du Châtelet moves immediately to assert that by the addition of "force" to the essence of matter we ensure that PSR is satisfied.

The argument for this conclusion is extremely compressed, but seems to go something like this.[49] Suppose that matter were purely extension. Suppose that the parts of a portion of this matter, no matter how small, were all at rest. Then they would be entirely similar. But by PSR, this cannot be the case. Therefore all the parts of matter, no matter how small, must be in different states of motion. Moreover, and crucially, in order to satisfy PSR the properties that differentiate one part of matter from all the others must be "in" that part of matter: there must be "a real difference between all the parts of Matter." The source of this real difference is an "internal force, or force tending toward motion" that is in all parts of matter. The introduction of a

---

[45] Keill's colorful diatribe against Malebranche's occasionalism includes such rhetorical gems as this: "At this rate one need not fear his headpiece tho' a Bomb were falling upon it with all the force that Powder can give it, for it could not so much as break his Skull, or singe his hair, if God did not take that occasion to do it" (1698, 8–9).

[46] The following material is based on Brading 2019; a more detailed discussion can be found there.

[47] See Brading 2019, chapters 1 and 2.

[48] Du Châtelet 2018b, 8.139.

[49] Du Châtelet 2018b, 8.139.

force in all parts of matter seems to be a bit of a leap, but we can interpret it modestly as being whatever is needed in order for a part of matter to be always in its own distinct (though perhaps changing) state of motion. Neatly, this is also going to be what provides a body with the power to act. If we accept that all changes in bodies are re-arrangements of parts of matter, then the power of one body to act on others is its power to rearrange them (or their parts). Du Châtelet identifies the force of a body tending toward motion, introduced in order to satisfy PSR, with the force by which it is able to act on others. We will not pause here to analyze whether this argument is any good, for our goal is to outline the main contours of her position. Her purpose is to provide an account of the nature of bodies such that they are capable of acting on one another. She argues that PSR requires enhancing the essence of body with something beyond extension, and so adds force of acting, which she calls "motive force."[50]

The addition of motive force as an essential property of matter does not complete the account of bodies, and of how they act on one another. Du Châtelet argues, on the basis of both reason and experiment, that matter has also a passive force.[51] For, how can one body act on a second unless the second resists?

Du Châtelet maintains that all changes happening in bodies can be explained by appeal to extension, active (or motive) force, and passive (or resisting) force, and that these three principles are mutually independent and jointly necessary and sufficient for an account of the nature of body.[52] Moreover, just as the extension of bodies is a phenomenon, arising from how we experience the simples (see section 5.2), so too are the active and passive forces of bodies: they are phenomena arising from the active and passive force of simples, through how we experience those simples. Referencing Leibniz, Du Châtelet distinguishes between primitive force, associated with simples, and derivative force, associated with bodies.[53] In arriving at this account of Nature, Du Châtelet adopted PSR (Principle) as a powerful tool, while also appealing to general empirical considerations (Evidence). And she is now in a position to address Action, thereby completing her solution to BODY. For this, we turn our attention to PCOL, and to her philosophical mechanics of collisions.

[50] Du Châtelet 2018b, 8.141.
[51] Du Châtelet 2018b, 8.142.
[52] Du Châtelet 2018b, 8.145–149.
[53] Du Châtelet 2018b, 8.155–158.

(ii) Du Châtelet on collisions

Du Châtelet asserts: "The active force and the passive force of Bodies is modified in their collision, according to certain laws that can be reduced to three principles."[54] And she then states her version of Newton's laws of motion.[55]

Important differences in wording from Newton's formulations notwithstanding, once Du Châtelet has Newton's laws, she can solve all the problems that fall under Newton's laws of motion, including the set of collision problems soluble using those laws.

What remains is to use these resources to provide a *causal-explanatory* account of the changes in motions of the bodies, consistent with the laws of motion. This is to be done using the theory of active and passive forces. Du Châtelet's account of the collision process proceeds as follows. At the end of chapter 11, she writes:

> When a moving body encounters an obstacle, it strives to displace this obstacle; if this effort is destroyed by an invincible resistance, the force of this body is a *force morte*, that is to say, it does not produce any effect, but it only tends to produce one. If the resistance is not invincible, the force then is *force vive*, for it produces a real effect, and this effect is called the effect of the force of this body. (Du Châtelet 2009, 11.268)

We have seen that all bodies have associated with them an active force, by which they move (or strive to move), and a passive force, by which they resist motion (or changes in motion). Du Châtelet here explains that the active force of a body manifests itself in two ways: as dead force (*force morte*), when the body strives to move but fails (due to an obstacle), and as living force (*force vive*), when the body is in motion. Du Châtelet further develops her account of the forces of bodies in chapters 20 and 21, and we arrive at the following account. When two bodies collide, during the time when they are in contact, they press upon one another by means of dead force, and in so doing they impress potential speed, or a tendency to motion, into one another. Eventually, all the active force of one body is used up, and it can no longer counteract the pressure of the other through its own dead force. The other body continues to impress active force, and the first body begins to

---

54 Du Châtelet 2009, 11.229.
55 See again Du Châtelet 2009, 11.229.

move—with living force—back in the direction of the impressed force. If it happens that the active forces are equal, then they destroy each other and the bodies remain at rest. If it happens that the two bodies are moving initially in the same direction, the dead force impressed by the faster body on the slower becomes a living force, and the slower body increases its speed. If it happens that the line of impact is oblique, the outcome of the collision is calculated by the components of force acting along the line of impact.

Our primary focus here is Du Châtelet's attempt to render the collision process intelligible by appeal to a complex theory of active and passive, dead and living, forces. Qualitatively, her account may seem promising. However, from the perspective of philosophical mechanics, this causal explanation succeeds only if it can be integrated with the rules of collision from mechanics. Again, her account looks promising because of her attempt to unify her theory of forces with a version of Newton's laws of motion. Ultimately, however, Du Châtelet's attempt fails.[56] This is partly, though not fundamentally, because it is incomplete: like her predecessors, she distinguished bodies into hard, soft, and elastic, but she did not explicitly develop her account for all three.[57] More importantly, however, her defense of *vis viva* in the concluding chapter of the *Foundations* leads her to disconnect her qualitative story from her quantitative account, so that the former is no longer explanatory with respect to the latter. This is what undermines her attempted unification.

By the middle of the 18th century, hers was the most developed attempt to provide a philosophical mechanics of bodily action, and yet it failed. With the benefit of over 250 years of hindsight, we can hope to pinpoint the philosophical reasons as to why, but in the 1740s the debate remained live.[58]

[56] For analysis, see Brading 2019, 95–7.

[57] See Du Châtelet 2009, 11.267: "Bodies receiving or communicating motion can be either completely hard, that is to say, incapable of compression, or completely soft, that is to say, incapable of reconstitution after the compression of their particles, or again elastic, that is to say, capable of regaining their original shape after the compression."

[58] The two most important open questions, for our purposes, were these: (1) Is an account of bodies in terms of active and passive forces necessary and/or sufficient for explaining bodily action? (2) Is an account of the forces of bodies as derivative, deriving from an underlying theory of non-extended simples, necessary and/or sufficient for explaining bodily action?

5.5 Euler and Action

Euler is an important figure in our story for his work explicitly spans all of philosophical mechanics. While the majority of his work lies primarily in rational mechanics (about which more in later chapters), he had a long-standing interest in philosophical physics: in the nature and properties of bodies, and the search for causes.[59] In 1741, Euler read Du Châtelet's *Foundations*. He wrote to her from Berlin, saying that he greatly admired her chapter on hypotheses, for hypotheses are the means by which we can arrive at *certain knowledge of physical causes*.[60] The letter continues with Euler announcing that he will offer his thoughts on the topic of bodies and their motions. All indications are that he wrote an extensive discussion of this topic, directly engaging with Du Châtelet's account of bodies, but this portion of the letter is, sadly, lost.[61]

Central to the Berlin Academy monad dispute[62] is Euler's rejection of the Leibnizian theory of simples with their active and passive forces, as advocated both by Wolff and his followers, and by Du Châtelet.[63] Beginning with his paper "Thoughts on the elements of bodies, in which the theory of simple things and monads is examined and the true essence of bodies is discovered," Euler argued for an alternative account of the "true essence of bodies."[64] He used collisions as a test of adequacy for his account, and claimed that we do not need the Leibnizian theory of active and passive forces in order to explain them. In the decade or so that followed, Euler developed and refined his account, and in the process he recast physics and its relationship to mechanics.[65]

---

[59] See especially his *Letters to a German Princess* and his *Natural Philosophy*. The literature on Euler's natural philosophy is small. Of particular relevance to our topic are Gaukroger 1982 and Watkins 1997, who discuss Euler's treatment of bodies, force, and impenetrability, especially in relation to Descartes, Newton, and Wolff. A general systematic treatment of Euler's natural philosophy has yet to appear.

[60] Du Châtelet 2018a, letter 380.

[61] He writes: "I begin with the first principle of mechanics, that all bodies by themselves remain in their state of rest or of motion. To this property one can well give the name force, when one does not say that all force is a tendency to change state, as does Mr. Wolff. All body is thus for you . . . " (Du Châtelet 2018a, 93). Frustratingly, the extant manuscript of the letter ends here, just as it was getting interesting for our present purposes.

[62] See section 5.2 and references therein.

[63] For Euler's involvement in the ongoing "monad dispute", see Calinger 2016, chapter 8.

[64] See Euler 1746b.

[65] See below and Chapter 6. We thank those present in KB's graduate seminar, fall 2018, in which the group worked on and discussed the natural philosophy of Euler.

(i) Euler on the force of bodies

In 1746, Euler argued that empirical investigations of the phenomena of
bodily motions show that bodies remain in the same state unless acted upon
externally. In the spirit of Newton's *Principia*, he recognized "*vis inertiae*" as
an essential property of bodies:

> The force of all bodies to remain in their state is called in the theory of mo-
> tion *vis inertiae*, and it is as general a property of bodies as extension, such
> that a body without this force would cease to be a body. (Euler 1746b, part
> II, §16)

Euler then argues for the identity of this force of continuing in the same state
with the force of resistance to change of state:[66]

> 17. It is not possible to imagine this force that enables bodies to remain in
> their state, without at the same time ascribing to them a force to resist all
> changes. For were a body to undergo all changes without resisting them in
> the slightest, one could not say that it is endowed with a force to remain in
> its state.
> 18. Since these two forces are of necessity connected with each other, and
> cannot be separated, it is the same force through which a body remains in
> its state, and through which it resists all change.
> 19. From this is clear that if the bodies were deprived of this force, they
> would have to undergo all changes without any resistance and there would
> be no impact and quite generally no resistance in the world; in consequence
> it would be as if the bodies could freely interpenetrate each other, and the
> concept of impenetrability, which is as important a property of bodies as
> extension, would cease to apply. (Euler 1746b, part II)

For Euler, bodies are extended, have a force of inertia, and are impenetrable: these three properties of bodies are needed for collisions between
bodies to be possible. This is his answer to Nature.

Euler argues that these three are sufficient for the explanation of
collisions: given impenetrability, the force of inertia is sufficient to yield

---

[66] In his 1746b, Euler refers to inertia as a force, but he soon modifies this language (see below). Cf.
also McMullin 1978, 36.

changes in the state of bodies on impact, and no further active force need be postulated. His argument is as follows.[67] Consider two bodies, one at rest and the other moving toward the first along a line of impact. In such a situation, it is impossible for both bodies to remain in their same state: in order for the former to remain in its state of rest, the latter would also have to come to rest; in order for the latter to remain in the same state of motion, the former would have to be moved out of the way by the impact. The reason why these changes of state occur, the very *cause* of these changes, Euler writes, is nothing other than *vis inertiae*. The same argument holds for any number of bodies in any states of motion where the lines of motion intersect. Having postulated *vis inertiae*, we have given the reason and the cause for the changes in the states of motion undergone by colliding bodies.

Contrary to those pursuing a Leibnizian approach to bodies, such as Wolff and Du Châtelet, Euler maintains that no further active force need be postulated in order to explain collisions. Indeed, he argues that postulation of an active force in bodies contradicts the principle of inertia, by which bodies are essentially passive. Euler retained this rejection of active force throughout the later developments in his account of bodies.[68]

Euler revisits and revises his 1746 treatment of bodies in his 1752 paper "Recherches sur l'origine des forces." There are three important differences concerning: (1) the status and conceptualization of inertia; (2) the role of impenetrability in the account of body and collisions; and (3) the desiderata for an adequate account of impact, and the role of physics and mechanics therein.

The first change that Euler makes concerns inertia. Euler makes the important move of denying that inertia is a force and reserving the term "force" for any cause that is capable of *changing* the state of a body.[69] Inertia, in contrast to force, is that property of a body by which it remains in the *same* state.[70]

---

[67] See Euler 1746b, §§23–28.

[68] See, for example, chapter 4 of Euler 1862. He used this general style of argument to reject not just Leibnizian active force but also gravity as an active principle in matter, and thought as a property of matter (Euler 1746b). The latter shows the importance of this general argument for Euler, beyond mechanics and physics. In his 1746d Euler argues that, to show that matter cannot think, it is sufficient to show the incompatibility of the property of thought with another accepted property of body. Since inertia is a force that preserves the state of a body (and is accepted to be so by all parties in the thinking matter dispute), and thought is a force contrary to inertia, bodies cannot possess both inertia and thought, or so Euler argues. (In this paper, Euler refers not only to Wolff, but also to Martin Knutzen, who is most famous today for having been Kant's teacher. Euler and Knutzen maintained an extensive correspondence, but the latter never mentioned Kant.)

[69] Euler 1752b, §8.

[70] In his *Principia*, Newton equated "*vis inertiae*" with "*vis insita*" in definition 3. He also introduced "*vis impressa*" in definition 4, and "*vis centripeta*" in definition 5. Euler's clarification

Calling inertia a force misleadingly implies some kind of action or activity in the body, whereas inertia is a property that *opposes* all changes of state. Forces that change the state of a body are to be sought *outside* the body that undergoes the change. Research in mechanics should seek the changes that are produced in the motions of bodies by a given force, on the one hand, and the forces that are responsible for given changes in the motions of bodies, on the other.[71]

The importance of this for natural philosophy, and for the conceptual foundations of physics, is perhaps not evident in the 1752 paper, but Euler's unpublished *Natural Philosophy*, written around a decade later, makes it clear. In chapter 1, entitled "On Natural Philosophy in General," he sets out the aim and scope of natural philosophy, placing himself squarely within the conception of physics dominant in the early 18th century (see Chapter 4). He writes:

1. Natural philosophy is a science [*Wissenschaft*] that aims to explain the causes [*Ursachen*] of changes that occur in bodies.
2. All changes occurring in bodies must have their reason [*Grund*] in the very essence and properties of bodies.
3. Hence, before all else, we must strive to investigate the essence and properties of bodies. (Euler 1862, 449f.)

The subject-matter of natural philosophy is therefore bodies and their properties, with a particular emphasis on causes. This much is familiar (again, see Chapter 4).

Like Du Châtelet, Euler selects the PSR as a methodological tool for the adequacy of meeting the aim expressed in 1, but he uses it differently.[72] Immediately following his statement of the aim of natural philosophy, he writes:

Wherever there is a change, there must be a cause that brings it about, for it is certain that nothing can happen without a sufficient reason. Whoever can point to the reason [*Grund*] why a change has occurred, has found its

gives us the separation of inertia from external or impressed forces, as is now familiar in our current versions of classical mechanics.

[71] Compare Newton 1999, 382.
[72] See Euler 1960, letter 128, for his criticism of the use of PSR by the advocates of monads.

cause [*Ursache*], and thus fulfills the ultimate aim of natural philosophy. (Euler 1862, 449)

However, he next emphasizes a very interesting difference between his conception of natural philosophy and the dominant one from the early 18th century. He insists that the aim is to provide reasons for *changes* only:

Thus natural science aims to study changes only. For, while an object remains in the same state, we cannot point to any other reason for that but the absence of causes that could bring about any changes. But as soon as a change occurs, we are entitled to ask for its cause. (Euler 1862, 449)

This has two consequences. First, it means that the search for the properties of bodies is subservient to the search for the *causes of changes* in bodies: what is required of natural philosophy is to determine *all and only* those properties of bodies relevant to the *changes* that bodies undergo. Second, it means that we are not required to explain those properties themselves: we do not need to say *how* it is that a body is extended, impenetrable, and mobile. Thus, Euler envisages a rather different application of PSR in natural philosophy than Du Châtelet. Where Du Châtelet sought an explanation of the extension of bodies (see section 5.2), for Euler a list of the properties of bodies is sufficient so long as those properties can be used to explain the changes that such bodies undergo. Though it may seem innocuous, in fact this is a very different conception of the goals and scope of physics. From the perspective of our analytical framework, it is a transformation of Evidence: with the change in goals and scope of physics, the justificatory task with respect to Nature and Action is recast. The consequences are far-reaching, as later chapters will show.

This account of force and inertia is the first of three important changes in his position that Euler makes in his 1752 paper. With this one in place, Euler turns to his main topic: the origin of the force by which a body is compelled to change its state when it collides with another body. Here, the two further important differences between the 1746 and 1752 papers emerge.

(ii) Euler's physics of collisions

Euler begins his 1752 discussion of collisions with the general case discussed in his 1746 paper, in which one body is on a collision course with a second body, and from which we can already conclude that a change in state of at least one body must occur.[73] He notes that this qualitative conclusion is completely general, following regardless of whether the bodies are elastic or completely inelastic, and argues that the crucial property of bodies that makes collisions possible is impenetrability.[74] Without this, bodies would pass through one another: it is impenetrability that prevents them from doing so. Moreover, this is what prevents bodies from remaining in the same state of motion, and therefore what requires them to change their state. Impenetrability, Euler claims, is an essential property of bodies, in the same sense as extension and inertia: without these properties, bodies would not be bodies.[75] If there were perfectly penetrable bodies, Euler goes on, these would be just like bits of empty space: for a body to occupy a place is for it to exclude other bodies from that place. Moreover, the impenetrability of bodies does not admit of degrees, for if it is not completely impenetrable, then it is penetrable. Therefore, the impenetrability of bodies is absolute.[76]

The conclusion of this first part of Euler's argument is that impenetrability is the reason why bodies change their state when their trajectories intersect. Euler then asks: what is the origin of the force that brings about these changes in state?[77] First, Euler asks whether this force is necessarily found with impenetrability, or whether the two are separable. His answer is that this force is necessarily coextensive with impenetrability and that they are absolutely inseparable. For, without this identification one could have impenetrability without any changes in motion, but that is not possible; and one could have changes in motion without impenetrability, but that is not possible either. We must therefore conclude that impenetrability is the origin of the force by which bodies undergo changes in their states of rest or motion, he avers.

This is the second important development from the 1746 paper. In that early piece, Euler had appealed to inertia in giving his causal account of the collision process. In 1752, Euler denies that inertia is a force and turns to

---

[73] Euler 1752b, §11.
[74] Euler 1752b, §12.
[75] Euler 1752b, §15. Euler there points out that without impenetrability, bodies would not be sensible. Cf. Newton's remarks in his manuscript "De Gravitatione" (Newton 2014, 42).
[76] Euler 1752b, §§16–17.
[77] Euler 1752b, §18.

impenetrability for the origin of the force of collisions. Moreover, he argues that impenetrability is *sufficient* for a physics of impact, as we will now see.[78]

To proceed, Euler notes that the force associated with impenetrability acts only when there is a threat of penetration and does not otherwise manifest itself. Hence, this force does not act continually (contrary to those—such as Wolff and Du Châtelet—who say that a force is a continual effort to change state).[79] The change of state that it produces in the second body is an indirect effect, arising as a consequence of the first body maintaining its impenetrability. However, it is a mistake to think of one body acting on the other, rather than of a mutual action. The force that arises from impenetrability when two bodies collide arises not from one body, but from the two bodies jointly. For, if one body was penetrable, then there would be no need for a change of state of either body, and no need for a force to operate. So it is the impenetrability of both bodies jointly that results in the appearance of a force. Moreover, the force will be of exactly the magnitude and direction needed to prevent penetration: no more and no less.[80] Euler concludes that when two bodies A and B run into each other, penetration is prevented by A and B acting on one another with equal and opposite force. In other words, Newton's third law of motion (equal and opposite action and reaction) for the case of collisions follows necessarily from the nature of impenetrability. This result enables Euler to conclude that impenetrability alone is the source of the force that acts to change the states of bodies in collisions.[81] Moreover, it follows from this that there is no need for the active and motive forces of bodies used by Wolff and Du Châtelet in their accounts of the collision process, and Euler rejects them. The appeal to impenetrability remains a constant in his account of bodies going forward.[82]

These results do not enable a quantitative treatment of impact, nor the recovery of collision rules. In the 1746 paper discussed above,[83] Euler did not investigate the collision process and its relationship to rules of collision in any detail. He turned his attention to this in a second paper of that year,[84] but

[78] Nothing in the above commits Euler to any specific notion of contact action as an account of collisions. Rather, whatever is the process whereby penetration is avoided, this is the notion of collision at work. For Euler on impenetrability and force, see also Gaukroger 1982 and Watkins 1997, 329f.
[79] Euler 1752b, §§20–21.
[80] Euler 1752b, §§25–27.
[81] Euler 1752b, §§30–2.
[82] See Gaukroger 1982, 134–7.
[83] Euler 1746b.
[84] Euler 1746c offers an intervention in the so-called *vis viva* dispute. First, he challenged the idea that there are two competing measures of the force of bodies in motion. Contrary to the Leibnizians, Euler argued that these are not utterly different in kind, but instead both pressure and percussion

it is the 1752 paper that is of greater interest for our purposes. Here, Euler takes a new approach, and thereby makes a critical move for the philosophical mechanics of collisions. This is the third important difference between the 1746 and 1752 treatments. In 1752, according to Euler, the *qualitative causal account* of collisions—in terms of impenetrability—offers a *complete physics*. That is: we *do not need* a *quantitative* account of the causal process of collisions, or any micro-level causal story of this process, in order to complete our *physics* of collisions. In what follows, we unpack and examine this claim.

(iii) Euler's philosophical mechanics of collisions

For Euler, the quality of bodies relevant to the physics of collisions is impenetrability: it is the origin of the force by which bodies act on one another during impact. Impenetrability is absolute; it is all or nothing: it is not a quantity. Nevertheless, impenetrability is sufficient for our causal story of bodily collision, as described above, and that story is qualitative. This is all we need for physics.

Mechanics, however, is quantitative, and the quality of bodies relevant to the *quantitative* treatment of collisions is hardness. Hardness, Euler writes, concerns the extent to which a body can be compressed, and therefore how much it deforms when acted upon by another body.[85] It is a property that comes in quantitative degrees. If determining the change of state that comes about as a result of collisions required perfect knowledge of the deformation, such knowledge would perhaps be impossible, since the outcome would be different for every degree of hardness. But happily, Euler argues, regardless of how much a body deforms during collision, we need pay attention only to whether or not this deformation remains after the collision is complete. This is enough to provide a quantitative treatment of collision outcomes.

forces are species of the same kind. This follows Newton. Second, Euler challenged the view that the dispute can be resolved empirically by looking at the effects of the force of bodies in motion, showing the complexity involved in quantitative treatments of the collision process and the deformations undergone by bodies during collision. He endorsed inertia as adequate for an account of all the changes in the motions of bodies that we see around us, and rejected the existence of hard bodies as "repugnant to the order of nature" via an appeal to continuity. He also comments (in line with d'Alembert 1743) that the *vis viva* dispute is largely a dispute about words; see Euler 1746c, 31–7. We will not pursue these interesting remarks here. For more on the *vis viva* dispute, see Chapter 3 and references therein.

[85] Euler 1752b, §§33–34.

As a result, we need to distinguish between only two principal kinds of bodies: those that keep their deformation after impact (inelastic), and those that return to their original shape exactly (elastic).

Euler's quantitative treatment of collisions is as follows. For both elastic and inelastic impact, the first stage of the analysis is the same. We begin by considering the period during which the bodies are acting on one another, and in particular an element of this time dt during which the speed increments of the two bodies are dv (for body A) and du (for body B) along the line of collision. Requiring that the force acting during any interval of the collision is equal and opposite, Euler concludes from

$$m_A dv = -P dt \quad \text{and} \quad m_B du = P dt \tag{1}$$

that

$$m_A dv + m_B du = 0 \tag{2}$$

where m is the mass of the relevant body,[86] and the force P is defined as rate of change of momentum $m\, dv/dt$. From whence he concludes:

$$m_A v + m_B u = \text{constant.} \tag{3}$$

This result, Euler says, shows that momentum is conserved throughout the collision process, and therefore

$$m_A v + m_B u = m_A a + m_B b \tag{4}$$

where a and b are the initial pre-impact speeds of A and B, respectively, along the line of collision.

Next, Euler turns to the value of the force P, by which bodies are deformed through impact.[87] This is measured by the change in the distance between

---

[86] For the problems with "elusive mass," and Euler on mass, see Chapter 7.
[87] Euler 1752b, §§41ff.

the mass-centers of the two bodies, dz, which is in turn composed of the distance traveled by the center-of-mass of body A (let this be dx) during dt, and that traveled by the center-of-mass of body B (say dy) in dt. Then, since $dx = v\, dt$ and $dy = u\, dt$, we have, from (1):

$$m_A v\, dv = -P\, dx \quad \text{and} \quad m_B u\, du = P\, dy \tag{5}$$

Therefore,[88]

$$m_A v\, dv + m_B u\, du = -P\, dz. \tag{6}$$

Integrating, we have:

$$\tfrac{1}{2} m_A v^2 + \tfrac{1}{2} m_B u^2 = K - \int P\, dz \tag{7}$$

where K is a constant of integration. Now consider the special case of the first instant of collision, in order to determine K in (7). At this moment, $v = a$ and $u = b$. Moreover, at this instant, the deformation is zero, and hence $\int P\, dz = 0$. From this we conclude that[89]

$$m_A v^2 + m_B u^2 = m_A a^2 + \tfrac{1}{2} m_B b^2 - 2 \int P\, dz \tag{8}$$

Having established (4) and (8), which hold for all types of bodies, Euler next differentiates between elastic and inelastic bodies.

Bodies that are entirely without elasticity, he writes, cease to act on one another the moment that they acquire the same speed along the line of impact, regardless of whether they deform one another; $v = u$. Therefore we can solve for v and u in terms of the masses and initial speeds (along the line of impact) of the bodies.[90] He concludes:

---

[88] Euler 1752b, §42.
[89] Euler 1752b, §42.
[90] Euler 1752b, §44.

As a result it is clear that the change of state that inelastic bodies undergo in collision is solely caused by the force of impenetrability of these bodies, and that we do not need to look elsewhere for the cause. In addition, notice that this rule does not depend in any way on the degree of hardness of the bodies, or on the quantity of forces with which these bodies act on one another during the collision; and as these forces depend principally on the degree of hardness, it is all the more remarkable that they always produce the same effect, however large or small they may be. (Euler, 1752b, §44)

Of course, this conclusion is somewhat circular. The justification for the claims that (i) the force of collision is equal and opposite (used in the derivation of (4)), and (ii) it ceases to act the moment that the threat of penetration ceases (i.e., when the bodies acquire the same speed, in the second part of the derivation), lies in Euler's treatment of impenetrability given above. Nevertheless, given this account of impenetrability, we are able to justify those claims and arrive at the accepted rules of collision for inelastic bodies. In this way, his physics of collisions plays a crucial role in his account of the rules of collision, en route to a philosophical mechanics of impact.

Notice also that Euler is able to remove a puzzle left unsolved by the accounts of collision considered in Chapters 2 and 3. Recall from Chapter 2 that a geometrically inelastic body, whether hard or soft, is one that may or may not deform on impact, but if it does, then it does not regain its shape afterward. A kinematically inelastic body is one that does not rebound. The difficulty was to explain why g-inelastic bodies are also k-inelastic (and similarly for elastic bodies, see below). Euler offers an explanation in terms of impenetrability. The impenetrability of bodies gives rise to only as much force as is necessary to prevent mutual penetration *and no more*, and that is why g-inelastic bodies move together after impact.

Elastic-body collision differs from inelastic because the bodies do not cease to act on one another once they acquire a common speed, but continue to do so until they have returned to their original shapes. Once again, impenetrability is the origin of this continued action: with the two bodies in contact, during the time that each body is recovering its original shape the threat of penetration (due to the changing shape) remains, and so the bodies continue to act on one another. As a result of this ongoing mutual action, the state of motion of each body continues to change, and so the final speeds of the two bodies may not be equal. This is the reason for rebound. Thus, g-elasticity leads to k-elasticity, via impenetrability as the origin of mutual forces between colliding bodies.

Since the final speeds may not be equal, an additional equation is needed for a quantitative solution to the collision of elastic bodies. For this, Euler turns to his result (8), above, from which he arrives at conservation of energy by a further appeal to impenetrability, as follows. Since impenetrability is the source of the force of collision, the additional force arising in elastic collision is just that required to remove the threat of penetration as the bodies regain their original shapes: it ceases to act once the bodies have regained their original shapes. Thus, Euler writes, the additional force is exactly that required to restore the distance between the centers-of-mass of the two bodies to that at the moment of impact (and prior to any compression). That is to say, the total deformation z is zero, and since there is no change in z, it follows that $\int P\,dz$ = 0. Thus, from (8) we arrive at conservation of energy for elastic collisions:

$$m_A v^2 + m_B u^2 = m_A a^2 + m_B b^2 \tag{9}$$

and this provides us with our second equation, using which we can solve for our two unknowns (the final speeds) in terms of the initial speeds and masses.

Euler concludes his discussion of collisions as follows: "There remains no doubt that, in all the collisions of bodies generally, the change of state is caused solely by the forces that result necessarily from impenetrability."[91]

At first sight, this seems a disingenuous conclusion. In elastic bodies, their elasticity plays a role in the change of state: where impenetrability yields bodies moving with a common speed after collision, elasticity leads to final speeds that may be unequal. An additional "force of elasticity" seems to be at work. But this is to misunderstand a division that Euler is making between impenetrability and elasticity. Impenetrability is a property of bodies as such: it is absolute (it does not come in degrees); it is the causal basis of the mutual action of bodies in collision; and it therefore lies in the domain of physics. Elasticity, by contrast, is an empirical property observed in the behavior of composite bodies: it comes in degrees, and is to be studied empirically. Hardness, too, is observed to come in degrees, and is subject to

---

[91] Euler 1752b, §46.

empirical study. Thus, hardness and elasticity have a very different status in Euler's natural philosophy than impenetrability.

This brings us back to the third important feature of Euler's 1752 paper. We saw above that impenetrability is sufficient to yield a physics of collisions: it provides us with a causal explanation of the collision process (see section 5.4). We have now seen that it is also sufficient for a philosophical mechanics of collisions, for it enables us to derive collision rules (his two equations by which we can treat collisions quantitatively). From there, we proceed on a secure foundation to elaborate our mechanics (by providing equations for special cases, deriving corollaries, and so forth) and to pursue our empirical investigations of particular kinds of bodies (in which we explore hardness and elasticity, and the behavior of bodies with varying degrees of each).[92]

Euler's rules of collision, as derived above, are not as general as they might appear from our presentation. They depend on two crucial assumptions: that the bodies in collision are spherical (so that the initial area of contact can be assumed to be a point), and that the density distribution within each sphere is spherically symmetric. Euler does not discuss this in the 1752 paper, but in a later piece, "Research on the knowledge of bodies from mechanics," he is clear that consideration of the effects of different shapes and mass distributions on the outcomes of collisions belongs to the domain of mechanics (Euler 1765a). So despite the appearance of a quantitative treatment arising from the physics of impenetrability, what we have is the quantitative treatment of a highly idealized case. This is the bridge, constructed by the provision of a philosophical mechanics, over which we pass from the qualitative causal treatment of physics to the specific quantitative cases treated in mechanics.

In our opinion, Euler's views offer a macro-causal account of collisions without a micro-causal account of the process. Specifically, impenetrability is a property of the whole simpliciter (in contrast to hardness and elasticity, which are properties arising from the composition of the parts), and it is impenetrability that we appeal to in giving a causal account of collisions in general. This suffices for our physics. The particular outcome of a given collision, depending on the hardness/elasticity of the bodies involved (as well as other properties, such as fragility), is an empirical matter (as it was for Newton in his *Principia*), a matter for empirical science, but knowledge of

[92] Euler's matter theory has proper and subtle matter, so he might have an account of elasticity in terms of these, along the lines of Malebranche (see Chapter 2). This is an open question about Euler's philosophy that requires further investigation.

these particular cases does not affect our knowledge of the general physics of collisions.

Unlike with the approach based on active and passive forces (as seen in Du Châtelet, for example), we do not need to attempt to say how the forces are used up during the process of collision. And, unlike those who appealed to subtle matter, we do not attempt to explain the causal process of collisions in terms of a micro-causal story. All of these attempts failed, as we have seen, and Euler does not adopt them. However, we also do better than those Newtonians who pulled the principles or rules needed to solve collision problems out of the air, or attempted to justify them empirically (but failed, see Chapter 3). Instead, we justify the needed principles by appeal to impenetrability. In our assessment, in the context of the approaches available at the time, Euler's approach to collisions appears to be a promising proposal. Nevertheless, it comes at a significant cost, for it defers a causal account of collisions at the microscale to some later date and removes responsibility for it from the domain of philosophical physics. We discuss the implications of Euler's reconceptualization of physics, and his reconfiguring of the domain of physics in relation to mechanics, in Chapter 7.

5.6   Conclusions

BODY persisted as a research problem in natural philosophy from the beginning of the 18th century to its end. By the 1730s, it would have been clear that the issue demanded urgent attention (as we saw in Chapters 2–4), and so it is not surprising that leading figures of the mid- to late 18th century engaged with it. For all their disagreements over the nature and properties of bodies, these philosophers agreed that: (1) bodies are spatially extended, impenetrable, and have a quantity of matter, and (2) an adequate philosophical physics would provide a causal account of how bodies act on one another.

In this chapter, we have examined BODY in the 1730s–50s, focusing on Du Châtelet and Euler. We paid close attention to their accounts of collision because, as we argued in Chapter 4, addressing PCOL is a necessary condition for solving BODY. In their work, we find attempts to begin from a philosophical physics of impact, and to work from there to an account that integrates the rules of collision, yielding a philosophical mechanics. In our view, their proposals are the best available options in the mid-18th century. Both face serious difficulties. Indeed, we think they take us to the end of the road for a philosophical mechanics of collisions based on a physics of extended, impenetrable, material bodies acting on one another by means of contact forces.

In the early 18th century it was Newtonian gravity, conceived of as action-at-a-distance, that was controversial as a means of body–body action. As the century progressed, it became increasingly clear that collisions were equally, if not more, problematic. New alternatives emerged in the work of Boscovich and Kant, each of whom offered a radical departure from physics as it was then understood. We investigate these in the next chapter.

6

Searching for a new physics:
Kant and Boscovich

6.1 Introduction

In the 18th century, solutions to the *Problem of Bodies* (BODY) were expected
to ascertain the properties, causal powers, and generic behaviors of bodies
(Nature); explain how, if at all, bodies act on one another (Action); and
provide justification for their results (Evidence and Principle).[1] In earlier
chapters, we have seen the central role of collision theory in addressing BODY,
and the widespread search for a causal-explanatory account integrating philosophical physics[2] with rational mechanics: a *philosophical mechanics* of
collisions. In Chapter 5, we showed just how difficult this task proved to be,
and we argued that the two main lines of approach available in the middle of
the 18th century, defended by Du Châtelet and Euler, respectively, arrived at
insurmountable obstacles. We argued that this was the end of the road for a
philosophical mechanics of collisions based on a physics of extended, impenetrable, material bodies acting on one another by means of contact forces. No
such project is viable.

It is here that Boscovich and Kant enter the story. What was needed was
a new way to construct the spatially extended, impenetrable, mobile, and
causally interacting bodies of a satisfactory philosophical physics. Kant and
Boscovich each attempted just this. More than that, on the basis of their new
accounts of body they sought to construct a philosophical mechanics that included a treatment of collisions. This chapter assesses those efforts.

Our main conclusions are two. First, the approaches of Kant and Boscovich
are transformative as to the goals of physics in two respects: (i) they limit the

---

[1] See Chapter 4 for BODY. For explication of all the shorthand terminology that we use in this
chapter, including BODY, PCOL, Nature, Action, Evidence, and Principle, as well as the terms "philosophical physics" and "philosophical mechanics," see Chapter 1.
[2] Recall that physics then was the sub-discipline of philosophy charged with the study of body in
general (see Chapter 4). We use the label "philosophical physics" now and then to remind the reader
of this.

Philosophical Mechanics in the Age of Reason. Katherine Brading and Marius Stan, Oxford University Press.
© Oxford University Press 2023. DOI: 10.1093/oso/9780197678954.003.0006

causal account that we can expect from physics; and (ii) they problematize the objects of physics. We explain precisely what we mean by these claims.

Second, their treatments of collisions are at once both radical and disappointing. On the one hand, there is a dramatic reconceptualization of what contact action might amount to. On the other, we show that many of the same problems in handling collisions persist in these new accounts. The promise of a new physics opening a path to a successful philosophical mechanics of collisions is unrealized, or so we argue.

Despite these shortcomings, all is not lost. Boscovich's theory is forward-looking in a way that Kant's is not, for Boscovich attempts to incorporate not just collision theory but the latest developments in rational mechanics more generally. These developments are our focus in later chapters of this book; we return briefly to Boscovich in Chapter 9.

We begin with the accounts of bodies offered by Kant and Boscovich (section 6.2). We next turn our attention to philosophical mechanics, in first Kant (section 6.3) and then Boscovich (section 6.4). We summarize our conclusions in section 6.5.

6.2 The physics of bodies in Kant and Boscovich

In Chapter 5, we saw Euler pose a challenge to those who, following Leibniz, attempted to construct bodies from monads or non-extended simples of any kind:

> The proponents of simple things argue that the simple things, that constitute a body, are at a distance from each other, and because of this distance have extent. However, if all these simple things were at a distance from each other, with nothing in between them, there would be nothing to prevent them from being driven together so closely that no distance between them remained. (Euler 1862, §13)

The challenge is explicit: what prevents non-extended simples from coming together so closely that no extension results from their composition?[3]

---

[3] Though the above statement comes from the *Natural Philosophy*, published only posthumously, Euler's concerns over how non-extended monads could compose an extended body were well-known: see Chapter 5. The analogous issue of whether points can compose a line was also a familiar topic at the time. We can distinguish two questions: (1) What keeps simples from coming so close together that they all collapse into a point? (2) What keeps them from being so tightly packed that

Boscovich and Kant share a common strategy: they combine non-extended simples with action-at-a-distance forces to arrive at extended bodies.[4] Boscovich introduced his proposal in 1745 in the context of the ongoing debates over collision, and developed it in A *Theory of Natural Philosophy* of 1758.[5] Kant offered his 1756 *Physical Monadology* explicitly as a solution to the problem of identifying the ultimate parts of bodies, and the means whereby these parts yield corporeal extension.[6] The differences between their accounts of bodies are important, but so too are the commonalities in their solutions to the problem of the extension and divisibility of bodies. Their theories have been studied in depth by philosophers. Our purpose here is to assess their proposals as contributions to BODY.

(i) Kant on bodies

His earliest account is in his *Physical Monadology*, a paper of 1756 whose full title reads, "The employment in natural philosophy of metaphysics combined with geometry, of which sample *I* contains the physical monadology." Kant there aligns himself with those who, following Newton, advocated the use of experience and geometry together as a means of avoiding error in natural philosophy. But he then moves to distance himself from some, such as Musschenbroek and 's Gravesande, for whom this method places the laws of nature at the epistemic limit of our inquiries (see Chapter 4). In a highly revealing passage, he writes:

if we follow this sound path, we can exhibit the laws of nature though not the origin and causes of these laws. For those who only hunt out the phenomena of nature are always that far removed from the deeper understanding of the first causes. Nor will they ever attain knowledge of the nature itself of bodies. (Kant 1992, 51)

they occupy no finite interval (though they may take up an infinitesimal stretch *ds*). To (1), both Baumgarten and young Kant gave solutions. For Baumgarten, see Watkins 2006; for the young Kant, see Stan forthcoming. To (2), only Kant did, but he required action-at-a-distance forces to do so, as we shall see.

[4] If the former are considered "Leibnizian" in origin, and the latter "Newtonian," then each drew on both traditions to offer a novel way forward. Boscovich 1922 is explicit about this: see part I, §2. Whether Boscovich's "force" is best understood as such is discussed below (section 6.4).

[5] See Boscovich 1745 and 1922, respectively.

[6] See Kant 1910 and 1992, respectively.

Kant's concern is that we will lack a complete knowledge of causes, and of the nature of bodies. We return to this below when we discuss the consequences of his view for physics. Kant assigns the search for "first causes" and the nature of bodies to metaphysics: "Metaphysics, therefore, which many say may be properly absent from physics is, in fact, its only support." More specifically, we require metaphysics to address the following problem:

> For bodies consist of parts; it is certainly of no little importance that it be clearly established of which parts; and in what way they are combined together, and whether they fill space merely by the co-presence of their primitive parts or by the reciprocal conflict of their forces. (Kant 1992, 51)

According to Kant, tackling this will require the union of metaphysics with geometry, or (equivalently, it seems) of transcendental philosophy with geometry.[7]

Difficulties arise immediately, Kant says: (1) geometry asserts that space is infinitely divisible, while metaphysics denies this; (2) geometry asserts that "empty space is necessary for free motion," while metaphysics rejects empty space; (3) geometry endorses universal attraction by means of forces inherent in bodies acting at a distance, whereas metaphysics rejects this in favor of mechanical causes. The first of these difficulties, more specifically stated, is how to reconcile the geometrical infinite divisibility of space with the physical extension of bodies. The bodies of physics are presumed to be extended, but how is this extension possible?[8]

Toward addressing this question, Kant opens Section I of his *Physical Monadology* by arguing as follows:

1. Bodies consist of parts, each of which separately has an enduring existence.
2. The composition of parts is nothing but a relation that can cease to exist without the parts ceasing to exist.
3. When all composition relations are abolished, the parts that are left are non-composite.
4. Therefore, all bodies consist of non-composite parts.

---

[7] See this famous statement: "But how, in this business, can metaphysics be married to geometry, when it seems easier to mate griffins with horses than to unite transcendental philosophy with geometry?" (Kant 1992, 51).

[8] Kant 1992, 53. See Chapter 5 for earlier attempts to address this issue.

Thereby Kant introduces his non-extended physical monad, located in space at a (non-extended) point. He refers to this point as the "place" taken up by the "mere positing of a substance," viz. by simply asserting it to exist, without regard to its causal activities.[9] Kant's monads are capable of motion in space, over time.

Next, he introduces additional properties. For now, the relevant ones are two forces, attractive and repulsive, each with its origin at the location of the monad. This enables him to associate an extended spatial region with each monad: its sphere of activity. Specifically, each monad has associated with it a finite-sized spherical volume whose bounding surface (call it S) lies where the monad's repulsive and attractive forces balance each other. At any point below this surface, there is net repulsion. That is, any other monad must expend momentum to reach below S, and when it comes to rest there it will be scattered back. At the center of S, the repulsive force becomes infinite.

This innovation—distinguishing between the *place* occupied by the *substance* and the *space* occupied by its *activity*—ingeniously allows Kant to evade the threat of contradiction arising from the divisibility of extension. To see this, we must distinguish two ways to claim divisibility. Both are conceptual, but it turns out that they differ importantly. In the first, we imagine slicing through our volume with a blade that has no width. Call this "slicing." In the second, we not only slice through our volume, but separate the resulting parts from one another spatially. Call this "breaking." More precisely, the two types of divisibility are these:

s-divisibility; slicing: to conceive a mental plane intersecting a material volume. Mutatis mutandis for material surfaces or lines.[10]

b-divisibility; breaking: to conceive of a separation of parts divided by slicing, either by spatial separation or by annihilation of one part.

---

[9] See Kant 1910, 483. As it happens, Newton's *Principia* makes possible the denial of premise 1 and, as we saw in Chapter 5, Euler explicitly denies it. However, Kant adopts it as a premise, and it would have been similarly accepted by many of his contemporaries. Kant notes explicitly that he is not using the Principle of Sufficient Reason in his derivation of simples, contra Wolff (and Du Châtelet).

[10] Imagine a loaf of bread and you driving a wide, thin blade through it. Leave the blade in, then do two mental operations: imagine the blade's thickness shrunk to zero (suppose it to be as thick as a line in Euclidean geometry); and abstract completely from its specific material, mass, temperature, etc. If the material is a plane, let the blade's analogue be a mental line; and for a material line, let it be a (mental) point. Soon after Kant, this approach became the canonical way for research in the mechanics of continuous media: we posit a mental plane—nowadays named an "Euler Cut"—to run through a control volume of material, and we study the stresses and torques crossing that plane in any given direction.

Here is what the distinction entails. The substance of the physical monad is itself indivisible in both senses: it can be neither sliced nor broken.[11] And, while the space it fills (its sphere of activity) is divisible, it is so in a safe sense: it can be sliced but it cannot be broken.

The extended bodies of our experience are composed of physical monads: they are aggregates of physical monads in an equilibrium configuration. For any two monads J and K, there is a number r such that, if K is placed at r units of distance away from J, the attraction and repulsion from J (on K) cancel out; thus, K will remain at rest there. (Evidently, there are many such locations; placing K anywhere on the bounding surface of J's sphere of activity will do it.) Vice versa for J in respect to K. So, having the monads J and K be at relative rest and mutual distance r is to have a composite body JK. And so on, for any finite number of physical monads; except that the equilibrium distance r_n (that separates a component monad from its nearest neighbors) will vary with the number n of monads in that particular composite body. In modern terms, a composite body is a lattice of mass points on which the net force is individually zero.

The material substances of Kant's physical monadology are both simple (i.e., part-less) and extended (albeit in a peculiar sense, as we have seen). Kant calls these substances *physical* monads, to signal that, unlike Leibniz's monads, they are not mind-like. Moreover, unlike the simples of either Wolff or Du Châtelet, there is no lack of clarity about their metaphysical status: they are physical and they are located in space and time. Kant's physical monads are assumed to be mobile, endowed with momentum, and able to respond to—as well as exert—impressed force. Any individual monad, with its sphere of activity, is a body. From here, it is a short step to the bodies of our experience: perceptible bodies are simply a subset of all composite bodies. They are composed of physical monads.

(ii) Boscovich on bodies

In *A Theory of Natural Philosophy*, first published in 1758, Boscovich offered a similar account of the extension of bodies.[12] He accepts extension as a

---

[11] Bear in mind that "slicing" requires mentally producing a *lesser*-dimensional object inside the thing to be "sliced" (i.e., a plane slices a volume, a line slices a plane, etc.). But a point is a zero-dimensional object. So, "slicing" it is not *conceptually* possible. Neither is "breaking" it, since a point has no proper parts to separate or annihilate. For further explanation, see Stan (forthcoming, part II).

[12] See Boscovich 1922.

corporeal property, and accounts for it by appeal to non-extended physical, or material, points that stand in distance relations to one another as a result of the second primitive of Boscovich's matter theory, force. For him, each such point has an associated *single* force that alternates with distance—between being repulsive and attractive—and when the forces between a collection of points are in equilibrium, the upshot is an extended body (see below for details).

In Book III, Boscovich raises the issue of the divisibility of extension, and tackles it with ease. For, while geometrical space is infinitely divisible, extended matter is not: "as soon as we reach intervals that are less than the distance between two points, further sections will cut these empty intervals and not matter."[13] The upshot is physical extension that can be sliced, but not broken into spatially extended parts. In other words, we can continue to slice the space occupied by the physical points and their fields of force, but such conceptual slicing does not divide the matter present in that region into *parts of matter*. Like Kant, Boscovich begins with non-extended simples that cannot be divided, and then the extended bodies of our experience arise from composites of simples standing in force relations to one another. A Boscovichian body arises from a finite number of simples, and thus allows for a finite number of divisions before further attempts at division no longer yield bodies as the resulting parts.[14] The differences between Kant's and Boscovich's accounts of bodies are important, but so too are the above commonalities in their solutions to the difficulties associated with the extension and divisibility of bodies.

Besides being extended, the bodies of physics were presumed to be mobile and impenetrable. The issues associated with mobility divide into two groups. First, there is the metaphysical question about the nature of motion, most famously expressed in the debate over absolute versus relative motion. What is it for a body to move? Change of place in absolute space? Change of distance relations with respect to other bodies? This question remained a matter of dispute in the late 18th century and beyond, and we touch on Kant's involvement briefly, later.[15] The second is the ability of bodies to act on one another through motion, including resisting the actions of other bodies

[13] Boscovich 1922, §§391–3.

[14] This was also Du Châtelet's (1740) solution, but she does not posit forces between her simples.

[15] Boscovich's discussion of space and time can be found in Boscovich 1922, supplements I and II. See also Guzzardi 2020 and references therein. Kant's contributions to the absolute versus relative space, time, and motion debate have been widely discussed in the literature. See, for example, Friedman 2013 and references therein.

when their mutual motions are impeded. For this, collisions are of course the primary testing ground, and so we turn to the problem of collisions (PCOL) below.

For all of the figures studied in the preceding chapters, impenetrability was accepted as a primitive property of bodies. In the theories of Boscovich and Kant this is no longer the case. Scott writes that "Boscovich used his own conception of 'impenetrability' as the theme" whereby to achieve a middle way "between the theories of the Newtonians and the Leibnizians."[16] Having constructed bodies from physical points endowed with a force, Boscovich states:

> The Impenetrability of bodies comes naturally from my *Theory*. For, if re-pulsive forces act at very small distances, and these forces increase indefi-nitely as the distances decrease, so that they are capable of destroying any velocity however large; then there never can be any finite force, or velocity, that can make the distance between two points vanish, as is required for compenetration. (Boscovich 1922, §360)

In other words, the force associated with the monad becomes repulsive at short distances, exponentially so with decreasing distance. One body excludes another from the space that it occupies in virtue of this repulsive force. This is a dynamical, rather than geometric, conception of impenetra-bility.[17] Though Kant does not dwell on it, a similar view of impenetrability follows naturally from his physical monadology.

This revision to impenetrability will be important for Boscovich's and Kant's theories of collisions. Scott says that for Boscovich impenetrability is accounted for in terms of forces rather than matter, but this seems to un-derplay how radical the proposal is.[18] For both Boscovich and Kant, the im-penetrability of extended bodies is *derivative*, an appearance manifest in the motions of bodies that arises due to the forces acting between non-extended material points. Every particle interacts with every other particle by means of action at a distance; contact action—the corollary of impenetrability—never takes place. Jammer puts the point well:

---

[16] Scott 1970, 65.
[17] Cf. discussion of "quantity of matter," section 6.4.
[18] Scott 1970, 65.

it is the same to assert that contact never takes place or to say that it always takes place, since two bodies have always a dynamic connection that depends on their relative distance alone.... Boscovich's theory thus reduces contact phenomena to actions at a distance and consequently eliminates impact as a fundamental concept of mechanics. (Jammer 1957, 157)

That is, impact—which a century earlier counted as a primitive, and as the *only* intelligible means whereby one body might act on another—now requires detailed theorizing in terms of action-at-a-distance forces.[19] We turn to Kant's and Boscovich's attempts to address PCOL shortly (see sections 6.3 and 6.4).

(iii) Assessment

First, however, it is appropriate to assess their attempts at a constructive account of bodies from the perspective of philosophical physics. Recall from Chapters 1–5, physics was expected to provide a causal account of bodies—of their natures, properties, and actions. To what extent do they achieve this goal?

We have seen that, for Boscovich and Kant, Nature (specifically, extension and impenetrability) is intimately tied to Action: contact action is replaced by forces emanating from physical points; the points act on one another at a distance, and the interaction of bodies derives from this point–point interaction. This picture seems comfortingly familiar in light of 20th century particle physics (at least as presented in books of popular science), thought of in terms of point particles and forces. That familiarity should not be allowed to obscure our view of what has happened: ostensibly, for both Boscovich and Kant, extended bodies remain the primary objects of physics, yet their solution to the problem of divisibility undermines that enterprise. Extended bodies are now composites, arising from extensionless physical points and spheres of activity or forces. This transforms not just the account of Nature, but the conception of physics itself. Their transformation in physics is twofold.

First, such physical monadologies leave natural philosophers with a dilemma. On the one hand, we might retain bodies as the subject-matter of

---

[19] For more on the complexities of contact action, see Chapter 7.

physics, but concede that these objects are derivative, constituted from a different kind of object whose study lies outside the domain of physics. A complete account of bodies would then require resources from beyond physics. Earlier in the 18th century, physics was the area of philosophy tasked with a complete account of bodies (their nature, properties, causes, effects, behaviors, etc.). Moreover, where such bodies were constituted from an underlying theory of matter, they were regions or parts of that matter, with matter and bodies sharing the same properties and investigable by the same methods. But now it turns out that we need to appeal to resources from *metaphysics*—such as simples—that lie forever beyond the reach of physics (i.e., beyond the means of inquiry available to physics). This approach, familiar from Leibniz, is clear in Kant. Physics cannot stand on its own: it requires a "metaphysical foundation." On the other hand, we might adopt physical monads and the forces between them as the proper subject-matter of physics. Bodies, constructed out of these monads and forces, can be studied by physics, but they are no longer physics' primary objects. This approach can be reached by considering Boscovich's position for, unlike Leibniz and Kant, his simples are in principle empirically accessible by means of investigating the force law.[20] This offers a route to the physical simples as objects of physics. Such simples (and their forces) are then the primary objects of physics, with bodies being a derivative ontology. Either way, the enterprise of physics is transformed: in the latter case, bodies are no longer its primary subject-matter; in the former, they remain so, but a complete account of bodies is now beyond its scope. This is one of the ways in which physics as the study of extended bodies disappears from existence as a viable philosophical enterprise by 1800. Insofar as bodies are extended, mobile, and impenetrable, they are no longer viable as the basic object of physics.

There is a second respect in which the general approach to BODY found in Boscovich and Kant transforms physics. We saw above Kant's concern that the Newtonians' methods are limited when it comes to causal knowledge:

if we follow this sound path, we can exhibit the laws of nature though not the origin and causes of these laws. For those who only hunt out the phenomena of nature are always that far removed from the deeper understanding of the first causes. Nor will they ever attain knowledge of the nature itself of bodies. (Kant 1992, 51)

This means that we will lack a philosophical physics (see Chapter 4). Yet this is not how Kant expresses the issue, for instead he assigns the search for "first causes" to *metaphysics*.[21] So, Kant is placing limits on the causal account that we can expect from physics. To complete our causal account of the natural world we must appeal to resources outside those accessible by the methods of physics. What this means, and how we go about it, is something Kant seeks to explicate.

For Boscovich, by contrast, forces are causes, and our *knowledge* of causes is limited to that of forces.[22] As a result, since forces are physical, a metaphysics of causation is beyond our epistemic reach. Boscovich is explicit that his ontology is compatible with all metaphysical views of causation then on offer, such as occasionalism and so forth.[23] He says that deciding between these accounts of causation is something that he does not seek to do, and that it cannot be done "from the phenomena, which are the same in all these theories." In this he is representative of the late 18th century trend in which discussions of the metaphysics of causation are widely judged to be irrelevant to discussions of causation in physics. Where this leaves the status of physics' claims to causal knowledge is an open question.[24]

And so, whereas in the early 18th century physics was the discipline responsible for obtaining causal knowledge of the natural world, and this was largely seen to be unproblematic and continuous with general metaphysical accounts of causation, in Kant and Boscovich we see attempts to either completely reconfigure this relationship (Kant) or walk away from it entirely

---

[21] See also Chapter 4 for a discussion of causation in early modern philosophy.

[22] That the forces are to be understood as causes is clear from Boscovich (1922, §519), where he discusses the possibility of other points, related by different force laws, present in the same space, and how they would be "perfectly independent" of one another and "could never acquire any indication of the existence of the other": they would be causally unconnected, despite living in the same space. It is in this same paragraph that Boscovich introduces the idea of another space, bearing no relations (including no spatial relations) to our own. This is a natural consequence of his position that (1) space is relational, (2) relations arise from the force law, (3) there may be different force laws relating non-overlapping sets of things. See Guzzardi (2020, 49) for a distinction between powers (the ontological seat of forces) and the forces themselves, which Guzzardi regards as mathematical. See the surrounding pages for Guzzardi's discussion of Boscovich on causation.

[23] Boscovich 1922, §§516–518.

[24] The emerging disconnect between the metaphysics of causation and causation within physics was discussed in Chapter 4.

(Boscovich). Thus, in different ways, both Boscovich and Kant limit the causal aspirations of physics.

And so, Boscovich's and Kant's proposals offer a radical approach to Nature, one that is intimately tied to their accounts of Action. The question is whether the general ideas that we have gestured to above can be turned into a successful solution to BODY when spelled out in detail. For example, can either Kant or Boscovich provide bodies that survive interaction with one another? Can they meet the demands of PCOL? To address this, we turn our attention to their philosophical mechanics.

6.3 Kant's philosophical mechanics

Both Boscovich and Kant are fruitfully read as offering a philosophical mechanics. The term is ours, and we use it to label the many attempts at addressing BODY that sought to integrate philosophical physics with rational mechanics, where the physics is responsible for a constructive account of the nature and properties of bodies, yielding a causal explanation of their behaviors, while the mechanics is responsible for a mathematical treatment of their motions.

As earlier chapters have shown, PCOL was central to attempts to address BODY, by providing a test case for the viability of any purported solution.

PCOL: What is the Nature of bodies such that they can undergo collisions?

The account of Nature and Action should be such as to allow a causal explanation of the collision process between bodies. Moreover, the explanation should yield (or at least be consistent with) the rules of collision: we should obtain a philosophical mechanics of collisions. How do Kant's and Boscovich's accounts of bodies cope with PCOL? As we will see, both addressed this issue explicitly.

Within the mid-century discussion of BODY and PCOL, the debate over *vis viva* was given new impetus by Du Châtelet in 1740. In 1745, Boscovich published his dissertation *On Living Forces* in which he reviewed the issues as they then stood, especially the conflict between hard-body collisions and the principle of conservation of *vis viva*.[25] Kant's first publication was also on this topic,[26] and his interest in it is part of the wider enthusiasm for natural

---

[25] Boscovich was a close contemporary of Du Châtelet and Euler. Like them, he wrestled with the inheritance from Leibniz and Newton, and with the search for an adequate account of bodies and of their collisions.

[26] His 1749 *Thoughts on the True Estimation of Living Forces* (Kant 2012, 1–155).

philosophy that he exhibited throughout his pre-Critical years. We begin with Kant, and then turn to Boscovich's contemporaneous proposal (section 6.4). Each offers an attempt to address PCOL via a constructive account of bodies, and each faces similar difficulties, familiar from the first half of the century.

In 1758, not long after *Physical Monadology*, Kant published a treatment of collisions.[27] This paper has received very little attention, but its significance stands out if we read it against the backdrop of our discussion so far. First, Kant aimed to contribute to the Leibniz–Malebranche agenda we have described in earlier chapters. He does not say so explicitly, but his exposition unmistakably reveals his problem as their conundrum: to explain the rules of collision by deriving them from a notion of body equipped with causal agency codified in dynamical laws. That derivation is his crowning result in the paper. We describe the derivation below, and assess its merits. Second, Kant confronted the question of whether collision processes are as intelligible as many 17th and early 18th century natural philosophers had insisted (see Chapter 5). More specifically, the process of collisions was commonly described as a "communication of motion." Kant wondered, as some before him had done, what this phrase really denoted. For it cannot be meant literally: everyone agreed that motion is a property (of individual moving bodies), and properties do not travel across substances. All attributes—thus motion too—always inhere in their respective substance, at all times. So, it is metaphysically absurd to suppose that motion (qua property) could be literally communicated, in collision. Then what *does* impact amount to, if it is not a case of substances transferring attributes?[28]

To achieve his first aim, Kant resorts to a three-step strategy. First, he crafts a new account of true motion; hence the title for his paper. Second, from this he claims to derive dynamical laws (of contact action) by a priori argument. And finally, with his laws in place, he explains briefly how to derive the rules of impact.

In the first step, Kant offers his key to a sound collision mechanics: a theory of true motion, *wahrhafte Bewegung*, which he construes as a peculiar version of relationism.[29] He begins with some very terse reasoning that leads him to

---

[27] Kant 1905. Translated as "New doctrine of motion and rest" in Kant 2012, 396–408. For context and analysis, see Stan 2009.

[28] See Chapter 4. See Watkins 2005 for the physical influx approach in Kant's immediate predecessors. On the early modern causality debate more generally, see Clatterbaugh 1999. For Du Châtelet's novel solution to this problem, see Chapter 5, and discussion in Brading 2019, chapter 4.

[29] True motion was that which bodies have in *re vera* (as Descartes put it), that is, independent of their *apparent* motion or rest as manifested to terrestrial observers. For instance, the motion and rest

infer that, in impact, the two bodies have a privileged quantity of motion, namely their velocity relative to the mass center of their collision. Quietly, he assumes that any moving body has a "moving force" equal to its "true" quantity of motion, mv, relative to their common center of mass.[30] It follows trivially that, in his mechanics, any two bodies *always* collide with equal force, and so neither counts as "weaker," "passive," "stronger" and the like.

From his analysis, Kant immediately infers two theses that function as his dynamical laws. The first says that, in any direct collision (even when one body is apparently at rest) both bodies are *truly* in motion: "it is impossible for a body to approach another one absolutely at rest." The second says that, "in impact, action and reaction are always equal." Kant claims that with these resources in hand he can "explain the laws of impact according to the new doctrine of [true] motion" he had defended above.[31]

Now we arrive at the third step, and here is how Kant's explanation of impact goes. Suppose that body A is taken to be at rest with body B moving toward it, in a straight line at some constant speed u. For Kant, these are just *apparent* states. Their *true* inertial states fall out of his theory of true motion above, as follows. Take their relative speed—in this case, u—and divide it into two individual body speeds, a and b, such that $a/b = B/A$, where A and B are the masses of the two bodies.[32] Obviously, $Ba = Ab$, and since mass times speed measures "force of motion," it follows that the bodies collide with equal and opposite forces: "both are in motion toward each other, . . . the one with the same *force* as the other. . . . and it is really with these forces [*Kräfte*] that the two bodies will act on each other in impact."[33]

At this key point Kant declares, casually and without any argument, that equal-force bodies always come to mutual rest after impact: "because of the

at issue in the Copernican controversy; or the (true) states of motion and rest asserted by the Law of Inertia. A theory of motion is a metaphysical account of the *nature* of true motion; an answer to the question, What does true motion *consist* in? From Descartes to our times, the two leading contenders have been absolutism—the view that true motion *is* change of place in absolute space, an immaterial, rigid, unbounded container-like entity metaphysically distinct from body; and relationism—that true motion consists in a (kinematic) relation to some distinguished body or material setup. Kant was a relationist, both early (in the 1750s) and late, after the so-called Critical Turn.

[30] This puts him close to the later Wolff, but quite far away from Du Châtelet. The center-of-mass frame is sometimes called the "zero-momentum frame," because relative to it colliding bodies have equal and opposite momenta. If that quantity measures their "force," then evidently they meet with equal and opposite "forces of motion."
[31] Kant 1905, 19ff.
[32] It is clear then that a and b, the bodies' "true" speeds, are with respect to the mass-center frame of their collision.
[33] Kant 1905, 18–9.

equality of their contrary forces, the bodies come to relative rest." Evidently, this description—and the causal-dynamical story it allows—holds only for the center-of-mass frame; only there are their "forces" equal, so that mutual rest ensues. His conclusion: this account yields "the rules of the relation that colliding bodies *enter into*, with respect to each other."[34]

To predict the *apparent* motions that the observer will see, we need only re-describe the impact's outcome relative to that observer (taken to be at rest).[35] Kant illustrates his point with a particular case, and then reminds the reader that such descriptions are not really explanatory: they are "just the outer phe-nomenon of what occurred immediately between the [colliding bodies]; and it is the *latter* that one needs to know," he insists.[36] Namely, the causal-process story he has given and defended above.

Thus Kant accounts for the quantitative patterns of apparent motion that an observer would see as they watch any two bodies collide head-on. In sum, the young Kant's natural-philosophical basis for collision mechanics is: a relationist account of true motion; two dynamical laws; and the presumption that moving bodies have "force of motion" equal to mass times true speed, not the Leibnizians' *vis viva*. Thereby, his paper is a late installment in the Leibniz–Malebranche agenda to explain why *those* patterns obtain, viz. to "give the reason" for the "rules of motion."

Recall that Leibniz and William Neile had first demanded a causal-explanatory basis for collision theory in the 1660s and '70s. Thus, the struggle to make sense of contact action had by now entangled sharp phil-osophical minds for well over a century. And yet, little progress has been made. First, Kant helps himself to a notion of mass, without justification.[37] Second, he asserts that bodies of equal "force" will come to rest as a result of head-on collision. This is to simply presume perfectly inelastic collisions, and Kant did not justify why the "force of motion" in each body should be annihilated in a head-on collision. These puzzles, present at the beginning of

[34] Kant 1905, 23.
[35] Kant (1905) helps himself to some key assumptions, though without discussion. Namely, that the bodies are in uniform motion, are non-rotating, and collide head-on; hence that their interaction is described from a proper inertial frame.
[36] Kant 1905, 24; our emphasis.
[37] Newton's second law of motion quantifies the relationship between force, mass, and accelera-tion. In *Physical Monadology*, Kant recognizes that this dynamical conception of mass involves both resistance and impulse. His physical monad is a point-sized entity endowed with just such a dual property, and the property is additive: physical monads aggregate into extended bodies, and *their* aggregate inertia is the sum of the component monads' mass. And, it equals the body's quantity of matter. See Kant 1910, 485–6. See also Chapter 7 for a discussion of "elusive mass."

the century, remained unaddressed in Kant's 1758 paper. As a result, the account of Nature, Action, Evidence, and Principle offered in his philosophical mechanics fails to meet the challenge posed by PCOL, and thereby to provide an adequate solution to BODY.

In the "Critical" years, Kant offered a new attempt in his *Metaphysical Foundations of Natural Science* (*MFNS*), a pithy tract of 1786 where he expounds a "metaphysics of material nature" for the sake of obtaining a "concept of matter ready for application" to natural science.[38] Though different in detail, the approach is again constructive; and again impact lies at the heart of *MFNS*. Kant first goes to enormous lengths to construct a notion of body. He then places collision center stage, in his chapter "Metaphysical foundations of mechanics," leaving the reader keen to see how he solves PCOL. In the end, however, nothing new emerges, and familiar problems resurface.[39] Moreover, *MFNS* is a *late* 18th century work; by the time Kant wrote it rational mechanics had moved on, displacing impact from its central role in philosophical mechanics.[40] Thus, while this book serves to emphasize the continuing importance of BODY through the end of the century, it is otherwise of little interest for our purposes: from the perspective of philosophical mechanics, *MFNS* was out of date before it was even written.

6.4 Boscovich's philosophical mechanics

We have already seen in general terms the constructive account of body offered by Boscovich in his physics. But in his *Theory* he was after a bigger prize, attempting exactly the type of project we call a philosophical mechanics. He says so explicitly in the opening paragraph of his Preface.[41] *Theory* consists of three parts: Part I outlines the theory as a whole (the

---

[38] Kant 1911. English translation Kant 2004.

[39] See Stan (forthcoming) for details. The gist is this. Kant's mature theory of collisions treats bodies as having real states of motion relative to their center of mass (or rather, relative to an inertial frame in which their mass center is at rest), and then any two colliding bodies run into each other with equal amounts of "moving force" which "cancel each other out" on impact, bringing the bodies to rest. Already one can suspect numerous assumptions in here, and closer inspection bears this out. Kant's account leaves several important questions unresolved. For extensive discussion of Kant's laws and their relevance to impact mechanics, see also Friedman 2013 (chapter 3) and Watkins 2019 (chapters 4–6).

[40] These developments in rational mechanics and their significance for philosophical mechanics are the subject of the remaining chapters of this book.

[41] Boscovich 1922, 13.

unified philosophical mechanics, in our terminology); Part II sets out his mechanics; and Part III develops his physics.

By physics, Boscovich means the study of bodies, their properties, and the changes they undergo. He opens Part III saying that he will address "the general properties of bodies," the "different species of bodies," and the "changes, alterations & transformations that happen to them."[42] By mechanics, Boscovich means rational mechanics along the lines conceived by Newton: the mathematical treatment of motions under forces. Specifically, in Part II Boscovich asserts that mechanics boils down to two problems: to find the motions given the force, and to find the force given the motions. And so, he sets out to investigate the "general laws of equilibrium and motion."[43] On his own terms, his physics is to be understood in the context of the goals of the overall project. And so, our question is this: To what extent does Boscovich succeed in his goal of producing an integrated theory of mechanics and physics—a philosophical mechanics—capable of solving BODY? To answer this, we begin with his mechanics, and then explore his attempted integration, adding more details from his physics as needed.

(i) Boscovich's mechanics

The core of Boscovich's mechanics concerns the motions of "points" under "forces." Each point has associated with it a "force of inertia": an "inherent propensity to remain in the same state of rest or of uniform motion in a straight line."[44] The points act on each other via an action-at-a-distance given by a single force law that varies with distance. Famously, Boscovich's theory begins from two fundamental assumptions, impenetrability and continuity. Impenetrability is the requirement that no two of the points can be spatially co-located. Continuity, the principle enunciated by Leibniz, is satisfied by the Boscovich force.[45]

Over various finite intervals, the force either repels or attracts other points within that interval (see Fig. 6.1). The first interval, from A to E, is one of repulsion; in the next interval, from E to G, the force is attractive. There follows another interval, G to I, of repulsion; and so on, alternately, up to distance

---

[42] Boscovich 1922, §358.
[43] Boscovich 1922, §204.
[44] Boscovich 1922, §8.
[45] Child 1922, xi.

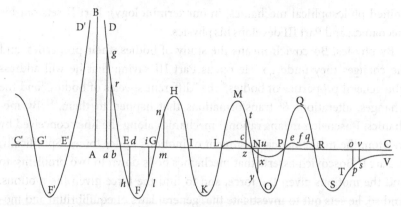

Fig. 6.1. Boscovich force. The curve represents the strength of the force. A is the location of the physical point, and so AC is an ordinate axis. The AB axis is oriented. Above A, the force is repulsive; below A, it is attractive. To infer the strength of the force at any location, project that location (for instance, *n*, *l*, or *t*) onto the AB axis, and you get the local strength.

R from the point, where the force turns attractive and trails off to zero at infinity. Moreover, within each interval but two, the force increases gradually from zero to a maximum strength, and then decreases to zero, at the same rate as it increased (e.g., from L to M, and then to N, respectively). Boscovich represented all these conjectures about his force in a famous graph.

"Limit points" are those where the force transitions from attractive to repulsive, or vice versa. Given a system of two points (say *a* and *b*), one situated at the origin and the other at one of these limit points, there will be no mutual force between them. So, if they are at mutual rest, they will remain at mutual rest: they will be in equilibrium. Following a disturbance to this equilibrium that momentarily increases the distance between the points, there are two possible outcomes. In the first case, the force on *b* may be attractive and the points will return to their original equilibrium configuration. In the second, the force on *b* is repulsive and the points will continue to move apart. The former kind of equilibrium is *cohesive*, and the latter non-cohesive.[46] In fact, a Boscovich central force originates from *each* of the points, so the *net* force is the critical dynamical player. The forces add as vectors to produce a single net force in which the pairwise motions of the points are mutual.[47] These are the

---

[46] See Boscovich 1922, §§179–180. In modern terms, these configurations correspond to stable and unstable equilibrium, respectively.

[47] This is important when interpreting his force-distance diagrams; see Boscovich 1922, §177.

crucial features of Boscovich's forces on which his mechanics depends: "force of inertia"; central forces; mutual pairwise motions; vector addition of forces; and equilibrium.

From these assumptions, Boscovich derives the main principles of his mechanics. First, he shows that every collection of points has a unique "center-of-gravity". This follows from the assumption of central forces, mutual pairwise motions, and the vector addition of forces.[48]

Second, he argues that the center-of-gravity of any isolated collection of points is at rest or moves uniformly. He begins by showing this holds for the simple case of points moving by "the force of inertia" alone, and then considers interacting points. In the latter scenario, since we have already postulated central forces and mutual pairwise motions, the result follows straightforwardly. It is, as he notes, "a most elegant theorem of Newton."[49] From this principle, Boscovich draws an immediate corollary of his own:

> The quantity of motion in the Universe is maintained always the same, so long as it is computed in some given direction in such a way that motion in the opposite direction is considered negative, & the sum of the contrary motions is subtracted from the sum of the direct motions. (Boscovich 1922, §264; italics in original)

This is a conservation principle for the total quantity of motion in the universe, though "quantity of motion" remains an undefined term thus far.

Next, Boscovich turns his attention from points to bodies. Since his bodies are sets of points in cohesive equilibrium, we can apply the above results to them. Such bodies act on one another when they (as two equilibrium configurations) are not in mutual equilibrium *jointly*, for then there will be a net force between them. His first conclusion is that there is equality of action and reaction between all bodies. Thus, Boscovich takes Newton's third law of motion to be a *consequence* of the assumptions he makes in setting up his mechanics.

To assess his mechanics fully, we would need to take stock of terms we have so far passed over, including "points," "forces," "center-of-gravity," "force

---

[48] The alert reader will be suspicious that an implicit assumption about mass has been made. We come back to this, and to the meaning of the term "center-of-gravity," in our discussion of "elusive mass," below.

[49] Boscovich 1922, §260. He means Corollary IV to the laws of motion in Newton's *Principia* (see Newton 1999, 421).

of inertia," "quantity of motion," and "mass." However, we can defer much of this until later. Our interest for now is in his account of bodies in relation to his treatment of collisions. Boscovich's approach consists of two general steps: first, the construction of the necessary extended bodies; and second, a demonstration that impact can be treated within his theory. We tackle each in turn.

(ii) The construction of extended bodies in rational mechanics

Boscovich constructs solid bodies by means of cohesive equilibrium occurring in regular tetrahedra. First, he discusses the motion of a point around an ellipse that has an additional point at each focus. He argues that cohesive limit points occur at the intercepts of the major and minor axes, and from there that cohesive equilibrium among three non-collinear points occurs only when the three points form an equilateral triangle (in the absence of an external force).[50] From collections of these triangles he builds tetrahedra, which he calls "particles."[51] From four of these particles we form a larger pyramid (a "particle of the second kind"), and so on until we arrive at large collections of points forming masses: the familiar solid bodies of physics, including practically rigid rods, as well as flexible rods, including those that oscillate appreciably (and thereby exhibit elasticity).[52]

As we have seen in earlier chapters, there was a standard three-way taxonomy of bodies in terms of their shape behaviors into soft, hard, and elastic. Boscovich explains how to think of this taxonomy within the context of his theory. Of soft bodies, he writes:

[50] See Boscovich 1922, §§225–239. He first considers equilibrium for three collinear points where the cohesion is extremely strong, so that it is very difficult to disturb the points from the line along which they lie, and once disturbed they return to that line in a very short time. Then he turns to non-collinear points, and argues for equilateral triangles exhibiting strong cohesion, and for these as providing us with "an idea of a certain solidity." He also considers four coplanar points, giving us "a more precise idea than hitherto has been possible of a solid rod," before arriving at tetrahedra. His discussion reinforces the interpretation that each point is equal in mass with the same force function being associated with every point. The differences in behaviors of bodies arise from the different arrangements of the points making up the body, and the consequent net force associated with that body. At large distances, these differences become negligible; this is why gravitational behavior then dominates.

[51] Boscovich 1922, §239.

[52] Boscovich 1922, §226.

By soft bodies are to be understood those, which resist deformation of their shapes, or compression; but which, when compressed, exert no force tending to restore shape; such as wax or tallow. (Boscovich 1922, §266)

Of elastic bodies:

Elastic bodies are those that endeavor to recover the shape they have lost; & if the force tending to restore shape is equal to that tending to prevent loss of shape, the bodies are termed perfectly elastic; and just as there are no perfectly soft bodies, there are none that are perfectly elastic, according to my thinking, in Nature. Lastly, they are imperfectly elastic, if the force exerted against losing shape bears to the force exerted to restore it some given ratio. (Boscovich 1922, §266)

And finally of hard bodies:

It is usual to add a third class of bodies, namely, such as we call hard; and these never alter their shape at all; but these also, even according to general opinion, never occur in Nature; still less can they exist in my Theory. Yet, if anyone wishes to take account of such bodies, they could consider them as soft bodies which are compressed less and less, until the compression finally becomes evanescent; that is, take the compressible soft body and reduce the compressibility towards zero to arrive at an idealization that we call a hard body. (Boscovich 1922, §266)

This is how we are to think of such bodies for the purposes of mechanics. As for "what are the causes of soft or elastic bodies, I do not investigate at present; this relates to the third part" (i.e., to his physics).[53]

For the purposes of mechanics, when we appeal to "rigid bodies," "inextensible threads," "rigid planes," and so on, we are working with idealizations never realized in nature, not even at its smallest parts; there are no tiny, rigid atoms in Boscovich's natural philosophy. Nor are there any perfectly elastic parts of matter, out of which the macroscopic bodies of our experience are

---

[53] The causes lie in the different arrangements that lead to different forces among the parts, yielding different locations for the limit points as well as different locations and strengths for cohesive equilibrium.

constructed, and of which the theorems of mechanics are exactly true. And so, Boscovich writes,

> Since, to my idea, there are no such things as continuous spheres or contin-
> uous planes, many of the things that have been said are only true as far as
> we can observe, & only very approximately and not accurately. (Boscovich
> 1922, §296)

This raises questions concerning the evidential status of mechanics, and of Boscovich's theory of points and forces out of which the bodies of mechanics are made.

(iii) Boscovich and PCOL

Boscovich treats collisions explicitly, appealing to both his mechanics and his physics, and so PCOL offers the opportunity for an exemplary window into his philosophical mechanics. The results, as we shall see, are disappointing.

Boscovich begins by saying that "to make the matter easier, I consider spheres, and these too homogeneous round about the center."[54] This allows him to treat the trajectories via centroids (that is, by a representative point at the center of the sphere). The principle on which he rests his collision mechanics is equality of action and reaction (cf. Newton's third law of motion) as a consequence of which "quantity of motion" is conserved. As already noted, Boscovich is explicit that "quantity of motion" is the product of mass and velocity.[55]

One equation is insufficient to determine the two unknowns (that is, the exit speeds of the two spheres post-collision). A second equation is needed, and for that Boscovich reaches outside his mechanics to his physics, begin-ning with soft bodies. His canonical situation is catch-up impact, and he says that when the faster approaches the slower closely enough,

> the parts at the back of the slower body, and the ones at the front of the
> faster, get compressed by a mutual repulsive force. And this compression

---

[54] Boscovich 1922, §266.
[55] Boscovich 1922, §268: "the velocity multiplied by the number of points represents the sum of the motions of all the points, i.e., the quantity of motion." Recall from earlier that the notion of mass is an open interpretive question.

increases until the bodies reach the same speed. At that point, the catch-up body stops approaching the other one any further, and so they stop compressing each other as well. As the bodies are soft, they exert on each other no other force beyond the above compression. Rather, they just keep moving at the same speed. (Boscovich 1763, §268)

This descriptive physics allows Boscovich to infer the second condition needed for solving the soft-body collisions: following impact, the bodies move together. Inferring the relevant "general formula" for their outcome is then routine: the final speed is $v = (m_a u_a + m_b u_b)/(m_a + m_a)$, for masses a and b, with initial and final velocities u and v.[56]

Notice that this conclusion is "routine" only so long as Boscovich has the Newtonian concept of mass. For, while one body *acts* on another in proportion to the number of points, how each body *responds* to the action of the other requires specification—such as via Newton's second law of motion. However, Boscovich, like many before him, faces a problem of "elusive mass": it is not at all clear that he has the requisite notion available (see below). His treatment of collisions seems to demand that he make use of it, dishearteningly committing Boscovich to principles and concepts not explicitly stated in setting out his theory.

Moving on to his treatment of elastic impact, the process takes place as above until the stage of maximum mutual compression. At this point, elastic bodies do not cease to act on each other, but instead each one's "force of recovering shape" leads it to return to its initial shape. As a result, the bodies "keep acting on each other, until they come back to their initial shapes; and their action will double the above effect."[57] By this pithy phrase, Boscovich leaps to conclude that the "double" action will result in the bodies rebounding at the same relative speed they had before impact. This provides him with the second condition he needs for deriving the outcome of elastic collision.[58]

Boscovich's treatment of collisions is disappointing. Though gesturing qualitatively at his new theory, in fact he simply repeats the results from decades before, his new theory playing no substantive role in determining

---

[56] This is Boscovich's algebraic expression in our notation.

[57] Boscovich 1922, §270.

[58] Boscovich also treats oblique impact, again from a centroid perspective, and he introduces two further conditions in order to derive his results. One of these is that in oblique collision, the force of impact is central, viz. acting along the line between the bodies' centers of gravity, a result that remains to be demonstrated within his mechanics and that allows him to transform oblique impact into cases that can be handled via the equations for direct collision.

collision outcomes. Boscovich makes no attempt to mathematize dynamics at the point of contact, nor even to connect his physical story of collision with his diagram of Boscovichian force. He treats the colliding bodies via their centroids, ignoring their extension in the quantitative treatment. Moreover, his treatment relies on Newtonian dynamical mass without explicit acknowledgment. A coherent mechanics of collision—one that successfully combines a notion of body, the forces at work in collision, and their kinematic outcomes—eluded Boscovich just as it had escaped all his predecessors.

Moreover, if we were to attempt a treatment that explicitly uses his force function, we would find that his account does even worse. The combination of Boscovich's collision theory with his doctrine of body entails that bodies are fundamentally and incurably *unstable*: Boscovich bodies oscillate without end, and nothing is available to restore and keep them in equilibrium configurations. The root cause of this problem is the absence of any dissipative forces—or any physical basis that can dampen oscillations and end vibrations.[59] This problem reverberates negatively into his very *notion* of body: if a body is a collection of points remaining at mutual rest in a stable equilibrium configuration, but *any* collision results in endless vibrations, then it follows that there are no Boscovich bodies in the actual world that are capable of undergoing significant interactions with other bodies. No body survives a collision.

If we are looking to Boscovich for progress and innovation in philosophical mechanics, we will not find it in his treatment of collisions. However, by the late 18th century collisions had been moved downstream, away from the foundations of mechanics, as a complex problem too difficult to solve at that time.[60]

---

[59] Consider the two "physical points" colliding head on. The outcome is elastic rebound (their mutual repulsive forces will drive them back from each other, to the same extent). As they recede, they reach a band of relative distance in which their mutual *attractive* forces dominate. So, under mutual attraction, their elastic rebound will slow down to an end, and turn into mutual approach. That will bring them close enough for their mutual repulsive forces—more exactly, for some repulsive regime within their force law—to take over, thus inducing mutual rebound, again; which will remove them far enough from each other that again some mutual-attractive regime of force becomes dominant, leading to relative approach; and so on, to no end. But, Boscovich bodies are just sets of interacting physical points; as bodies, they have no sui-generis forces over and above the single, alternative force of Boscovich's matter points. And so, head-on impact between bodies is bound to result in endless vibrations *within* each body, as the collision makes its constituent points oscillate relative to each other. Incidentally, Kant's physical monadology has the same problem; Smith 2013 first noticed that.

[60] Indeed, to best understand the ambitious goals of Boscovich's *Theory*, as a philosophical mechanics, we think it should be read in light of d'Alembert's *Treatise*; see Chapter 9.

(iv) Boscovich and "elusive mass"

Boscovich is of particular interest when it comes to "elusive mass." This is because he makes no explicit appeal to mass when setting up his mechanics. Two interpretative options present themselves.

Interpretation 1

In the first, we assume that this was an oversight: we interpret Boscovich's mechanics as closely to Newtonian point particle mechanics as possible, associating an equal quantity of Newtonian mass with each Boscovichian point and treating the mass of a body as the additive sum of the masses of the constitutive points. At first sight, this interpretation seems to align with what Boscovich says later, when he turns his attention to physics in Part III. Here, he writes that the "mass of a body is the total quantity of matter pertaining to that body; & in my Theory this is precisely the same thing as the number of points that go to form the body."[61] Moreover, this reading has several advantages. First, it makes sense of Boscovich's term "center-of-gravity," which he says in reality "does not depend on gravity, but rather is related to masses." "Center-of-mass" indeed seems to be a more appropriate term for the quantity that he calculates. Second, his proof that every collection of points has a unique "center-of-gravity" goes through straightforwardly if (a) he means center-of-mass and (b) every point is implicitly assumed to have an equal mass. Third, if his points are point masses, and if (crucially) he is willing to appeal to the necessary principles, he will be able to derive all the results then available for point-mass mechanics.

Doubts about Interpretation 1

That things are not so straightforward is clear from Boscovich's very next remark: "Here now we have a certain indefiniteness, or at least the greatest difficulty, in forming a definite idea of mass; & that, not only in my theory, but in the usual theory as well."[62] Whereas the first quotation may look like an endorsement of Newtonian mass, this second remark gives us pause: Boscovich holds "mass" to be indefinite, and an unhelpful notion as a result.

In the ensuing discussion, it becomes clear what he is worried about. As we saw in Chapters 2–5, in the early 18th century many relied on a distinction

---

[61] Boscovich 1922, §378.
[62] Boscovich 1922, §378.

between proper and foreign matter. Boscovich's concern is that this distinction cannot be made in a principled way, leaving the mass of a body indefinite. Specifically, the quantity of matter associated with a body—understood as the *volume of its proper matter*—is indefinite. Now, measuring mass by its geometric volume (viz. the "geometric" notion of mass) is to be distinguished from the *dynamical* notion expressed in Newton's second law of motion (see Chapter 7 for details), where the relevant aspect is a body's response to impressed force. It seems to be the geometric conception that Boscovich finds problematic, but it is the dynamical conception that is crucial for mechanics. So, we can take his criticism of "mass" to target *geometric* mass, and continue to maintain our interpretation that each Boscovichian point has a well-defined quantity of *dynamical* mass.

Thinking of Boscovich's "mass" as picking out geometric rather than dynamical mass is consistent with Boscovich's use of the term to denote *collections* of multiple points (and never in connection with a single point). Moreover, though Boscovich never uses the term "mass" in setting up his mechanics, he does assert that his points have a "force of inertia" associated with them, and that as a consequence of this bodies do too.[63] It is striking that Boscovich separates "mass" (for which he offers a geometric conception, as we have just seen) from "force of inertia," and that he never uses the term "mass" for the latter at all. In his use of these labels, he perhaps makes the distinction we have made between geometric and dynamical mass. If Boscovich's "force of inertia" is correctly interpreted as dynamical mass, then Interpretation 1 stands.

According to him, "the force of inertia consists in a propensity for staying in the same state of rest or of maintaining a uniform state of motion in a straight line, unless some external force compels a change of this state."[64] Statements along these lines are familiar from Newton's first law of motion, and it is tempting to think of this as expressing dynamical mass. However, Newton's first law tells us nothing about what changes in motion a body undergoes when a force is impressed: it gives us only "kinematic" mass (see Chapter 7). Lacking Newton's second law (or some analogue thereof), we have no information about the relationship between the mass of a body and its response to impressed force. And so, despite the initial plausibility of attributing Newtonian dynamical mass to Boscovichian points, pursuing

---

[63] Boscovich 1922, §8.
[64] Boscovich 1922, §382.

this line of interpretation leads to trouble: it requires attributing to Boscovich a concept he never explicitly endorses, and for which he lacks the requisite resources.

Interpretation 2

There is a second interpretative line that we might take instead. Having noticed that Boscovich makes no reference to mass in setting up his account of the motions of points, we might wonder what he means by "force." If his "forces" determine the motions of points without any appeal to mass, then they are perhaps better thought of as acceleration fields. The word "force" was still an ambiguous term with many senses then, as we have seen. Hence, while "acceleration field" may sound anachronistic, we think it is a helpful use of later terminological clarifications: Boscovich lacked a term for an acceleration field at the time, so though we are applying this terminology with hindsight, it is not unreasonable to test out the concept against the uses Boscovich makes of his term "force."[65]

The "acceleration field" interpretation of "force" has four advantages. First, Boscovich explicitly presents the area under his "force" curve as representing the change in the square of the velocity of the point. This is the case when acceleration is plotted as a function of distance, so the "force" curve is naturally thought of as an acceleration field in one spatial dimension.[66]

Second, in an acceleration field all bodies placed at the same location undergo the same acceleration, independent of their mass. So if Boscovich's "force" is an acceleration field, then all points would accelerate the same way, and he would not need to appeal to mass. As a result, our interpretation is not only consistent with his seeming failure to mention mass, it does better: it explains and justifies it. If his forces are acceleration fields, then mass indeed plays no role in the motion of his points.

The third advantage is that it ensures consistency with Boscovich's assertion that he is introducing a *single* force. This requires a slightly longer explanation. Recall first that in Boscovich's theory there is only *one* force, given by the function graphed above (Fig. 6.1). He opens his "Preface to the Reader"

---

[65] Whyte (1961, 284) follows Child (1922) in noting that "Boscovich's *vis* and *massa* did not mean what "force" and "mass" meant for Newton and all mechanical theories . . . Boscovich's theory is still the only example of a pure kinematic theory." See also Guzzardi (2020), who respects Boscovich's terminology in arguing for a mutual "determination" interpretation of Boscovich's force.

[66] Boscovich 1922, §176. The area under an acceleration–displacement graph is $(v^2 - u^2)/2$, where v is the final and u the initial velocity. See also Child (1922, xiii), who long ago pointed out that Boscovich's force has the dimensions of acceleration.

thus: "Dear Reader, you have before you a Theory of Natural Philosophy deduced from a single law of Force."[67] It is from this force that gravitational phenomena arise (at appropriately large scales). Next, notice that Boscovich's *only* explicit example of this force is gravity. Now, in a gravitational field all bodies fall at the same rate, whether a piece of gold or a feather, as he explicitly points out.[68] Putting all this together, we may infer: it is not just gravity that satisfies the equivalence principle (i.e., that all bodies fall at the same rate), but rather throughout nature Boscovich's force produces mass-independent accelerations, just like gravity. There is only *one* force in Boscovich's theory and, since gravity has this property, so does Boscovich's force quite generally. Boscovich's "force" is an acceleration field in which all points (and therefore all bodies) "fall freely" (to use terminology from general relativity). So our third advantage is this: it allows for a unified treatment of Boscovich's "force," whereas appeal to "dynamical mass" would require a fragmentation, differentiating the domain of gravity (for which the equivalence principle holds) from other regimes of his "force" (where the equivalence principle fails). Boscovich is explicit that he is introducing a single force and he never discusses any such differentiation of it.[69]

This interpretation of his force directly yields the fourth advantage. It makes sense of our earlier observations about Boscovich's use of the term "mass," including: his use of it to denote *only* a geometrical (and not a dynamical) conception of mass; his separation of "force of inertia" from his use of "mass"; and his lack of a dynamical concept of mass. According to our new approach, Boscovich lacks an analogue of Newton's second law and a notion of dynamical mass, not because he makes an unfortunate omission, but because he believes he can do without them. The place where he introduces "force of inertia" supports this reading, for he writes that a point existing by itself will be at rest or move uniformly, but:

if there are also other points anywhere, there is an inherent propensity to compound . . . the preceding motion with the motion which is determined

[67] Boscovich 1922, 13; translation amended.
[68] See Boscovich (1922, §213) for his discussion of how, at appropriate distances, behavior in accordance with Newtonian universal gravitation arises. The inverse-square dependence on distance comes from the shape of the force law. The acceleration field's dependence on the mass of the attracting body arises from the number of points combining to form the source of the acceleration field. The acceleration of the attracted body depends only on these, so that the "mass" of the attracted body drops out of consideration, as discussed. Nowhere is a dynamical conception of mass needed.
[69] Guzzardi (2020, 56ff.) suggests that multiple *powers* may underlie the single force law, but we do not think this contradicts the claim that the result is a single force.

by the mutual forces that I admit to act between any two of them, depending on the distances & changing, as the distances change, according to a certain law common to them all. (Boscovich 1922, §8)

At this juncture, were there any mass-dependence of the changes in motion resulting from the mutual forces (as Newton's second law dictates), Boscovich would have told us. He does not. And the most natural conclusion is that this is because he didn't think he needed it.

Doubts about Interpretation 2

There are some disadvantages, however. First, whereas our first interpretation opened the road to Boscovich incorporating all of Newtonian point-mass mechanics into his system, the second offers no such easy path. His results will have to be re-examined to see whether they go through.

Second, and relatedly, his key concept "center-of-gravity" loses its ready understanding as our familiar notion of center-of-mass. Specifically, if Boscovichian points do not have mass, then center-of-gravity seems to become a merely geometric notion concerning the spatial distribution of these points, lacking any of the dynamical import we are used to associating with it via Newtonian inertial mass. Moreover, if Boscovich arrives at his results on the motion of the center-of-gravity via the "force of inertia" (viz. kinematic mass) associated with each point, his proofs of these results will need to be reworked using only acceleration fields, to prove that they go through.

Third, when it comes to developing his mechanics there are remarks that are most naturally read as invoking dynamical mass. For example, he writes that "the common center of two masses lies in the straight line joining the centers of each of the masses, & that the distances of the masses from this point will be reciprocally proportional to the masses themselves."[70] If we deny ourselves Newtonian mass, then we will need to do some heavy lifting to see whether this can be made sense of. Moreover, recall that Boscovich introduces the term "quantity of motion" without explaining what he means by it. Later, he claims: "the velocity multiplied by the number of points represents the sum of the motions of all the points, i.e., the quantity of motion."[71] This quantity is conserved, he says, due to equality of action and reaction, so it plays a crucial role in his mechanics. If each point has Newtonian

[70] Boscovich 1922, §253.
[71] Boscovich 1922, §268.

mass associated with it, then "quantity of motion" is straightforwardly understood as linear momentum. If not, then we are faced with some hard work to understand whether and how our second interpretation can incorporate an adequate concept of "quantity of motion."

Finally, even if Boscovich works only with "force of inertia," the *measure* of kinematic mass in Newtonian mechanics is its acceleration under an impressed force: dynamical mass. So, if "force of inertia" plays a quantitative role in his mechanics, he will need some measure of it, and this may require Newton's second law (or some analogue thereof). As we will see in the coming chapters, 18th century theorists sought alternative dynamical principles to Newton's second law, many through attempting to generalize from equilibrium (a notion in statics) to a principle adequate for dynamics. Boscovich places equilibrium at the heart of his mechanics. On the one hand, this looks promising because weight is a property that is empirically accessible and tracks mass. On the other hand, in Boscovich's theory this weight behavior must arise from the points moving in the acceleration field and his theory—interpreted as positing a single force in which all points are in free fall—lacks the resources for adequately theorizing the balance.

To resolve these challenges to the second interpretation, and to determine the extent to which either of the interpretations is viable, we would need to look at the details of Boscovich's mechanics. We defer this for another occasion, for either way there is an important conclusion already available: there is no straightforward interpretation of Boscovich's theory that avoids the challenges associated with Newtonian mass. This adds yet another difficulty for his account of collisions, because in our discussion of impact we granted him a notion of mass. Therefore, despite Boscovich's fine attempt to produce a physics adequate to the needs of mechanics, obstacles on this road to a philosophical mechanics of collisions remain.

6.5 Conclusions

Thus far in this book we have argued for the centrality of PCOL to 18th century philosophy, and for its pressing importance. If philosophical physics cannot address PCOL, then it cannot solve BODY, and if there is no solution to BODY, then no area of philosophy that includes bodies (human or otherwise) among its objects escapes. Knowing this, it is not surprising to find that BODY persisted as a research problem in natural philosophy throughout the century, and that a great deal was learned about the possibilities (and other-wise) for a solution.

We have focused our attention so far on constructive solutions: those that begin from the qualities and properties of matter and seek to build an account of bodies from there.[72] The 18th century opened with two options for matter theory: the continuously divisible matter of Descartes, and any theory in which there are smallest parts of indivisible matter, such as atomism. Mathematically, rational mechanics at that time made use of a third option, namely, point particles endowed with mass and force. By the end of the century there were, thanks to Boscovich and Kant, candidate matter theories to go with it. Philosophical mechanics provides a tool by which to analyze the extent to which the resources offered by Boscovich and Kant are adequate for addressing BODY. This matters because each of them explicitly set out to do so.

Boscovich and Kant both begin with non-extended physical "points" endowed with forces and attempt a constructive account of bodies; one which recovers the Nature of bodies as extended, impenetrable, and mobile, and which explains the Action of one body on another.[73] The consequences for physics are twofold. First, the move to points and forces puts pressure on the conception of physics as the study of bodies; it seems that the fun-damental ontology of the natural world is something very different in-deed, and it is unclear whether this ontology is going to be investigable by the methods of physics. Second, both Boscovich and Kant in their different ways limit the scope of physics in delivering knowledge of causes. At the be-ginning of the century, knowledge of bodies and causes lay at the heart of physics; in Boscovich and Kant that ambition is transformed, and in some ways curtailed. This might have been a price worth paying if a solution to PCOL, and thereby to BODY, had been forthcoming. But it was not. As we saw, many of the old problems in handling collisions reappear in the accounts of Boscovich and Kant. Almost a century after the Royal Society discussions of impact, a satisfactory solution to PCOL remained elusive.

Notwithstanding all the difficulties we have encountered, in the late 18th century extended bodies persisted as important objects of study, not just in physics but also in rational mechanics. It is easy to see why: they are the

---

[72] See Chapter 1 for the constructive and principle approaches to *Body*.

[73] Again, we have interpreted Kant above as endorsing a mass-point picture of matter. But, it is not the only reading compatible with his *Physical Monadology*, which is often opaque and ambiguous. The excellent Smith 2013 argues for a different picture that is closer to our modern view of a deform-able continuum.

objects of our sensory experience, with the human body among them. One way or another, philosophy requires that we recover an account of extended bodies in order to proceed. Qualitatively, the attempt to construct bodies from matter points endowed with forces seems promising. But philosophical mechanics demands that this promise be made good quantitatively, via integration with rational mechanics. The constructive accounts of Kant and Boscovich make the difficulties vivid, and serve to demonstrate that efforts to achieve a philosophical mechanics continued throughout the 18th century. Moreover, as we will see in the coming chapters, concurrent developments in rational mechanics changed the landscape for philosophical mechanics, presenting philosophers with new challenges as well as new opportunities.

7

Shifting sands in philosophical mechanics

7.1 Introduction

Throughout the 18th century, the dominant strategy whereby philosophers sought to tackle the *Problem of Bodies* (*BODY*) involved two steps. First, develop a material account of bodies. Then, integrate the rules of collision, augmenting the account of bodies as necessary (the problem of collisions, or PCOL).[1] Despite widespread and extensive efforts to bring this strategy to fruition, none succeeded (see Chapters 2–6). And yet, a coherent account of bodies remained crucial and indispensable for philosophers and rational mechanicians alike. The need for a solution to *BODY* did not go away.

With hindsight, we can see that an important shift occurred mid-century. Rational mechanics, the province of mathematicians, became the research field most relevant to *BODY*. This chapter explains why, and with what consequences. Most importantly, a divide emerged between philosophical physics and pertinent advances in rational mechanics. After 1750 most professional philosophers failed to absorb the new insights of mechanics into their doctrines of body. Consequently, we cannot turn to philosophers of the period to unpack the significance of contemporaneous developments in mechanics for *BODY*. Instead, we must examine rational mechanics for ourselves.

Doing so reveals two things. First, the relevant area of rational mechanics is the theory of constraints. As a result, PCOL is displaced as the problem to be solved by philosophical mechanics, replaced by what we call "the problem of constrained motions," or PCON. Second, with the shift in locus to rational mechanics, *BODY* is no longer the primary problem engaging our protagonists. From Descartes onward, *BODY* had been an explicit problem among philosophers, self-consciously acknowledged and grappled with (see Chapters 1–6). This allowed us to study it with the tools of contextualist

---

[1] For a longer explanation of *BODY*, and of the requirements on any satisfactory solution, including PCOL, see Chapter 1.

Philosophical Mechanics in the Age of Reason. Katherine Brading and Marius Stan, Oxford University Press.
© Oxford University Press 2023. DOI: 10.1093/oso/9780197678954.003.0007

historiography. The move to rational mechanics as our domain of study means that new methodological tools will be required.

We begin by addressing this methodological challenge explicitly (section 7.2). We then turn to the shift, behind which were two factors. First, conceptual difficulties with three notions that philosophers had relied on— mass, contact action, and extended-body motion. Rational mechanics, it turned out, was best equipped to clarify these (sections 7.3–7.5). Second, developments in rational mechanics itself that were largely undigested by philosophers. The chief causes were a lack of relevant training, along with some institutional barriers (section 7.6). As a result: the rift, and our new task. In section 7.7, we explain the relevant problem to be solved in mechanics (namely, the mechanics of constrained motion, MCON) along with its correlate in philosophical mechanics (PCON). We offer a brief introduction to 18th century rational mechanics (section 7.8), by way of an on-ramp to Chapters 8–11, where we take up the story in detail. Section 7.9 summarizes our conclusions for this chapter.

7.2 Methodology

In the earlier chapters of this book, our actors were themselves explicitly engaged with the problems and questions we call BODY. In contrast, for those who worked in rational mechanics, this was seldom the case. Yet, bodies continued to be presupposed across philosophy and rational mechanics. Indeed, the claims of rational mechanics at the time to be offering a science of bodies in motion means its protagonists were obliged to say *something* about what those bodies are, especially since philosophy had so far failed to provide a notion of body adequate for their purposes. Euler explicitly asserted that *mechanics* has authority over the account of body in general, and not philosophy (section 7.6). So, rational mechanics is not completely silent; one thing we can do is examine the explicit claims found in the texts of the period, and see how they fare with respect to BODY.

A second thing we can do is uncover and analyze the philosophical significance of the moves made by those rational mechanicians, *independent* of the motivations and purposes of the protagonists at the time. That is, we can look into the problems solved, the presuppositions and assumptions made, and the means whereby solutions were achieved (or not), to see how these developments impact BODY.

In what follows, we do both, by means of the three methodological heuristics articulated in Chapter 1 and employed in this book. First, we seek to recover meaning from use, where by "use" we mean: in philosophical argumentation and in theoretical problem-solving. Second, we strive to use anachronism judiciously, developing our arguments just as a premier authority, fully au courant with the state of the art then, could have done (with the proviso that some of the words we use have changed their meaning since that time). Finally, we explicitly recognize our own authorship: BODY is *our* problem, and we chose it because it interests us. It was also, as we have seen in the preceding chapters, often *their* problem too, a live problem of the day. A little more articulation of this last heuristic will be helpful here. Whenever we discuss BODY, it is always someone's problem; ours, theirs, or both. Our methodology prohibits treating BODY as floating free, as having a disembodied life of its own (excuse the pun). This point is especially important in the chapters that follow where, very often, we will be addressing BODY using work in rational mechanics that was carried out by people whose primary motivation lay elsewhere. In these instances, it is not that our protagonists were somehow mistaken about their own projects, engaged in working on BODY without knowing it; they are not sleepwalkers. Rather, it is *we* who bring BODY to their work, and we who use their work to examine the specific form, contours, and development that this problem takes in the context of the projects that they were engaged with at the time. Moreover, as we have emphasized throughout, an adequate solution to BODY requires a *philosophical mechanics*: it demands resources beyond those of either philosophical physics or rational mechanics alone. So, in our investigations into rational mechanics we are on the lookout for two things: developments that pertain to rational mechanics itself (*actors'* category), and those that pertain to philosophical mechanics (*our* category of philosophical analysis).

So now we see our task. Looking back from our present-day vantage point, we seek to (1) identify what developments were taking place in rational mechanics relevant to BODY, and (2) spell out the philosophical significance of these developments for BODY. This is the work of philosophers of physics. The task is demanding, and it will take the remaining chapters of this book to see it through.

7.3 Elusive mass

We begin our argument by identifying three strands in the standard strategy for tackling BODY that became increasingly problematic as the 18th century unfolded. The first of these is "quantity of matter," or mass, and the second is contact action. Both hide complex conceptual issues that, as of the 18th century, remained unresolved. The third is the theory of motion for extended bodies, and this is where enormous progress was made. We treat them in turn.

In the preceding chapters we have encountered multiple philosophers who introduced a notion of mass, either explicitly or implicitly, and we have seen that it was frequently unsatisfactory, lacking justification within the metaphysics of the system being developed and/or failing to meet the demands of collision theory. In Chapter 2, we labeled this issue "elusive mass." We are now in a position to articulate it in more detail, and to show its consequences. It shifts attention concerning the general properties of bodies from philosophical physics to rational mechanics—at least for mass, the single most important general property of all.

In Descartes' physics, the "quantity of matter" associated with a body is simply its volume. This is what makes his notion intelligible according to his epistemology. Newton in his *Principia* introduced the term "mass" to designate a body's "quantity of matter," and he included the volume of a body in its definition. Yet this similarity between Cartesian "quantity of matter" and Newtonian mass is deceptive, for the two concepts are radically different. Unlike Descartes, Newton associates an "inherent force" with matter—a "power of resisting," a "force of inertia"—and uses his second law of motion to relate this quantitatively to changes in motion in response to external forces. Developed through considerable conceptual labor, the result is a complex innovation whose importance was under-appreciated by many, and often not recognized at all.[2] Indeed, while all the rules of collision we have considered thus far invoke some notion of mass, this apparent ubiquity hides a surprising proliferation of ideas and lack of conceptual clarity. If we claim that mass is among the general properties of bodies, then what exactly do we mean by this?

[2] To see just how much effort it took Newton to obtain his mass concept, see Fox 2016 and Smith 2019.

(i) Geometric mass

Consider first the idea that "quantity of matter" is measured by the volume of a body. An immediate problem is that two bodies of equal size may behave differently in important test cases. In particular, both collisions and the balance scale may seem to indicate a difference in mass among equal volumes. Take, for example, one ball made of balsa wood and another, equal in size, made of lead: an impact from the latter feels rather different, and they do not balance one another when placed on a scale. A solution is readily to hand. Introducing a distinction between the apparent volume of a body (given by its bounding surface, such as the spherical size of our wooden ball) and its true volume (the apparent volume minus the volume of any "pores" inside the body) addresses this while preserving the volumetric—or geometric—conception of mass. There is very little discussion of mass in 18th century natural philosophy, and one explanation for this is that both atomism and plenism could treat mass geometrically by means of this distinction. Atomists could choose to assume that all atoms are of uniform density, and account for the varying masses of equal-sized objects by variation in the number of atoms present in a given volume: a piece of lead of a given volume is heavier than a piece of wood (of equal volume) because it contains a greater quantity of matter; it contains more atoms per unit volume. Those who adopted continuous matter theory were typically plenists, and their response was to distinguish between proper and foreign matter. Only the former contributes to the quantity of matter associated with a body, and constitutes the "true volume" of the body. All true volumes of equal size are equally dense. Foreign matter, though present in the overall volume occupied by the body and thereby contributing to its apparent volume, does not belong to the body proper: it is simply passing through "pores" in the proper matter. It does not contribute to its true volume, nor therefore to its mass: the equal-sized pieces of lead and wood, though having the same apparent volume, differ in true volume, and so differ in mass. This latter approach is found in Du Châtelet, as we saw in Chapter 5.

(ii) Kinematic and dynamical mass

In addition to volume (and the geometric conception of mass), there is the persistence of a body in its state of motion as described in Newton's first law of motion (call this "kinematic mass"), and there is the weight of a body. Together, these are the three most common ideas associated with the term "mass" in the figures we have considered so far.

Yet none of these captures the dynamical conception of mass expressed in Newton's second law of motion. This law quantifies the change in motion that a body undergoes as a result of an impressed force, and that change in motion depends upon the mass of the body: $F = ma$, in its most familiar formulation, where F is the force acting on the body, m is the mass of the body, and a is the resulting acceleration.[3] Given a known impressed force and a measured acceleration, we can calculate the mass of the body. Notice that this is *independent* of any consideration of its volume. In contrast to the geometric conception, Newton's second law concerns *dynamical* mass.[4] And notice that Newton's first law by itself is insufficient for this dynamical mass concept, because that law tells us nothing about how a body responds to an impressed force; it gives us *kinematic* mass only.

It is not obvious that the above threads in the mass concept will come together into a coherent package. To see this, consider that a body has parts, and ascribing to it a quantity requires a sum of the magnitudes of those parts. But what is being added, and what is the mathematical form of that addition? Here, indeterminacy intrudes for, by 1750, three different matter theories had emerged, and each gave a different answer. As a consequence, quantity of matter, and hence mass, is indeterminate unless coupled to a preferred theory of matter. To see this more clearly, consider what it means for a perceptible body K to have a "quantity of matter." Three exact formulations are available:

$$m_K = \sum_i m_i \tag{1}$$

and also

$$m_K = \int_V \rho_K dV \tag{2}$$

---

[3] This formulation is not found in Newton's work. It is due to Euler 1752a.
[4] The distinction between "geometrical" and "dynamical" mass is due to Biener & Smeenk 2012. See McMullin, 1978, chapter 2, for a taxonomy of the different roles that mass plays in Newton's *Principia*.

and, finally

$$m_K = \Sigma_i \rho_i V_i \qquad (3)$$

The first formulation is entailed by matter theories in which the mass point counts as the unit body. In words, it says that the material mass of a (macroscopic) body is the sum of the masses associated with its component points, or "physical monads." The second formula falls out of the theory that matter is a physical continuum. It says that a body's quantity of matter is the integral (over the body's volume) of its density ρ.[5] The third concept goes with the picture of matter as made up of discrete atoms, that is, of finite rigid volumes filled with mass density, and situated at finite distances from each other. In words, the quantity of matter for a body of this kind is the sum of its individual masses, for each constituent atom. Importantly for our purposes, deciding among these will depend on detailed work on the motions of extended bodies, and this falls within the domain of rational mechanics.[6]

(iii) Mass in rational mechanics

When it comes to the motions of bodies, *dynamical mass* proves to be the most important mass concept. It provides a measure of quantity of matter that has multiple virtues: it is empirically accessible via the body's kinematic response to impressed force; it is intimately tied (via gravitation theory) to weight, itself empirically accessible; and it is fruitful for problem-solving in rational mechanics generally. And, it is those prioritizing *mechanics* who have most need of the dynamical concept, and most at stake in clarifying the concept of mass.

This clarification took time. Newton himself was not always clear about the different aspects of his mass concept that he wrapped together into a single notion. Euler was the first, so far as we know, to attempt a purely dynamical

[5] In turn, the notion of (mass) density requires that density be describable by a continuous, differentiable function ρ non-negative everywhere in space, endowed with the structure of a Lebesgue measure (so that mass density at a point can be additive over the extended volume taken up by the body).

[6] Crucially, even the *kinematics*—the exact description of motion—for an extended body depends on the choice of its basic-level constituents (from among the three candidates above). For a trenchant statement of that important lesson, see Wilson 2009.

theorization of mass, freed from an associated geometric conception.[7] We see this explicitly in *Mechanica*, where he set out a research program for a complete mechanics.[8] The first step of Euler's program considers those extended bodies that can be treated as point particles; that is, the extended mass distribution of the body plays no role in the trajectory of that body, and so its motion can be treated by means of a representative point. This is important, for these point particles may differ in their quantity of matter (or mass). Euler introduces his concept of mass in the middle of chapter 2. Consider two point particles a and b, acted upon by two external forces p and q, respectively. The two points will "acquire the same acceleration" just in case $q/p = b/a$. This equality, he explains, is "the foundation of measuring the force of inertia ... whereby matter or mass in mechanics must be considered." In virtue of it, a point particle counts as having greater or lesser mass than another, by a quantified amount.

Euler's demonstration proceeds as follows. Let $q = np$. Then, from $q/p = b/a$, we have that $b = na$. Question: In this case, will the force q on the point particle b have the same effect as the force p on the particle a? The na parts of b are acted upon by force $q = np$, so each part a is acted upon by each force p (i.e., we can treat them separately as acted upon by separate "elements" of the force); therefore $b = na$ is acted upon by force $q = np$ as a is acted upon by the force p. So, the point particles are related in magnitude to one another not by the summation of geometrical parts, but dynamically. The equal division of b into n parts is made not in the sense that the elements are equally small (the geometric notion of mass), but in the sense that a given force yields an equal effect on each element (dynamical notion). Moreover, these parts are not prior to the whole: we can divide b into however many parts we need.[9]

Euler follows Newton in asserting that the force of resisting changes in the state of motion or rest is the same as the force of continuing in the same state. This latter is inertia which, for Euler, is an essential property of bodies rather than a force: it is the property that bodies have for conserving their state.[10] Nevertheless, he continues to talk of the "inertial force" of a body, and the "matter, or mass" of a body equals the sum of the inertial force of its component point particles:

---

[7] Nevertheless, he was committed to bodies as extended (as we have seen). More work is needed to spell out the relationship between mass dynamically conceived and bodies geometrically conceived in Euler's philosophy.

[8] Euler 1736.

[9] See Chapter 5 for Euler on extension in relation to the divisibility of bodies.

[10] Euler, 1736, Preface.

For every body, its force of inertia is proportional to the quantity of matter of which the body consists. ... The body's force of inertia must be estimated from the force, or power, required to put the body out of its state of rest or uniform rectilinear motion. Now two different bodies will be equally changed out of their state by two powers that are respectively as the quantities of matter contained in those bodies. Thus, their forces of inertia are proportional to those powers; hence also to their quantities of matter. (Euler 1736, 57)

So, as Euler says explicitly, the mass of a body is measured by the change in the state of motion arising due to a disturbing force, or the acceleration due to an impressed force, and the upshot is that the inertia of a body is measured by its acceleration in response to a given impressed force. Quantity of matter, or mass, is measured dynamically.

We have said that dynamical mass measures inertia of *linear* motion. That qualification is critical for understanding mass correctly. As became clear by the mid-1700s, bodies also have *rotational* inertia, not just inertia of translation; that is, they likewise resist changes to their circular motion. But, the measure of that species of resistance is more complex. To infer how much a body will resist torques (attempts to make it turn around an axis) it is not enough to know how much dynamical mass it has; we must also know exactly how that mass is arranged (the shape of the body and the mass distribution within that shape). That joint information—of the quantity of mass and its spatial distribution—is encoded in a parameter now called *moment of inertia*. In rational mechanics, Huygens first broached that notion, but it was Euler in the 1750s who formulated an exact theory of the bodies' moment of inertia, the rotational analogue of linear mass.

For our purposes, this is important because rotational inertia emerged as a concept in—and is critical for—the theory of the motions of extended bodies. This makes it important for BODY. Yet, it is a highly technical concept, forged and refined in rational mechanics. As a result, it is yet another factor pushing research pertinent to BODY out of philosophical physics and into the domain of rational mechanics.

Alongside those who worked with a notion of mass as central to their projects, there were those who attempted to construct mechanics without mass, eliminating it as a fundamental concept. We see this in d'Alembert's mechanics, and also in Boscovich's *Theory* (see Chapters 9 and 6, respectively). The examination of any such eliminative project requires us to identify *which* elements of the mass concept are the targets of elimination and which, if any, are being retained. The distinctions made above are useful for this purpose.

In sum, the concept of mass, and even the need for such a notion, continued to trouble 18th century philosophers and mathematicians alike. To some modern eyes, the schism in Enlightenment conceptions of mass may seem shocking, but it was not entirely unexpected. As we noted above mass was, from the beginning, a hard-won concept. Newton, its inventor, struggled to see it clearly, and when he turned to writing, he was not always clear and univocal about the exact meaning of his novel concept. The difficulties with mass persisted throughout the 18th century, and they affected everyone. The important point for our purposes is this: "mass" remained an elusive concept mid-century, and if attempts at clarification were to succeed then rational mechanics would have to play a central role.

7.4 Contact action

Mechanical philosophers of the 17th and early 18th centuries had this in common: they deemed contact to be the only intelligible means whereby one body acts on another. In the wake of the *Principia*, they—especially in continental Europe—challenged Newton's claim to have produced a physics of gravitation on the grounds that action-at-a-distance is unintelligible. The contrast, they alleged, was with contact action as the standard. Yet as we saw in Chapter 5, Maupertuis in 1732 threw down a challenge to his French compatriots: he argued that contact action is no more intelligible than Newtonian action at a distance. With the spotlight on the intelligibility of contact action, a new problem comes to the fore: what exactly *is* contact action; what does it amount to, and in what terms should it be described?

Two views emerged. The first is the "traditional" view, inherited from the 17th century claim that bodies are extended parts of matter, and maintained into the 18th century among Cartesian-, Newtonian-, and Leibnizian-inclined philosophers alike. On this view, each body occupies a space—it has a geometric shape—and bodies are in contact if these shapes "touch." We call this *geometric contact*, or *g-contact* for short. This notion was officially a part of Wolff's doctrine, and he in fact restricted all corporeal action to it.[11]

---

[11] "There can be no change in a body except from another body in contact with it" (Wolff, 1731, 112, §128).

However, the notion of "touching" behind this picture of contact remained problematically unclear, as we will see.

The second view was new in the 18th century. On this *dynamical* view, two bodies count as being in contact if their mutual repulsive *forces* go above zero. Call this *d-contact*. This view can be found in both Boscovich and Kant, as we have seen in Chapter 6. It remained a constant in Kant's thought, retained even after his drastic conversion to a continuum theory of matter. For him, "contact in the physical sense is the immediate action and reaction of *impenetrability*," which relies on a repulsive force for its obtaining.[12] In other words, geometric touching—the fact of two bodies' bounding surfaces merely overlapping at a common area—is not enough to count as contact *action* between bodies.[13]

Kant saw clearly the ambiguity in his century's understanding of contact-action. In 1786, he explained it, and expressed his preference for the d-contact picture:

> Contact in the mathematical sense is the common boundary of two spaces, which therefore [on account of being their boundary] is neither inside the one space nor inside the other. . . . Now a circle and a line, a circle and another circle—they make contact at a point; surfaces make contact over lines; and bodies over surfaces. Mathematical contact underpins physical contact, but there is more to the latter notion. For physical contact to obtain, we must also think a dynamical relation, namely, . . . a relation of repulsive forces, i.e. of impenetrability. Physical contact is the mutual action of the repulsive forces [built up] on the common boundary of two matters. (Kant 1911, 512)

These two notions are really distinct; they do not count as just two explications of the same phenomenon. In fact, they are not even coextensive—they do not pick out the same physical situations entirely. Two bodies can be in g-contact but not in d-contact, and vice versa (see Figs. 7.1 and 7.2).

Moreover, these notions are separately incompatible with certain theories of matter. Suppose that matter is really made of mass points: then g-contact cannot obtain, and so it counts as an empty concept; only d-contact is possible

---

[12] Kant 1911, 511.
[13] Boscovich similarly theorized the mutual action of bodies dynamically rather than geometrically; see Chapter 6.

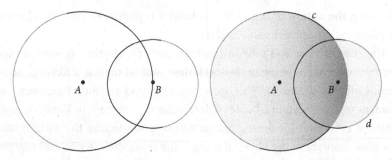

Fig. 7.1. Two point masses in contact. Left: *A* and *B* are in a type of geometric contact: two of their associated spaces (viz. the volumes over which they exert repulsive force) overlap. And yet, *A* and *B* are not in dynamical contact: the repulsion force on each (by the other) is zero. Right: Two mass points, *A* and *B*, in dynamical contact. Specifically, *B* is within *A*'s field of repulsive force (i.e., below the bounding surface *c* at which its repulsion drops to zero).

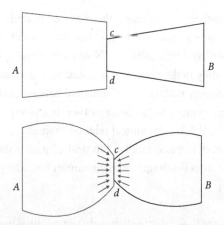

Fig. 7.2. Two deformable continua in contact. Top: Bodies *A* and *B* are in geometric contact: they overlap at the surface *cd*, which counts as common to both. However, neither of them is exerting any repulsive force on the other. Bottom: The same bodies, *A* and *B*, in dynamical contact. Upon their mutual compression, repulsion stresses build up below their common surface *cd*, forcing them away from each other.

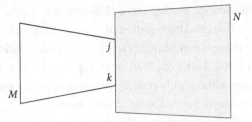

Fig. 7.3. Rigid bodies in contact. *M* and *N* are non-deformable under interactions. They overlap at the common surface *jk*, and so they count as being in geometric contact. However, there are no internal stresses in a rigid body, and so *M* and *N* cannot exert repulsive forces (built up by compression, on either side of *jk*) on each other.

in that theory.[14] Or, suppose that it is made of spatially extended, rigid atoms, lacking any action-at-a-distance forces: then at fundamental scales only g-contact is possible, and d-contact counts as an empty concept (see Fig. 7.3).

Really, of the three canonical theories of matter available in that century, just one—namely, the physical continuum—can accommodate both notions of contact. Unsurprisingly then, Kant, who had just converted to that matter theory by 1786, made the distinction official.

In any case, that contact action admitted of two distinct explanations must count as a further source of difficulty in philosophical mechanics after 1750. What had seemed an unproblematic notion less than a century before was now revealed to require extensive theorizing, with different options available for investigation. Any satisfactory resolution would, perforce, require detailed work in rational mechanics.

7.5 A general theory of bodies in motion

We turn now to the area of conceptual difficulty that mattered the most in the latter half of the 18th century. The strategy for tackling BODY we have focused on in earlier chapters takes the rules of collision as the principal resource

---

[14] No two point masses can overlap, nor can they share a common boundary—some point, line, or surface common to both. For a lucid and very helpful account of the various notions of contact and pictures of matter, we recommend Smith 2007 and 2013.

from rational mechanics needed for a philosophical mechanics.[15] The formulation of these rules had been settled in the mid-17th century, and so from this perspective there was nothing new for 18th century rational mechanics to contribute to BODY. However, these rules capture the motions of extended bodies only under some highly restrictive assumptions; a general theory of extended-body motions proved more troublesome.

For some two decades after the 1687 *Principia*, rational mechanics made no advances beyond Newton's results.[16] Most Europeans were busy quibbling with Newton's conceptual foundations or trying to reverse-engineer some of his key results from different assumptions.[17] It was the 1730s before significant new ground was broken. Two desiderata emerged that would shape rational mechanics going forward. The first is a quest for generality. Protagonists sought principles that were *demonstrably general*, covering all problems previously solved along with new classes of problems too. We explain this in more detail momentarily. Looming large as a stimulant in the quest for general principles was an obstacle. Mathematicians ran up against limitations in the scope of problems treatable by the techniques currently available, whether in Newton or elsewhere. Of particular importance for our purposes is the lack of adequate resources for treating the motions of extended bodies. This problem falls within a wider project constituting the second desideratum: a mechanics of constraints.

Upshot: a great deal more rational mechanics will be needed for a philosophical mechanics of bodies in motion; or, more simply, for an adequate solution to BODY. Since much of the work was incomplete as of mid-century, this engendered a shift from philosophical physics to rational mechanics as the primary locus of research pertaining to BODY.

---

[15] The laws of motion had a dual status, as laws of nature as well as axioms of rational mechanics (see Brading forthcoming), and their formulation was a matter of continued debate during the 18th century.

[16] Newton, in Book I of his *Principia*, had pursued with great success an approach to motion that he called "rational mechanics" (see Chapter 1). In his time, "mechanics" normally denoted a science of *equilibrium* (viz. the statics of some select rigid bodies) or also the engineering dynamics of the five ancient-Greek "simple machines" plus the inclined plane. Newton's rational mechanics is different: he assumed some dynamical parameters as initial conditions, and inferred their kinematic consequences, viz. predicted the ensuing motions; or he assumed certain trajectories as given, and inferred the forces necessary or sufficient for producing them. Newton put it as the "Science of the Motions that ensue from any forces whatsoever; and of the forces required for any motion whatever, precisely [*accurate*] stated and demonstrated." A conceptual innovation that Christoph Pfautz signaled for attention in his account of the book for the Germans; cf. Newton (1687, Preface) and [Pfautz] (1688, 304). Euler in 1736 repurposed the old term "mechanics" to denote the science of *motion*, not equilibrium.

[17] Lucid accounts of how the Leibnizians did that are Bertoloni Meli 1993 and Aiton 1972, 123–51.

The calls for generality from leading natural philosophers began barely a decade after Newton died. This baffles us moderns because—by a prejudice that Ernst Mach instilled in us—we expect everyone then to have seen right away that *Newton's* three laws were truly general principles of mechanics.[18] And yet, no one then building up mechanical theory thought that. Consider these pronouncements. One is by Daniel Bernoulli, who had just solved Leibniz's old problem—a case of motion under rheonomic constraints—of a bead moving outward in a tube that rotates in a plane around a fixed point:

> One would be wrong to regard this sort of problem as just an isolated research whose sole fruit is the pleasure to have solved it. . . . Such mechanical problems never fail to enlighten us: often they yield *new* principles, and teach us *new general laws* that nature follows in all her operations. . . . The wonderful System of the world, whose important discovery was reserved for the great Newton, is simply a consequence of Mechanics. And, what is left to discover in this System depends solely on achieving a greater perfection—in my opinion, not so much by improving the calculus as by getting closer to the *general laws* of motion, which are *still hidden* to us. (D. Bernoulli 1746, 55; our emphasis)

Euler expressed the same sentiment as he struggled with the motion of an *extended* body around a fixed axis:

> These principles are of no use in the study of motion, unless the bodies are infinitely small, hence *the size of a point*—or at least we can *regard* them as such without much error: which happens when the direction of the soliciting power passes *through the center of gravity*. . . . But if it does not pass through that center, we *cannot* determine the entire effect of these powers. That is all the more so if the body to be moved is *not free*, viz. is constrained by some obstacle, depending on its structure. (1745, §17; our emphasis)

And again, years later, still struggling with the motion of extended bodies:

---

[18] Mach claimed that "Newton's principles suffice for solving *every* mechanical problem we encounter in practice, be it in statics or dynamics. We need not appeal to any *new principle* for that. If we run into obstacles, they are always just mathematical, i.e. formal. Not difficulties with the *principles*" (Mach 1883, 239; our emphasis).

I must note that the principles of Mechanics so far established suffice only for the case when the rotation is around a steady axis. . . . But if the axis of rotation changes, the principles of Mechanics we know so far *are insuffi-cient* to determine the motion. Thus we must find and prove *new principles*, suited for solving the problem above. (Euler 1752a, 188-9; our emphasis)

Implicit in their lament above was the denial that Newton's laws had the generality they required. So much for Mach's prejudice.[19]

Examined more closely, the calls for generality faced in two directions. One was *forward-looking*, so to say, and essentially open-ended. Namely, it called for the discovery of laws general in the sense of being applicable to mechanical systems *as yet untamed*. At the time, the epitome of these untamed systems was the rigid body moving around a fixed point, under the action of an external force.[20] But there were others, too; for instance, the motion of elastic membranes and of thick plates. The other was *retrospective*, as it were, because it called for the *unification of known* solutions (under a single law or a small set of connected laws). To be precise, by 1740 many mechanical setups had been solved from "local" theories. For these systems, equilibrium conditions or equations of motion had been derived from laws known to hold for *some*, but far from all, mechanical processes.[21] The retrospective calls for generality demanded these known solutions be *re-derived* from a single, unifying law. Thereby, that law would be shown to be a general

---

[19] Notice the *epistemic* importance of this lack of generality, in relation to Evidence. Suppose one believed that Newton's laws are complete in principle (that is, they cover all physical phenomena). How would one gather evidence for this? One would do so by using those laws to construct theories of motion for physical systems. But the lack of generality means that we have systems for which Newton's laws are no help. So, the epistemic barrier to treating these systems by Newton's laws becomes a problem of evidence for the claim that Newton's laws are complete. Absent demonstrable generality, we are not justified in making that claim.

[20] This setup needed treating because it was the key to modeling mathematically a long-known, but as yet intractable, celestial phenomenon, viz. the earth's precession of the equinoxes. As a reminder, Newton's treatment (of that precession) was entirely *qualitative*: he gave no equations of motion—let alone one derived from his laws—that would account for the known precession of the earth's axis. The need to solve rigid-body motion became even more pressing in 1748, when a new celestial phenomenon—terrestrial nutation—was discovered; it too must be modeled as an additional, second-order precession of a rigid body rotating around a fixed point in space. In fact, it was the discovery of nutation that spurred d'Alembert and Euler separately into treating, from different dynamical laws, the kinematics of forced rigid-body motion, in 1749-50; see Chapters 9 and 10, respectively.

[21] Among such local laws were familiar ideas such as Conservation of *Vis Viva*, the Parallelogram Rule, the Law of the Lever, Conservation of the Gravity Center, the so-called Pendulum Condition, and similar statements of limited explanatory scope. For some examples of how these limited-scope laws worked in practice, see Maltese 1992a, and Clairaut's aptly titled booklet of 1742, "On a few principles of dynamics that yield solutions to a great many problems" (Clairaut 1745).

principle of mechanics, superior to the 17th century laws known to hold only for some specific, limited conditions or configurations.

To sum up: in the early 18th century, philosophers debated the correct formulation of the laws of motion (revising and tweaking Newton's laws, for example) and suggested alternative principles (such as Leibniz's conservation of *vis viva*), but the presumption then was that such principles would be sufficient for all cases. The debates in rational mechanics show that, at that time, there was not sufficient epistemic warrant to support this presumption. Rather, it was an open question which principles are truly general, because it was an open question which principles would allow the treatment of the kinds of systems listed above. Resolving this required detailed work in mechanics. Moreover, the unsolved problems of interest frequently concerned bodies moving subject to constraints, such as when a rigid sphere moves around a fixed point. In fact, treating constrained motion lies at the heart of developing a general theory of the motions of extended bodies. And so, insofar as solving BODY demands that we have bodies capable of motion, and of undergoing changes in motion, the rational mechanics of constrained motion is moved center stage philosophically.

7.6 Shifting sands

Throughout the first half of the 18th century, the dominant strategy for tackling BODY took philosophical physics as its starting point and sought a constructive solution. Despite decades of widespread and concerted effort, by the middle of the century this strategy had yet to deliver (see Chapters 2–6). So far in this chapter we have argued for the importance of rational mechanics for further progress in addressing BODY, and this suggests a different strategy, one that takes rational mechanics as its point of departure: develop a rational mechanics of the motions of extended bodies, and show that this is consistent with a philosophically viable theory of matter.

Euler's physics represents a transition between the two strategies (see Chapter 5). For Euler, the search for the properties of bodies is subservient to the search for the *causes of changes* in bodies: what is required of natural philosophy is to determine all and only those properties of bodies relevant to the changes that bodies undergo. Moreover, we are not required to explain those properties themselves: so long as they can be used to explain the *changes*, this is all we need in order to admit them. Now, since it is *mechanics* that identifies which properties are necessary and sufficient for this task, it is to mechanics rather than matter theory that we must turn for identification of the properties of bodies in general.

Notice the shift in domains of authority that this transition engenders. By 1758, Euler had a clear tripartite division for the study of bodies, making explicit where the new boundaries lie. In the opening paragraph of "The knowledge of bodies from mechanics," he states that there are three domains of knowledge of bodies: geometry, mechanics, and physics.[22] Geometry treats bodies only insofar as they are extended. Mechanics treats bodies *as material*, without regard for any other properties with which they are endowed, while physics treats all remaining properties and qualities of bodies. In order to treat bodies as material, mechanics requires the quantity of matter (or "mass") associated with any body, and how that quantity of matter is distributed within the volume of the body.[23] Euler emphasizes that this knowledge is "absolutely necessary" when the motion of bodies is at stake; hence knowledge of quantity of matter and its distribution falls within the remit of mechanics.

Traditionally, it had been the role of philosophy (and specifically physics) to provide the bodies that are the subject-matter of rational mechanics, specifying bodies' nature, properties, and causal powers (see Chapter 5). In Euler's reconceptualization, rational mechanics has authority to determine those properties of body relevant to its materiality: it is mechanics that will tell us, for example, that mass is the relevant quantity and what is meant by the concept.[24]

And so we have two strategies for tackling BODY, distinguished by taking either philosophical physics or rational mechanics as the starting point. In principle, philosophers might adopt either of these two approaches, pursuing philosophical mechanics in parallel with the rational mechanics of the mathematicians. In practice, by the late 1750s the two groups of practitioners had begun to drift apart. Developments in rational mechanics outstripped the mathematical training of most philosophers at the time, and meanwhile institutional factors exacerbated the rift.

---

[22] Euler 1765a, 131.
[23] See discussion of "elusive mass" in section 7.2.
[24] Again, see section 7.2.

(i) An emerging rift

It comes as no surprise today to find laws of physics expressed mathematically as differential expressions. But consider the situation in the early 18th century. Not only Descartes' laws of nature, from 1644, but also Newton's laws of motion in his *Principia*, were expressed in words alone.

In sharp contrast, after 1710 mathematicians began to benefit from a breakthrough that let them expand the scope of rational mechanics to an unprecedented width: they invented the differential equation. Applied to matter, that invention yields a new way of describing its behavior, viz. the equation of motion, which quantifies infinitesimal changes at every point in an extended body.[25]

This application had three consequences. It turned collective attention away from causal processes between whole bodies—toward the kinematics of change at a point. It transformed rational mechanics in a way that favored nomic unity—by generating the collective task of inferring new equations of motion (for as yet untreated systems) from dynamical laws. And, it put rational mechanics—its chief results and methods—even further beyond the reach of professional philosophers. Next, we explain how these aspects deepened the rift between mathematicians and philosophers.

Equations of motion
The task of quantifying kinematic change over infinitesimal neighborhoods— by deriving equations of motion for that particular type of body—set mathematicians apart from those who pursued a traditional approach to BODY. The latter persisted after 1750 in the German-speaking lands, where its supporters advocated for three closely-knit priorities. One, that Nature— the essential properties and powers of body—be given prominence by placing it explanatorily upstream from the more local, specific pictures and processes described in mechanics. Two, that the basic laws of philosophical mechanics be tightly connected to matter theory—preferably by having the laws inferred from Nature. And three, that properly discharging Action must result in causal explanations of a privileged kind, namely, ones that incorporate broad descriptions of whole-body changes over finite intervals of time, from premises about the causal powers of body in general (or at least of its

---

[25] For the historical stages that led to this breakthrough, and some of its philosophical implications, see Stan 2022.

broadest species, e.g., elastic, solid, or fluid bodies). With its new objective, rational mechanics now diverged from this in all three respects: it prioritized mechanics; it prioritized generality of the principles over commitment to a particular theory of matter; and it endorsed equations of motion (expressing the kinematics of infinitesimal change) over qualitative explanation of whole-body processes from a causal dynamics embedded in matter theory via loose reliance on finite quantities.

Ontic and nomic unity

A second, and related, task becomes the pursuit of nomic unity: mathematicians aimed to integrate the various parts of the theory of rational mechanics into a whole, by showing that they are connected to one another *deductively*. Specifically: show that all equations of motion can be inferred from a single dynamical law, itself stated as a differential expression. They meant inferability in a strong sense that mixed purely deductive reasoning with explicit analytic rules (for the manipulation of variables and operators). Epitomes of this integrative endeavor are Lagrange's two theories of mechanics, which we discuss at length in Chapter 11. However, this endeavor likewise set them at variance with those who thought that philosophical mechanics must fall out of the metaphysics of body, via verbal deductions often couched in qualitative language. Whereas rational mechanicians favored *deductive nomic unity*, the traditional approach of the philosophers sought *explanatory ontic unity*: begin from ontological commitments to a theory of matter or of body, obtained from outside rational mechanics; and integrate therein the results of rational mechanics via *explanatory* rather than deductive relations. Viewed from one side of the chasm, the other likely looked irrelevant to the task at hand. Euler, the intellectual colossus of 18th century mathematics and mechanics, attempted both approaches; he stood alone in straddling the abyss.

Institutional changes

The fragmentation above correlated with an institutional development: the rise of the royal academies of Paris, St. Petersburg, and Berlin. These academies in effect were research institutes *avant la lettre*, providing secure employment for a meritocratic elite of mathematicians.[26] The leading

---

[26] This is in contrast with Britain, where the Royal Society remained a voluntary association, devoid of the royal funding dispensed at Paris or Berlin. Instead, the British had to rely on funding from

figures of rational mechanics found such institutions to afford the most appealing career prospects; which in turn tended to keep them from university campuses, where exposure to metaphysics of matter might have been more likely, as would engagement with it.[27] At the same time, employment on campuses—in the provinces, as professors of metaphysics at traditional universities—kept the supporters of matter-theoretic, ontic unity programs away from the royal academies, where they could have engaged with rational mechanics in the 18th century.

In addition, some contingent but very effective obstacles kept philosophers from immersing themselves more deeply into the advances of rational mechanics. One obstacle was linguistic. French was the official language of business at the Berlin Academy, but many local metaphysicians (with Kant as the most lamentable instance) had little French, and so they could not follow the mathematicians' new results and methods.

Another obstacle was really formidable. After 1740, the mathematical framework and techniques of rational mechanics became a good deal more complex than the geometric approaches of Huygens, Newton, and Hermann. In consequence, grasping the newer research required special training. However, such advanced training was not available at university, and self-study was an option only for the most mathematically-gifted minds of the century.[28] As a result, before the 1810s most professional philosophers simply could not follow the state of research in advanced rational mechanics.

Continental Europe had two chances to make philosophical mechanics into a central area of work in the Republic of Letters. Regrettably, it missed both, which sharpened the emerging split we have uncovered in this chapter. One chance was at the beginning of the Berlin Academy. Frederick II wished it to be a place where science *and* letters coexisted and collaborated—unlike at the Academy in Paris, which he otherwise sought to emulate. To that end, he would have the Berlin Academy led jointly by a mathematician

members' contributions. Inter alia, this disadvantage kept them from recruiting foreign talent away from France, Germany, or even Russia (which Euler chose in the last decades of his career).

[27] Johann Bernoulli and Jakob Hermann in the 1720s were virtually the century's last rational mechanicians employed in academia—at Groningen and at Padua, respectively. After that, only Lagrange would work in a teaching capacity, but at very atypical venues (viz. military engineering schools), which kept him from philosophical faculties at liberal arts universities.

[28] Prior to 1740, all of the century's premier mathematicians—Hermann, Daniel Bernoulli, Euler, Maupertuis, and Clairaut—had learned the "higher analysis" by private coaching from Johann Bernoulli, a prominent father of the integral calculus. D'Alembert and Lagrange were self-taught; on the latter's mathematical education, see Pepe 2008.

(Maupertuis) and a philosopher (Wolff). However, Maupertuis took six years to arrive, and Wolff refused outright. The latter had misgivings about the heavy French presence at Berlin, and little regard for Euler, whom he sometimes belittled, in private letters. And, he had reasons to mistrust the vagaries of royal power.[29] And so, Frederick's planned duumvirate came to nothing. Instead, Maupertuis would run the institution by himself, starting in 1746. Then Euler in 1747 challenged German metaphysicians to a debate, but they perceived it as a great slight, resolving to keep their collective distance from Berlin.[30] To these years we owe some of Euler's most important contributions to philosophical mechanics: a fresh defense of absolute space and time; his mature theory of matter; and an ontology of body and impressed force. But, none of the local philosophical talent took it up to engage with it at the time. So, overall, the Berlin institution missed a great opportunity for philosophical mechanics.

As a result, in Prussia by the late 1750s metaphysics and rational mechanics had grown into mutual indifference. Their estrangement deepened after 1765, with Lagrange doing solitary work on the frontiers of rational mechanics while his colleagues in the Academy's Division of Speculative Philosophy had turned to traditional metaphysics.[31] Philosophical mechanics thereby goes dormant in Germany, at least until Kant's solitary efforts in the 1780s, and then Gauss' renewal of rational mechanics in the 1820s.

The other chance was at the end of life for the Paris Academy of Sciences. From its inception in 1666, this institution had been walled off from philosophical inquiries. Its given mission was to advise the government on scientific matters, review intellectual-property claims, and do pure research in the sciences and mathematics broadly conceived.[32] Philosophy was left to universities and the Academy of Letters, a different institution. After 1789, the Academy of Sciences became a target of revolutionary hostility. Some had

[29] On these facts about Wolff, see Winter 1956, 22ff., and Wolff's correspondence with Manteuffel, in Ostertag 1910. Academic intrigue that spilled into politics at the royal court caused Wolff in 1723 to be banished from Halle, and to lose his job at the university there. He went to Marburg (outside Prussia), where he taught until 1740, when he was finally able to return to his old school.

[30] Euler there became embroiled, sometimes against his will, in various academic disputes—about monads; space and time; and physical action—that turned heated, and had some political undertones. A good account is Leduc 2015.

[31] Recent work on philosophy at the Berlin Academy in the 1700s is Leduc & Dumouchel 2015. Calinger 2016 recounts Euler's prolific activity there. The classic synopsis of research output at the Academy during the Euler years is Winter 1956; von Harnack 1900, volume 1A, is an older but useful narrative of the Academy in the 18th century during Frederick II's reign.

[32] In 18th century France, "mathematics" covered arithmetic and number theory, several species of geometry, analysis, theoretical astronomy, kinematic optics, and rational mechanics.

come to see it as an arm of the *ancien régime*, so they called for its termination. Academicians who had joined revolutionary legislative bodies sought to save it in some form. Among them was Condorcet, who drafted a proposal for reform; he called for the academy to be supplanted by a National Society of Sciences and Letters. It would employ mathematicians *and* philosophers alike—an institutional first, for France. If created, his Society would have given two previously separated groups the chance, place, and resources to collaborate on common problems in the foundations of science. Unfortunately, political headwinds kept Condorcet's proposal from becoming a reality. In 1793, the Royal Academy of Sciences was dismantled, with no successor institution to give philosophical mechanics a home to take hold and flourish.[33]

And so: the Age of Reason ended with two groups—philosophers and mathematicians—who were, for the most part, separated from one another by intellectual dispositions, disciplinary premises, and institutional barriers. To a regrettable extent, they were kept from working together on problems for whose adequate solution both groups were needed.

7.7 From rational to philosophical mechanics

The above rift means we cannot turn to *philosophers* then to understand the philosophical significance of contemporaneous developments in rational mechanics for BODY. Instead, we need to turn to mechanical theory itself. Here, we specify the central problem to be addressed by rational mechanics (MCON) as a prerequisite to solving BODY, and the corresponding necessary condition on any satisfactory solution (PCON).

(i) MCON

A general theory of the motions of extended bodies requires a treatment of *constraints*, for reasons we explain in a moment. Newton himself handled almost no constrained systems in his *Principia*. Canonically, he studied scenarios in which bodies can be treated as point masses, free to move

---

[33] Chapter IX of Hahn 1971 recounts the Academy's last years, and the efforts to save it. A more sophisticated account is in Gillispie 2004.

everywhere in empty space.[34] Mathematicians in the early 18th century had only just begun to struggle with constrained motion. We call this challenge MCON. For our purposes, we divide MCON into two. This will help us explain the importance of constrained motion for BODY.

In the first, the task is to build a mechanics of extended bodies that, throughout interactions, continue to move as *one* body, or single-shaped volume. We call it MCON1:

> **MCON1:** Given an extended body subject to *internal* constraints, how does it move?

More specifically, given an extended body whose parts are mutually constrained among themselves (i.e., held together to form one body), what is the motion of each of its parts? A solution to MCON1 would enable us to determine the motion of *every* part of an extended body, as the whole moves. We give an example below.

The second is the difficult case in which the bodies' possible motions are not free, on account of constraints that limit their mechanical behaviors. We call it MCON2:

> **MCON2:** Given an extended body subject to *external* constraints, how does it move?

Specifically, when a body's path is impeded by an obstacle, what is its resulting motion? A solution to MCON2 would enable us to determine the motion of a body when subject to external constraints, as we explain a little more below.

---

[34] More precisely: he did not handle any such systems from *his* own dynamical laws, by means of *mathematized* models of motion. Admittedly, Newton in Book I does discuss some extended bodies—the attraction of ellipsoids—but he does so in a static context or with the aim of *eliminating* their extension. That is, Newton takes those spheroids to be at rest; and he seeks to replace them with *point-sized* sources of gravity (on other particles passing nearby). In sum, Newton's theorizing there is not about extended bodies in motion. Another seeming exception is Newton's treatment of the compound pendulum (in Book I, §10). But, note what he does there. He *accepts as valid* Huygens' solution to the compound pendulum (from his 1673 *Horologium oscillatorium*, see Huygens 1986), and merely adapts it to the case of central gravity—viz. directed to the center of the earth, decreasing as the inverse-square distance from it, instead of the parallel constant gravity (which was Huygens' assumption in 1673). Plainly put, Newton just *recalibrated* numerically Huygens' result, which was proven from non-Newtonian principles (namely, from a *vis viva* assumption), and which consisted in *replacing* the extended body (i.e., the pendulum bob) with an equivalent *mass-point*. In effect, he used an eliminative strategy born out of the inability to treat extended-body constrained motion mathematically and from first principles. See also Chapter 8, as well as Stan 2022.

An illustrative example of early work on MCON1 is rigid-body motion. A rigid, extended body is one whose parts move together not just as one connected body, but in a fixed configuration: if I move one end of a rigid rod, the rest of the rod moves accordingly. The parts of the rod are *constrained* to move together. Another much-studied type of constraint was the motion of an incompressible fluid.[35] This alone should suffice for a glimpse of two things: first, the central place that rational mechanics comes to have in tackling BODY; and second, how hard that problem became in the 1700s, and how much effort it cost to tackle it.

However, extended bodies are just one species of constrained system. Articulating their mechanics qua extended is merely MCON1; the problem is broader than that. Its other half was the need to mathematize the motion of bodies when external constraints are present (MCON2). French theorists sometimes called them "invincible obstacles."[36] Put more exactly, external constraints are *kinematic impossibilities*, or restrictions on free motion. They are conditions (on the system) which entail that certain directions of motion, trajectories, configurations, or shapes are disallowed—so, they never obtain—because they are incompatible with the given conditions.

Kinematic constraints were a placeholder: they stipulated that a rod remains rigid; they did not give an account of the *forces* by which this is so. That the constraints they treated were just kinematic was a mixed blessing for 18th century theorists. The constraints being studied ultimately amounted to restrictions on the overall *shape* of a system or of its location in *space*; shape and spatial location were the common topic of various *geometries*; and the constraints were *holonomic*: so the makers of rational mechanics often resorted to exploiting the *geometry* of a system in order to gauge the effect of constraints on its motion.[37] Namely, they stated constraints as

---

[35] For this object, the constraint is that the motion be isochoric, that is, the fluid takes up an equal volume throughout the time it moves, even though it changes shape. For the other, the constraint requires *self-congruence*, that is, the body must keep the same shape and size throughout its motion.

[36] Two easy examples are the inclined plane and the mass-point pendulum. In the former, gravity urges bodies to move vertically downward, but the inclined plane *prevents* any body on it from so moving. As a result, that body is constrained to move in the direction that remains kinematically free, or unconstrained: namely, parallel to the inclined plane. Mutatis mutandis for the pendulum bob, which is prevented by the string (from which it hangs) from moving vertically down as gravity impels it to do.

[37] Constraints are *holonomic* (a 19th century term) when they can be stated as functional dependencies between just the positions of the constituent bodies in a mechanical system. If some constituents' motion depends on the *velocities* of some other part in the system, the constraints are anholonomic. Further, holonomic constraints come in two species. They can be *scleronomic*, when their expression is only a function of space-coordinates, not of time. (A famous 18th century example was the downward motion of a bead on a curved wire, under gravity.) Or, they can be *rheonomic*, when the constraints are a function of coordinates and also of time. (A famous rheonomic system

restrictions on the geometry of a system. That was a blessing, as it enabled rational mechanicians to draw on the immense trove of geometric knowledge accumulated by 1740, or even add to it, with a view to mechanical application.[38] At the same time, this implication required theorists to explain how these conditions on the system's geometry link up explanatorily with the dynamical laws and the forces—which are generally *unknown*—responsible for maintaining the constraints. However, these figures never explained this aspect. It is quite hard to explain constraints from mass- and force-distribution, and so they never broached this problem. That had an unforeseen effect: rational mechanics in the Enlightenment becomes unmoored from matter theory, and from physics more broadly. In modern terms, the mathematicians actively pursued physics avoidance, and so their mechanics suffers from the mystery of missing physics.[39] We see how this plays out in later chapters.

(ii) The problem of constrained motions: PCON

The role of rational mechanics is to provide solutions to MCON1 and MCON2. With this in place, we can hope to move to a philosophical mechanics, and to a solution to *BODY*.

From the perspective of *BODY*, the challenge is clear. The mechanics of constrained motions provides us with a theory of the motions of extended bodies. And so, just as the mechanics of collisions makes demands on our accounts of Nature and Action (hence the importance of PCOL, as we saw in earlier chapters), so too does the mechanics of constrained motions. That is to say, any purported solution to *BODY* must address this question:

PCON: What is the nature of bodies such that they can be the object of a general mechanics (that is, a general theory of the motions of such bodies)?

then was Leibniz's example of a bead moving outward in a tube that rotates in a horizontal plane. Daniel Bernoulli and Euler in the 1740s worked out its equation of motion.)

[38] This was especially the case in differential geometry. For instance, important results on geodesics (of curved surfaces) came out of Euler's researches of inertial particle motion on a rigid curved surface; and early notions of Gaussian curvature came out of efforts (from Daniel Bernoulli to Poisson and Sophie Germain) to quantify the strain energy of a point in a bent elastic plate, which thus curves in two dimensions, not just one.

[39] "Physics avoidance" and the "mystery of the missing physics" are apt phrases that come from Mark Wilson; see his 2009, and the recent collection of case studies, Wilson 2018.

Whereas PCOL required the integration of the rules of collision into a solution of *Body*, PCON requires the integration of the results of MCON1 and MCON2. Hence the importance of developments in rational mechanics in the latter half of the 18th century for *Body*.

7.8 Rational mechanics ascendant

A little-noticed paper by Jakob Bernoulli in 1703 made a contribution to rational mechanics that proved epoch-making in retrospect (we analyze it in Chapter 8). But it was not until the 1730s that rational mechanics began to emerge as a powerhouse of innovation and development. From 1740 on, it settled into a pattern of massive and steady growth, well past the terminus years of the period at issue in this book.

Very importantly, this new brand of rational mechanics stands apart from Newton's version in three respects. First, it incorporates *entirely new* principles, with no precedent in the 17th century. Even when, rarely and superficially, some of these principles *look* like one of Newton's laws, they differ from it quite significantly in real content, mathematical form, and manner of application. That is the case for Euler's laws of motion, which resemble Newton's principles superficially. Other post-Newtonian innovations were Maupertuis' "principle of least action," "d'Alembert's Principle," and a plethora of further basic laws that we explain in later chapters.

Second, the new, post-1738 approaches treated constrained-motion systems, whereas Newton had tackled exclusively kinematically *free* bodies and particles, as noted above. Indeed, the early Enlightenment had inherited next to nothing of help in tackling systems with constraints, but in 1703 Jakob Bernoulli found a heuristic that would prove immensely fruitful. With confidence increasing that constrained motion *was* tractable, after all, Clairaut found ways to mobilize known mechanical principles into new heuristics that allowed him to handle a broad spectrum of constrained motions. These advances spurred d'Alembert into prolific action, culminating in his great treatises on constrained particles, fluid motion, and rigid-body dynamics (1743–52). At the end of the century, Lagrange took up an insight he attributed to d'Alembert, in his ground-breaking *Mechanique analytique.*[40] For details on these breakthroughs in MCON1 and MCON2, see Chapters 8–12.

Finally, soon after 1742, rational mechanics moved to tackle MCON1, namely, the motion of *whole* bodies regarded as extended volumes of matter.[41] Johann Bernoulli had the initial breakthrough, as he sought a way to quantify the motion of inviscid fluids under gravity. His insight was to single out an arbitrary, very small volume (in the fluid), identify all the forces acting on it, and infer the net acceleration that they would induce; then repeat the procedure for every similar volume there. The fluid's overall motion will then be the time-integral of the accelerated volume elements in the extended fluid mass.[42] However, it was Euler who saw the real merit of Bernoulli's discovery. Armed with his old teacher's insight, Euler in the late 1740s began to develop it into a general recipe for building up the rational mechanics of extended bodies. The gist of his breakthrough was a recipe that we call the Euler Heuristic. This recipe allowed Euler to make immense strides toward a theory of rational mechanics truly broad in scope. In Chapter 10, we discuss it in greater depth; for now, suffice it to say that it came out of Johann Bernoulli's insight above, which he claimed to have had late in the 1730s.

7.9 Conclusions

In this chapter we have argued that, during the mid-part of the 18th century, rational mechanics emerged as the primary locus of research relevant to BODY. This has three broad consequences, whose implications we address in the chapters that follow. The first concerns the role of collisions in relation to BODY; the second concerns the importance of the mechanics of constrained motions for BODY; and the third concerns the very conception of physics itself—its goals, methods, and place within philosophy.

We have seen that impact was the explanatory gateway to everything else in philosophical physics. Yet in mechanics, collisions occupy no such privileged place. In Hermann's widely-read *Phoronomia*, collisions are treated in chapter 6, well behind other mechanical processes. Euler in *Mechanica* does not even include an account of collisions; his introduction there hints

---

[40] Lagrange 1788.
[41] In contrast, Huygens and Newton had treated them as particles, or point-sized bits of inert matter, devoid of extension; as had Leibniz's disciple, Jakob Hermann.
[42] For details, see Chapter 10.

at a reason why (see Chapter 10). As the 1700s wear on, canonical textbooks relegate impact to an even more peripheral place. By 1788, when Lagrange wrote his *Mechanique*, he did not even think it worth having a treatment of collisions at all.

Two circumstances precipitated this demotion of impact. First, the enormous growth in rational mechanics that took place from the mid-century onward put two new priorities at the forefront of research: the quest for general laws of motion, valid across mechanics broadly, not just for impact action; and the imperative to model mathematically new mechanical setups and systems, instead of retreading old knowledge (which the rules of collision were then). The other circumstance is the realization that impact is a complex problem in mechanics, and a good deal more work was needed before collision theory could be secured within the newly developing foundations of mechanics. Hence, PCOL loses its importance for *BODY* and, some thought, should be shifted downstream until the needed resources from mechanics were at hand.

Instead, we see PCON rise to prominence as a necessary condition on any satisfactory solution of *BODY*. By the mid-18th century, a general mechanics of extended bodies, whether free or constrained, was an explicit target of theorizing. The task was to build a rational mechanics that is *general*, and also consistent with a *single* matter theory; that is, to provide a generalized philosophical mechanics. Constraints emerged as the most pressing foundational problem facing mechanics in the early decades of the 18th century. Moreover, as the century progressed, the relationship of *BODY*—and the vulnerability of proposed solutions—to developments in mechanics became increasingly fraught with philosophical and conceptual difficulties. In the remainder of our book, we examine PCON and the associated attempts to solve *BODY*.

Our third major conclusion concerns the place of physics in relation to philosophy. At the beginning of the 18th century, matter theory and physics were under the umbrella of philosophy. Matter theory fed into philosophical physics, the study of material bodies. The bodies of philosophical physics, in turn, were fed into rational mechanics, as its material subject-matter. By the end of the century, rational mechanics had laid claim to authority over the general properties and principles of bodies and their motions. But, as we will see in the coming chapters, physics did not yield to this attempted take-over, and instead took steps toward independence, drawing resources from mechanics, but refusing to fall entirely under the dominion of either philosophy or mechanics.

So here is where we find ourselves: one hundred years after Descartes'
Principles of Philosophy, BODY remained unsolved. Physics, the area of philosophy charged with its solution, had yet to deliver. Mechanics, equipped
with new tools and new energy, was asserting itself as a source of knowledge
of the nature and properties of bodies, but any claim to be offering a solution to BODY was, mid-century, little more than a promissory note. Yet the
philosophical exigencies requiring a solution remained. The unfolding of
this story, and its consequences for philosophy, are the subject of the coming
chapters.

8
Early work in the rational mechanics
of constrained motion

8.1 Introduction

The first half of this book ended in disappointment: the figures involved failed
to solve the problem of bodies (*Body*). They thought the solution would
come out of making coherent sense of the problem of collision (PCOL). In
turn, that struggle with impact slowly taught them the need for an account
of whole-body motion, and led them to a broader realization: before anything else, a sound philosophical mechanics ought to solve the problem of
constrained motion (PCON). And, as we saw in Chapter 7, tackling PCON
shifts our attention away from philosophical physics and into rational mechanics. With the above backdrop in place, we now backtrack to revisit the
first half of the 18th century with *rational* mechanics center stage.

From Chapter 7 recall that PCON demands two contributions from rational mechanics. First, a mathematical account of extended bodies subject
to *internal* constraints—bodies that, throughout their motions, continue to
move as *one* thing, viz. as an extended and connected whole of adjoining parts
constrained to stay together as one throughout the motion. Second, a mechanics of bodies impeded by *external* constraints, viz. obstacles, trajectories,
or regions of space that are inaccessible—kinematic prohibitions. These are
MCON1 and MCON2, respectively (see Chapter 7).

Below, we present the first efforts to discharge both tasks, and we draw out
the implications for PCON. To elucidate those efforts, we focus on two key
episodes: the early treatments of the vibrating string, and of the compound
pendulum. Historically, these efforts went from 1690 to the mid-1730s. In
line with our philosophical, analytic interests in this book, we concentrate
on the inferential structure, warrant, and matter-theoretical assumptions behind the efforts at issue.

Our aim is threefold, and forward-looking. First, we seek to impress on the
reader just how difficult it was to mathematize the motion of extended matter.

Philosophical Mechanics in the Age of Reason. Katherine Brading and Marius Stan, Oxford University Press.
© Oxford University Press 2023. DOI: 10.1093/oso/9780197678954.003.0008

At these early stages, breakthroughs were modest. With them as a backdrop, it will come as no surprise that much of the 18th century was given to solving PCON; specifically, to creating a rational mechanics of *extended* bodies with constraints. Second, we aim to open a new vista. These early efforts uncovered some heuristics that proved enormously fruitful for the later theory of constrained systems. In retrospect, the Lagrangian synthesis of Chapter 11 will appear as the apogee of an insight discovered around 1700. Third, we seek to uncover some serious philosophical complications. Specifically, success in these new areas of rational mechanics often came at the expense of empirical warrant and clear links to a picture of body. In our terms, in these mathematizations it becomes unclear what roles Nature, Action, and Evidence play, and whether they are compatible with each other. As it turns out, these worries will have long-term implications for the chances of an internally coherent philosophical mechanics.

As backdrop, we begin with an outline of the theorists and centers driving the research in the half-century at issue here (section 8.2). Next, we study competing attempts to tackle one half of PCON: the motion of extended bodies. To illustrate their efforts, we choose the vibrating string (section 8.3). Next, we examine a major attempt to solve the other half of PCON, namely, external constraints. As illustration, we use their work on the compound pendulum (section 8.4). And, we complement it with French work on constrained particles (section 8.5). From these five decades of collective efforts, we draw some lessons for philosophical mechanics in that period (section 8.6).

8.2 Personnel and work sites

Much work in the first two research areas surveyed below came from a small group united by family ties and adherence to Leibniz's version of the calculus. Jakob Bernoulli had encountered it in the 1680s, taught it to his brother, Johann, and they went on to apply it to new problems in statics and dynamics. Jakob's research career unfolded in academia, at Basel, whereas Johann had to journey in search of stable employment—first to Paris, where he taught calculus to the marquis de l'Hôpital; then to the University of Groningen; and finally to Basel, where he outlived his brother by decades.

In various senses, the elder Bernoullis fathered the Basel School, by far the most important research group in mechanics during the century after

Newton. Its second generation counted their distant cousin, Jakob Hermann. Recommended by Leibniz, he took up Galileo's former chair at Padua, where he wrote *Phoronomia*. Daniel Bernoulli was Johann's son and unwilling antagonist; much of Daniel's later research took place at Basel, the family's hometown. The youngest and by their joint assent the brightest member of the Basel School, was Johann's student, Leonhard Euler, whose long career spanned half of Europe and most of the Enlightenment.

Johann Bernoulli mattered in another respect. At the time, training in "higher analysis" (viz. the new calculus and its physical applications) was informal and hard to come by. Many important figures—Euler, Clairáut, and Maupertuis—learned it from Johann Bernoulli, who coached them.

There were two institutional centers of gravity for advanced research in rational mechanics in the early 18th century. The first site was at St. Petersburg, where the ambitious Peter the Great had set up the Imperial Academy, a research institute designed to compete with France and Britain. And it did, largely thanks to its hiring the entire second-generation Basel School. Hermann went there first, then Daniel Bernoulli, and finally Euler, who stayed until 1740, when Frederick II enticed him to Berlin, to work at *his* academy.[1]

The second center was the Paris Academy, then in the process of becoming the world's powerhouse of research in mechanics for a good century. There Alexis Clairaut began his rapid, meteoric rise aided by an early training in analysis. Of interest for our topic here is a long paper he presented in 1742, "Sur quelques principes qui donnent la Solution d'un grand nombre de Problèmes de Dynamique."[2]

Finally, nearly all the figures we study did work in Newton's wake too (on celestial phenomena, that is), not just in the rational mechanics of sub-cosmic processes. However, we set aside their astronomical research. For one, the meager advances in celestial mechanics (during the five decades at issue here) did not make a difference to BODY. For another, it was the mechanics of *terrestrial*, sub-lunar motions that drove theorists to understand that new foundations were needed to solve BODY: new laws and concepts that go beyond Newton's resources.

---

[1] For the Bernoulli family, see Fleckenstein 1949; for Jakob Hermann, see Nagel 2005; Euler's definitive biography is Calinger 2016; good accounts of the Basel School's activity at St. Petersburg are Boss 1972, and Nagel & Verdun 2005.

[2] Clairaut 1745. On the rise of the Paris Academy of Sciences in the 18th century, see Hahn 1971, 58–251. A scientific biography of Clairaut is Brunet 1952.

8.3 New territory: oscillating systems

In this section and the next we survey and discuss some early attempts to create a rational mechanics of two oscillating systems: the vibrating string and the compound pendulum. The attempts to tackle the former were modest, cautiously aimed at inferring just one characteristic quantity of the motion of a thin string fixed at both ends. Many pursued it, and we will focus on just two attempts, by Brook Taylor and by Johann Bernoulli. Others have studied these pieces before, but our perspective is novel. We read them as episodes in philosophical mechanics: as further attempts to integrate mathematized models (of motion patterns) with a philosophical physics of body and its generic laws.[3]

From that vantage point, early work on the vibrating string appears broadly instructive about rational mechanics in the early 1700s. Relative to the theorizing that preceded them, Taylor's and Bernoulli's results epitomize four tendencies. First, they show a felt need to *extend* rational mechanics to motion patterns and mechanical processes not covered by the theories of the late 1600s. Second, they help us diagnose a certain *heuristic silence*, as it were: the principles bequeathed by Newton, Leibniz, and Huygens gave no clear, univocal guidance for how—or even whether—they apply to cases and setups not solved by these pioneers. Third, they suffer from *descriptive incompleteness*. Specifically, the several attempts to quantify the vibrating string have a shortcoming in common: they are unable to predict the motion of every point in the string. In modern terms, they merely infer an effective parameter of the string, not its equation of motion. Finally, they *obscure connections*: while they refer to material bodies, the papers above end up severing links to an underlying theory of matter; at the very least, they obscure those links considerably.

Moreover, the papers above matter to a foundational agenda that gained momentum as the century advanced. For one, they show, contrastively, what later figures meant by their calls for generality in mechanics. Namely, discovering principles that were general in the sense of provably entailing solutions to many different mechanical setups. The treatments of the vibrating string that we will present *lacked* this virtue. Second, these treatments attest

---

[3] These papers, and related work at the time, have been surveyed and discussed in Truesdell 1960, Maltese 1992b, Cannon & Dostrovsky 1981, Villaggio 2008, and Feigenbaum 1985. Except for Feigenbaum's work, the others share an anachronistic reliance on modern mathematics, to the detriment of physical and conceptual insight, and of understanding in context.

to a centrifugal tendency that accelerated throughout the century. It is the tendency of rational mechanics to *pull away* from matter theory, and to settle into outward trajectories that break free from linkages to any picture of the *bodies* whose motions these rational-mechanical models aim to represent. Now we move to present their content.

Taylor treated vibration in a 1713 paper entitled "On the motion of a taut string," and reprised his solution in 1717 so as to illustrate Newton's calculus of fluxions; we refer to these pieces as *Motus nervi* and *Methodus*, respectively. Johann Bernoulli followed up in 1728 with a paper that re-derived Taylor's results from different material assumptions, mathematical notions, and heuristic principles.[4]

Both men start from the same givens, or initial conditions. Let AB be the axis on which the string lies when at rest. Pinch it at the middle point C, and stretch it into the tensed *triangular* shape ACB, then release it. They assume that, post-release, the middle point will travel from C to C′ (on the other side of the rest axis) and back, periodically; and that the tensed string starts out as a triangle ACB but after release it rearranges itself into a *curve* of some sort (see also section 8.6, especially Fig. 8.6). Separately, Taylor and Bernoulli pursue the same question: what is the *frequency* of the moving string? Namely, what is the period of the point C of maximum elongation? Implicit in their reasoning are two restrictions: the string vibrates in the fundamental mode; and its vibrations are infinitesimal.[5] From the first restriction, they draw an easy corollary, nowadays called the "harmonic condition": as the string self-relaxes from the tensed position, *all* the particles in it cross the axis AB at the same time.

Against this backdrop, Taylor's study is a two-stage argument. First, he *assumes* the harmonic condition, and asks what type of force (acting on the string particles, driving them toward AB) would secure that condition, viz. cause the particles to cross AB simultaneously. He proves two results about it: the strength of the restoring force must be directly as the distance from the axis; and it must be as the local curvature.[6] Next, Taylor proves the reverse conditional. He assumes that the restoring force on any particle is as

---

[4] Bernoulli 1742b.

[5] They say that the vibrations must be very small, but for their reasoning to go through, the vibration must be *infinitesimal*. Large vibrations remained intractable until the late 20th century; cf. Antman 1980.

[6] That is, at any instant the restoring force on a particle C must be as its distance to the rest axis AB, and proportional to the radius of the osculating circle (tangent to the string curve) at the point C. See also section 8.6, especially Fig. 8.6. This shows why he must restrict the string to infinitesimal vibrations; if they were finite, the curvature CG and the distance CD would differ by a *finite* amount, and Taylor's proof would be fallacious. So, he must impose that both CG and CD are first-order

the distance to the axis AB; and infers that the force will cause "all the string points to arrive at the axis [AB] at the same time."[7] That is, the string is a simple harmonic oscillator.

Taylor then obtains a novel result. From the restoring force being as the distance (to the rest axis) he infers: there exists a simple pendulum that oscillates isochronously with the string AMB.[8] Finding the length of that pendulum gives him the frequency. And this equals the result he sought, viz. the frequency of the vibrating string itself: because the string and its equivalent pendulum oscillate isochronously *ex suppositione*, thus at the same rate.

In his treatment Taylor relied on two premises with dynamical import: that the restoring force is proportional to the local curvature; and that it is the resultant (as given by the Parallelogram Rule)[9] of the local tensions that pull away from each other; see also Fig. 8.6. This fact deserves notice, because soon we will see d'Alembert take up that dual assumption, and use it to infer the wave equation (Chapter 10).

Efforts to produce a rational mechanics of vibration continued throughout the period. The most visible was a 1728 piece by Johann Bernoulli, "Thoughts on vibrating strings."[10] The title is apt, for his paper is really a set of terse notes, not limpid argument. Still, the piece is remarkable for its striking similarity with Taylor's treatment above, which it aims to emulate. In particular, Bernoulli re-derives Taylor's key result—the frequency of the fundamental mode—but from different dynamics. Taylor's key assumption had been the Parallelogram of Forces (cf. section 8.6). In contrast, Johann Bernoulli relies on a different, and obscure, dynamical principle: "I too have obtained [Taylor's formula], but from the principle of live forces."[11] In effect, he seems to reason as follows.

We explain the case of one mass—see Fig. 8.1; for strings loaded with multiple masses, Bernoulli generalizes inductively. He aims to derive the time t

---

infinitesimals, and hence they differ by a second-order infinitesimal amount, which may be safely neglected.

[7] Taylor 1713, 28.
[8] Namely, the pendulum bob P and the string's middle point M would take the same time to reach their respective rest axes, if released at the same instant.
[9] We gloss here over the technical and conceptual difficulties involved in extending the addition rule for powers from simple cases (of time-independent forces acting on point particles) to complex cases involving extended matter.
[10] Reprinted as Bernoulli (1742b, 198–210). The full title is "Thoughts on strings that vibrate with little weights placed at equal intervals on them; in which, from the principle of live forces, it is sought the number of vibrations the string does during one oscillation of a pendulum of known length."
[11] Bernoulli 1742b, 208.

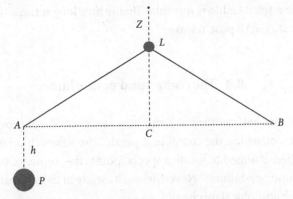

Fig. 8.1. Bernoulli's derivation of the string frequency from a *vis viva*
principle. *L* is a discrete mass that a tensed flexible string *AB* carries to *C*, where
it arrives with a velocity *v*. *z* is the height from which *L* would have to fall freely
so as to acquire *v* at *C*. The tension force in the string is equal to the weight *Ph*.

for the load *L* to arrive at *C* on the string's rest axis *AB*, its line of equilibrium.
Let *z* be the height from which, *if L fell freely*, it would arrive at *C* with the
same speed as it *actually* does when the vibrating string carries it to *C* after
release.

Bernoulli starts from the key assumption that "the *vis viva Lz* of the little
mass *L* [as it moves to *C* under the action of tension in the string] equals the
vis viva of *P*," a weight supposed to be equal to the tension.[12] This quick move
allows him to infer:[13]

$$Lz = Ph. \tag{1}$$

Three of the four quantities above are known; or are inferable from known
quantities. That yields an expression for the height *z*, which then yields the

[12] Bernoulli 1742b, 200. Why those two live forces are equal remains unclear, and Bernoulli leaves
it at that, without further light. Commentators are just as cryptic as he, on this point. One calls it the
"energy theorem" and the "balance of kinetic energy" (Villaggio 2008, 229, 230). Another glosses
it as "the potential energies of *P* and *L* are identical" (Maltese 1992b, 725). These are anachronistic
impositions, not explanations in context.

[13] Bernoulli's assumption is *laminar* gravity, hence the string particles obey Galileo's law of free
fall: if a body drops from a height *z*, its time of fall *t* and its speed *s* of arrival are equal to √z. Hence *z*
equals the speed squared, and so *Lz* equals its *vis viva* at the end of fall, just as Bernoulli has it in expression (1).

sought time t, from Galileo's formula. That is how long it takes the string to reach the rest axis AB post-release.[14]

8.4 The compound pendulum

Throughout the half-century at issue here, another object proved fruitful for rational mechanics: the compound pendulum. Research on it was fairly circumscribed: it aimed to locate a special point (the "center of oscillation") in a compound pendulum.[15] Nevertheless, its study in those decades matters greatly for philosophical mechanics.

First, their attempts to mathematize the compound pendulum illustrate the struggles to make rational-mechanical sense of objects that Newton was unable to handle. While work on the compound pendulum preceded Newton's *Principia*, his book just took it for granted, because that object is intractable from his Second Law.[16] Second, the treatments of this setup vividly illustrate the proliferation of principles in the period we survey here. Theorists then relied on many, distinct dynamical laws: virtual work principles, energy principles, equilibrium conditions, and novel force laws. As a result, the landscape of rational mechanics turns rich and diverse to an extent that we miss if we focus on the *Principia*. Third, and most significantly, in some of these treatments were the seeds of later theories that, in terms of generality of treatment and explanatory power, surpassed Newton's approach, and birthed modern analytic mechanics as we know it.

The compound pendulum is a body with *dual* constraints, internal and external. The internal constraint is that the bob's component particles keep the same distance to each other as the bob oscillates. And, the bob is presumed

---

[14] His reasoning is far from clear, but see a summary in Villaggio 2008, 227ff.

[15] A *compound*, or "physical," pendulum is a theoretical model in which the bob is extended, and attached to a rigid massless rod swinging vertically from a suspension point. In contrast, a *simple*, or "mathematical," pendulum models the bob as an unextended mass point attached to a massless rod. The center-of-oscillation problem is to locate in a physical bob the point H that swings *isochronously* with an equivalent mathematical pendulum. Specifically, let P and M be two equal-mass pendulums, compound and simple, respectively; and let M oscillate with frequency w. The center-of-oscillation problem is to locate in P the point H that oscillates with frequency w. (All other points in it oscillate with *different* frequencies; see below.)

[16] Newton in his *Principia* treated the compound pendulum in Book I, section 10. There he *took for granted* Huygens' Formula above—he did *not* seek to rederive it from his own laws. Instead, Newton just proved that Huygens' results (about the beat of a compound pendulum) also hold in a field of *central* gravity (not just in Galilean-laminar gravity, as Huygens had assumed in his treatment of 1673).

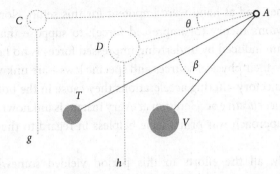

Fig. 8.2. Constrained motion. Where the weights would be (relative to each other) after an instant, if they were free. That is, if the constraint CAD was not rigid. *Cg, Dh* are the (equal) accelerations of gravity. NB: If they were free as above, their mutual angle β at the next instant would be greater than θ, their mutual angle at the *previous* instant.

rigidly attached to the swinging rod.[17] Together, these constraints make a great difference to the motion of component particles in the bob. Consider their behavior if they were *free*, that is, allowed to move without kinematic restrictions (see Fig. 8.2, where the particles C and D fall freely to g and h). Starting from rest, *all* free particles would fall at the *same* rate: so a and b would cover equal distances in equal times. Hence at every subsequent instant, we would find the particles at the *same* height above ground, falling in *straight* lines.[18]

But constraints alter these motions radically. The requirement that the bob stay attached to the rod entails that its particles must move in *circles*, not in straight lines. And, the rigidity conditions entail that the particles must fall at *unequal* rates, thereby crossing *unequal* distances in equal times (so that the pendulum's shape remains self-congruent through its fall and rise). Together, these constraints entail that, in a pendulum bob, some particles must accelerate and others decelerate—relative to their speed of free fall—so as to keep their mutual positions in the bob unchanged. But, no one then knew *how much* the individual particles' motion changes thereby; the dynamical

[17] Note that the rod itself is taken to be massless: it *kinematically* constrains the motion of the bob, but does not contribute to the *dynamics* of the scenario.

[18] These facts, and the particles' individual speeds at every second of their fall, had been known since Galileo's 1638 tract, *Two New Sciences* (Galilei 1974). The kinematics of free fall from rest is in the Third Day of that book. Huygens rederived Galileo's theory in Part I of his treatise, *Horologium oscillatorium* (Huygens 1986).

principles they had at hand were impotent for this setup. Nor could they rely on Newton's theory (of impressed force): to suppose that kinematic constraints are induced by underlying impressed forces is no help, because these forces—their physical sources and specific laws—are unknown, hence so are the trajectory-altering accelerations they cause in the bob. In consequence, the *net effective* acceleration at every instant is unknown, and so the Newtonian approach was prima facie helpless in regard to the compound pendulum.[19]

Admittedly, all the efforts in this period yielded somewhat limited results: no one was able to advance theory beyond the representative-point approach to this object. Just like the vibrating string above, the compound pendulum daunted all efforts to tackle it as an extended body (MCON1). Nevertheless, this work concentrated theorists' attention and energies on the motions of bodies subject to external constraints (MCON2), and it stimulated them to think about it more generally, beyond the specificities of pendulum motion.

At the time, the compound pendulum received attention from the Bernoullis, Brook Taylor, Jakob Hermann, and the young Euler. Space being at a premium, we explain in detail just one solution, by Jakob Bernoulli. It was the most significant: Bernoulli there originated a heuristic—for inferring actual motions from unknown forces—that later d'Alembert and Lagrange would put to very powerful uses (see Chapters 9 and 11). To understand his intervention, we need some background.

Huygens in 1673 had first treated the compound pendulum, by deriving a formula for a special point in the extended bob, viz. its center of oscillation. Specifically, he calculated its distance l from the suspension point (around which the physical bob swings in a vertical plane):[20]

$$l = \frac{\sum(m_i r_i^2)}{\sum m_i r_i} \tag{2}$$

We call it the Huygens Formula. He had inferred it from a key premise that some rejected then. They did not think that Huygens had enough *evidence*

---

[19] The "most widespread mistake about Newton's three laws of motion is that they alone sufficed for all problems in classical mechanics" (Smith 2008, §5).

[20] In the formula, m_i is the mass of the i-th particle in the compound bob, and r_i its distance from the point of suspension. For the reception of Huygens' pendulum work, see Howard 2003, chapter 3.

for it. Their objection led other theorists to show the formula was correct after all, by rederiving it from some principle they thought incontestable.[21]

Among them was Jakob Bernoulli. At first he studied the case of a compound pendulum made of two or more discrete particles on a rigid rod that swings around a fixed point. Though he succeeded, Bernoulli knew that his approach was too narrow: it worked only for the special case of collinear weights (all on a single straight rod, that is). So, a more general approach was needed. After much effort, Bernoulli found it in a brief but seminal paper of 1703. There, Bernoulli states a general principle, and then applies it to the compound-pendulum problem. In effect, these two moves amount to proposing a new *dynamical law* and a new *heuristic* for handling systems with rigid constraints.

Bernoulli claims to derive his solution from the "Lever Principle." His phrase is misleading, because it denotes a radically new insight. As an entry point into his novel idea, consider a *bent* lever—specifically, a massless rigid rod CAD, with A between C and D, and the angle CAD less than 180°. Let gravity be equal and parallel everywhere, and A be a fulcrum fixed in space. And, let two weights be placed on the lever at points C and D. Now ask yourself:

Under what conditions is the bent lever in static equilibrium?

Jakob Bernoulli first answered this question, two millennia after Archimedes. Bernoulli reasoned as follows. For each weight, "conceive" the acceleration of gravity "as being composed of two other ones," that is, resolve it into two components: one normal to the arm, viz. *tangential* to the weight's circular trajectory; and one *radial*, viz. along the arm (see Fig. 8.3). He continues: none of the radial components produce any effect, because they "disperse themselves entirely" along their arms, and so "they become completely lost." Hence, "only the [tangential] motions have their effect."[22]

---

[21] The premise was: "Let a pendulum made of several weights be released from rest, then cross some part of a whole oscillation. Let the constraint (holding the weights together) be dissolved. Suppose each weight converts the speed it has (when the constraint is removed) into motion upward, and rises as high as it can from that motion. Thereby, the compound bob's center of gravity will rise to the same height where it was before release." See Huygens 1673, 98. Among the first follow-up pieces to Huygens' derivation was Bernoulli 1691; for discussion, see Vilain 2000.

[22] Bernoulli 1703, 82.

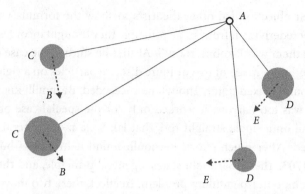

Fig. 8.3. Bernoulli's lever principle. He represented each weight at C, D twice. CB and DE are the tangential components of the acceleration of gravity on each weight.

Put in exact terms, his Lever Principle becomes:[23]

$$\Sigma(AC \cdot c \cdot CB) + \Sigma(AD \cdot d \cdot DE) = 0. \tag{3}$$

Namely: a bent lever is in equilibrium if the net sum of the arm times the mass times the tangential velocity increment is null.[24]

Now here is how Bernoulli applies his Lever Principle. Let CD be the simplest compound pendulum—two weights C and D rigidly connected to a suspension point A and to *each other*. Suppose that C and D were not mutually constrained, but in fact are *free* to move relative to each other. If they were so free, after an instant of fall they *would* have changed their mutual situation: in particular, they would arrive at T and V, respectively. In that case, their mutual angle would be greater (see Fig. 8.2). But a change of angular distance would violate the rigidity condition, viz. that the bob CD must retain its shape throughout the oscillations: the "weights C and D cannot then be at T and V, respectively." Hence, *instead* of arriving at T and V (as they

---

[23] c and d are the masses attached to the lever arms at the points C and D, respectively. CB and DE are tangential components of the acceleration of gravity.

[24] In modern terms, that is the net virtual work done by the (gravity-induced) torque on the system. Note that the individual products are signed quantities: plus if the velocity increments go one way (say, clockwise), and minus if they go the other way (counterclockwise).

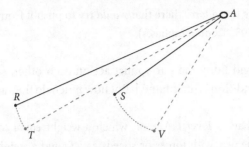

Fig. 8.4. Internal constraints and their action. The difference between where the weights C and D would be, versus where they actually are, after an instant (i.e., the difference between their falling free vs. falling mutually constrained). If the bob was not rigid, in an instant the two weights would fall to T and V, respectively. But the rigidity condition constrains them to move in fact to R and S, respectively.

would, were they free, not constrained), the weights must arrive respectively at R and S; see Fig. 8.4.[25]

Still, that leaves the exact location of R and S undetermined. How might one infer it? Without any notice, Bernoulli's reasoning now reaches a critical juncture, of great subtlety and immense theoretical value.

To arrive at R and S instead of T and V, the weights must act on each other *via the constraint internal* to the bob, whatever its physical underpinnings may be. Specifically, each weight must induce in the other an angular displacement RT and VS, respectively. Bernoulli's breakthrough was to see that the displacements are really the angular differences between the locations where C and D *would* have been, *were* they mutually free; and the locations where they *actually* are, *due to the constraint*. Nowadays, we call them *virtual* displacements induced by the constraints. He explains the dynamics associated with these virtual displacements:

> because of the isochronism [condition], the weights C and D cannot be at the locations T and V: in fact, they must be at R and S. . . . Consequently, the effort of gravity, which acts on the weight C, is not exhausted at R. Hence its leftover effort RT must be used to push the body D across VS. However, because body D must *resist as much as it is pushed*, it is *as if* D *were in S*

---

[25] Which trivially ensures that the angle CAD remains self-congruent throughout the displacement, hence the bob CD stays rigid.

and *as if there was a force* there that *would* try to push it from S back to V. (Bernoulli 1703, 82; our emphasis)

That is, in a rigid body the parts must act on each other, so that the self-congruence condition is maintained. But how much do they act? He answers:

And so, we have a lever CAD, on which a weight equal to C (pushing or pulling one way with forces or speeds as *RT*) and a weight equal to D (pushing or pulling the other way, with forces or speeds as *DV*) balance each other. (Bernoulli 1703, 82)

That is: to obey the constraint, the parts do virtual work—they exert torques on each other, which cause virtual displacements. Over the whole body, their net virtual work vanishes, hence rigid constraints are *workless*. That follows from applying his Lever Principle to the pendulum's bob:[26]

$$C \cdot m \cdot RT + D \cdot n \cdot SV = 0.$$

With this crucial insight in place, locating M, the center of oscillation—and thus obtaining Huygens' Formula—becomes a mere exercise in plane trigonometry.

In any case, his paper is not important for that formula, which had long been accepted as a fact. It really matters for the deep insight it contains: his Lever Principle for a bent lever. For the 18th century, the principle mattered in three ways. First, it broadened the notion of equilibrium beyond the narrow sense it had in Archimedean statics.[27] In Chapter 9 we will see d'Alembert extend that notion ever further. Second, it extended the nomic basis of statics well beyond the meager fundament it had inherited from Ancient Greece. So significant was his extension that in the 1700s some thought the laws of static equilibrium were enough to support *all* of mechanics (see section 8.6). Third, it led them to focus on virtual displacements and the work associated with

---

[26] Specifically, let the length of the arms *AC* and *AD* be *n* and *m*; and recall that the displacements *RT* and *SV* are directed magnitudes. Then Bernoulli's Lever Principle entails that, in the compound-pendulum bob, the quantities $C \cdot m \cdot RT$ and $D \cdot n \cdot (-SV)$ are equal.

[27] In statics before Bernoulli, the notion of equilibrium was based on the Archimedean (straight) lever: two bodies are in equilibrium if their respective distances to the fulcrum are inversely as their masses or weights.

them. That change of focus—on the virtual work of forces and torques on a body—would lead to a crucial breakthrough. It taught theorists that internal constraints are workless: their net virtual work vanishes; see Chapter 11.

8.5 From special problems to general principles

In the early 18th century, work on constrained motion had two features relevant to our analyses in Chapters 9–11. It was piecemeal—it amounted to individual solutions to local, specific problems. And, it was unsystematic: there was no unity of treatment, and no common heuristic; the dominant spirit was epistemic opportunism.

The mechanical setups most studied were the motions of several particles connected by flexible strings; the orbit of a particle moving outward in a rotating tube; and the motion of a particle falling down a curved incline that slips horizontally under the particle's weight.[28]

> This kind of setup is a Class of physico-mathematical Problems, so to say, whose common requirement is to find the motion of several bodies placed inside certain grooves . . . when these bodies are connected by strings, and so they influence each other's motions. (Clairaut 1739, 3)

In just a few years, those efforts lead to a deep change of mindset; a new awareness arises about the true import of these problems. Many begin to call for *generality* of treatment—for principles or approaches that apply to all these constraints uniformly, and neither depend on, nor assume, any particular facts about a limited class of bodies. "The fourth principle I use here . . . leads to solutions for *all* the problems" of determining the motion of bodies under mutual constraints, ventured Clairaut.[29] D'Alembert saw the chief merit of his own book as answering precisely their call for generality, with which he was familiar: "I will teach here how to solve all problems of Dynamics by a single Method, very simple and direct."[30]

---

[28] Many then grappled with the motion of a particle in a rotating tube (first mentioned by Leibniz, who had left it unsolved); Nakata 2002 explains the various approaches they used. The first treatment of fall on a mobile incline is Bernoulli 1742c. The calls for generality of treatment began at the Paris Academy around 1740; Schmit 2017 gives the context for them, and some of the first attempts to answer that call.

[29] Clairaut 1745, 3; added emphasis.

[30] D'Alembert 1743, xxiv. See Chapter 9 for explanation.

In those years, the chief landmark in the mechanics of constraints was a paper by Clairaut that matters here in two respects. First, it illustrates a nascent impulse of Enlightenment science that 17th century theories lacked: the drive for generality as we explained it above. Second, it shows the *heuristic* predicament of rational mechanics at this time. As they look into mechanical systems beyond the *Principia*, theorists realize that none of the laws they had inherited from the 1600s was truly general in the sense they required; nor did those laws come with clear instructions for how to apply them to novel setups. We elaborate these points in section 8.6. Now we move to give an overview of Clairaut's results and approaches in his "Quelques principes."[31]

The mathematics of his solution is a mixture of algebraic and geometric methods. He represents particle trajectories as polygons (with infinitesimal sides dx), and seeks to infer displacements dx or velocities dx/dt by diagrammatic reasoning, viz. by inferring ratios between infinitesimals based on the geometry of the constrained system at issue.[32] Once an expression for dx has been found, integrating it gives Clairaut an equation for the curve, or trajectory, represented analytically.

Now here is how Clairaut's mathematical methods matter to our topic. To infer the strength and direction of the constraints' action (on the moving bodies they constrain) he always needs to rely on the *particular makeup* of the system at issue. The diagrammatic representation of *that* specific setup discloses how the constraint changes the motion in *that* case. Thus, Clairaut's approach to constraints is *not general*. He does not have a recipe (for handling constraints) that is non-specific (i.e., applies uniformly) despite his talk of general principles. This fact—the indispensable reliance on graphic methods—is methodologically significant. It will be left to Lagrange to overcome this heuristic limitation, by devising an approach to constraints that is truly general, viz. does not require ad hoc reasoning from the specifics of the target system.

---

[31] Clairaut 1745. This is not Clairaut's sole major work in mechanics. In the 1740s, he also made very major contributions to gravitation theory from Newtonian, impressed-force foundations. But, his paper above was an important contribution to the rational mechanics of constrained motion, MCON2. Moreover, "Quelques principes" had an ancestor that began in 1735—a shorter piece in which he treated a smaller number of problems, by a single method (Clairaut 1739).

[32] Example: "To extract an equation from these conditions, I drop a segment *no* perpendicular to C*k*, and I note that *nko* and *CPH* are similar triangles. Which gives me $no = gdt^2 \sqrt{(1 - z^2)}$." See Clairaut 1745, 28 and passim.

The setups Clairaut treated are as follows. A body falling down a curved incline that slides horizontally under an arbitrary impulse. A particle sliding outward in a tube that rotates around an endpoint C fixed in a horizontal plane. Two particles M, N connected by a flexible string as they fall individually down two distinct curved inclines. Two particles M, N, each constrained to move in a given rigid groove, and connected to each other by a rigid rod. A particle K falling on a curved incline that moves under K's weight, and thereby pushes another particle, C. A rigid body with a component point constrained to move in a curved groove MN. An extended body moving under gravity while connected by strings to an arbitrary number of particles. Two particles attached to a rod—one rigidly, the other mobile alongside it—while the rod is impelled in a horizontal plane; or it falls vertically under gravity. The same setup, but now both particles can slide freely along the rod above. A tube loaded with an arbitrary number of mobile particles, moving horizontally from an initial impulse. Two particles rigidly attached to a rod constrained to pivot around a fixed point after an initial impulse. A particle M that falls down a mobile, curved incline; as it falls, M drags an extended body sliding horizontally at the top of the incline, which likewise moves sideways under M's weight.[33]

To solve the problems above, Clairaut relies on certain propositions that he denotes by the umbrella term "principle." They differ in role and function, as we explain in section 8.6. Two of his principles are really heuristics. They state recipes for solving a class of related problems—hence they count as stepwise instructions, not first-order knowledge claims.

Other principles *are* knowledge claims: premises that he invokes when he needs to reduce some unknown quantities in a master equation.[34] Some of these auxiliary principles are explicit (e.g., Conservation of *Vis Viva*). Others are tacit—he relies on their content without drawing attention to them (e.g., the two laws of the lever, Archimedean-straight and Bernoullian, or bent). Yet another principle is a knowledge claim, but it is equivocal. Clairaut calls it the "general principle of accelerative forces."[35] Sometimes, he means the

---

[33] Schmit (2017, §2) has a synopsis of Clairaut's various problems and their immediate antecedents and successors in rational mechanics on the Continent.

[34] That is, when he has more unknown quantities than equations involving them. In each case, Clairaut iterates the same approach to solutions: (i) set up a master differential equation between the sought quantity and other variables, known or unknown; (ii) use some auxiliary principle—some accepted general statement—to obtain an extra equation for an unknown in the master formula; (iii) repeat, until one has as many equations as unknowns in (i).

[35] Clairaut 1745, 6 and 38, for his use of lever laws.

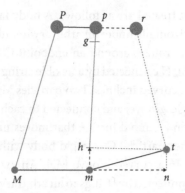

Fig. 8.5. Clairaut's "principle of accelerative forces" applied to constrained motion. *M* and *P* are mass points moving for two instants while constrained by a rigid rod. The task is to find the position of *M* at the end of the second instant, after it leaves place *m*. Suppose M was free, not constrained: then it would travel to *n*. And, suppose the action of the constraint would change its motion by a velocity increment *mh*. Then the actual path of the particle is the resultant of these two tendencies. That is the real meaning of Clairaut's principle. By the Parallelogram Rule, the actual path is the diagonal *mt* of the parallelogram *hmnt* that the two tendencies form, if represented geometrically.

Second Law: on a unit-mass particle, the net force equals its acceleration d^2x/dt^2. Some other times, he means the Parallelogram Rule for the composition of forces; see Fig. 8.5. With the benefit of retrospective, behind his account of this principle—and latent in the way he uses it—we recognize a deep insight:

> *From one instant to the next, the path of a constrained particle is the Parallelogram-Rule resultant of two motions: the velocity it would have, if it were free; and a velocity increment caused by the constraint.*

This insight lurks behind his instructions on how to apply his principle of accelerative forces—on how to think about the action of constraints at every point, and how they interfere with the exogenous, applied forces. As he explains (see Fig. 8.5):

> I begin by imagining the system in some position, and I draw each of the little straight lines that the bodies cross in an instant. Next, at the end of every such little line I draw the little lines that the bodies *would* describe in the second instant *if* they were *free*. Then, I draw the directions along which the strings, levers or other instruments act. On these directions, I mark the

little segments that denote the forces of these instruments. I determine the length of those segments as follows. For each body, these segments and the extension of the lines crossed in the first instant [the imaginary lines of counterfactual free motion] form a *parallelogram*. The endpoint of its diagonal is where the body *will* be at the end of the second instant—assuming the string or lever has neither stretched nor bent. By this *method*, we obtain any two consecutive sides of the respective curves that the bodies (in the given system) will cross. Then finding the equations of these determinate curves becomes a matter of mere calculus. (Clairaut 1745, 21; our emphasis)

This principle is worth our attention here, because it turned out to be a key step toward a general solution to MCON2. Clairaut did not have that solution—Lagrange first did, later—but it required the insight behind Clairaut's principle.

Still, the only type of constraints Clairaut tackled are *rigid*, whether directly or ultimately. In "Quelques principes," most constraints are rigid surfaces (*parois, lamelles, rainures, plans*), fixed or mobile. A few are strings (*fils*), but they are *flexible*, not elastic. This fact is key: as flexible, these strings bend isometrically—they can curve at any point, but their overall length is constant—whereas true elastics can change both shape *and* size, or length. Being merely flexible, Clairaut's strings can be analyzed away, as it were, into aggregates of microscopic *rigid*, inflexible rods connected by pins and joints.

8.6 Implications for philosophical mechanics

Now let us evaluate how the results above bear on the long-term project of a philosophical mechanics of extended matter in motion—in particular, on the problem of constraints, PCON.

(i) Faltering progress in rational mechanics

When it came to MCON1, progress was very difficult. The true aim was to mathematize the motion of *every* particle in an extended body—not just of a "representative point," or centroid. For vibration theory, however, that aim remained out of reach until d'Alembert's papers of 1747 and later. The work we surveyed above inferred just a modest result, viz. the frequency of one point in the string: the point of its maximum elongation from the rest axis. Taylor and Bernoulli express it as the "equivalent pendulum" (i.e., as the length of the *mass-point* pendulum that would beat at the same frequency as the vibrating string). In essence, then, both were forced to take the representative-point approach, and so their struggles with MCON1 fell short of real success. In the early 18th century many strove to "solve problems concerning systems with many degrees of freedom through the determination of mass-points considered 'equivalent' to the system."[36]

Clairaut too takes the centroid approach to solutions. That is, he does *not* treat the constrained body as an *extended* whole. Instead, he replaces the body with a point (which he endows with "mass," a notion left unexplained) attached to a weightless rigid rod whose other end always moves along the constraint.

In effect, this substitutive approach skirts around MCON1. Namely, it avoids giving an equation that tracks the motion of *every* single point in an extended body. That task remains too difficult for mechanics in 1742, and so the best Clairaut can do is solve for the motion of just one point—a centroid. An understandable heuristic, to be sure, given the expressive and computational limits of Clairaut's kinematics and dynamics. In taking the centroid approach, however, he does away with extension as a theoretically relevant, effective parameter in his mechanics. Eo ipso he does away with *bodies;* his mechanics is unable to account for the motion of a body as an extended whole.

(ii) Unstable explanations

Another aspect of the above studies on vibration illustrates our point that 17th century mechanical principles lacked heuristics for further application. That aspect is their proffered explanation for the string's *shape* as it vibrates. Consider Taylor's account of how shape evolves post-release from the tensed state (see also Fig. 8.6):

> Let a string stretch from A to B. Pinch it at z, and pull it to distance Cz from the axis AB. Because the string is bent only at C, upon release that point moves first. Then right away the string becomes bent at the locations D and d nearby, so these points begin to move as well; then E and e, and so on. Now at the beginning, point C moves very fast, because of the great

[36] Maltese 1993, 54.

bending [*flexura*] at place C. But then, as the curvature [*curvatura*] at the nearby places D, E, etc. goes up, these points also move ever more swiftly. By the same action [*eadem opera*], as the curvature of C goes down, that string point is accelerated less. And likewise everywhere, as the slower points get more accelerated and the faster ones less—eventually it comes to be that, the forces having regulated one another suitably [*inter se rite temperatis*], the motions of all the string points conspire, and they approach the axis together, then recede from it together, and so on to infinity. (Taylor 1713, 28)

Some have called it a "qualitative" account of the "redistribution of the curvature" as it propagates outward from the string's plucked center point.[37] That seems too charitable. The tensed string starts out as a triangle, isosceles with C as a vertex. There is *no* curvature at C. To say that the string curves as it travels away from C is to presume that curvature arises ex nihilo, then propagates outward. Even that miraculous birth appears ruled out—by Taylor's matter-theoretical premises (see below). Truesdell was right to call Taylor's account "obscure if not faulty."[38]

Bernoulli's explanation is hardly better. Recall, he concludes that the string's shape as it vibrates is *sinusoidal*. But, he cannot explain that from his matter-theoretic premises. He starts with a discrete loading: then the string's shape post-release must be a polygon (with a weight at every vertex). Bernoulli ought to explain how a polygon turns into a smooth curve—by what causal process, treated rational-mechanically. Much like Taylor, he is unable to explain that. In effect, both Taylor and Bernoulli exploit the geometry of their models so as to extract results in rational mechanics alone. But, in doing so, they sever the explanatory link to the *material properties* of the body they quantify: to the matter theory that underpins it. In short, these technical difficulties in MCON1 spell trouble for the general problem of constrained motion: they are barriers to using the results in rational mechanics as resources for a philosophical mechanics.

(iii) Elusive physics

Philosophical mechanics aimed to integrate smoothly a physics of body with a mathematized mechanics of its motions. The early-Enlightenment authors

---

[37] Maltese 1992b, 717.
[38] Truesdell 1960, 130.

we surveyed above ran into very serious difficulties on this front: they were unable to combine matter theory and rational mechanics *coherently*. They start from certain physical premises (about the makeup of their target body) and infer to mathematized conclusions (about its motion) that contradict their premises. Or, they start from physical assumptions that contradict each other. To illustrate our point, we turn the spotlight on Taylor's early account.

In *Motus nervi*, he starts from matter theory: he regards the string as made up of equal "rigid particles, infinitely small."[39] From that premise he seeks to derive a result in rational mechanics, viz. the strength of the restoring force at a point on the string. He reaches that result by two routes. One is to regard the restoring force as the *resultant* of two other forces, namely, the tension forces Bt and bt that pull the two *rigid* particles (flexibly joined at t) away from each other. For that conclusion he relies on the Composition of Forces, or Parallelogram Rule, whereby he infers that segment tr gives the size and direction of the sought resultant.

The other approach is to regard the restoring force as proportional to the *curvature* of the string at B and b. In turn, curvature varies as the radius of the osculating circle at those points. Geometrical reasoning (plus the small-vibrations assumption) then leads him to conclude that the curvature (and so the sought force) is as the distance to the rest axis (see Fig. 8.6).

But these two approaches contradict each other, because they rely on in-compatible matter theories. The official one takes a string to be a flexible composite of rigid bits of mass—infinitesimal parallelepipeds, really. When deformed, the string is always a *polygon*, with actual vertices: because rigid bodies do not curve, if they are straight. But—because he needs curvature as well, qua measure of the local restoring force—Taylor quietly assumes another *matter* theory, viz. the physical continuum. In this picture, a string is a deformable continuous line, and when it vibrates its shape is a *true curve*, not a polygon. It has no vertex, but it does have a well-defined degree of cur-vature at every point between the fixed ends A and B. In contrast, polygons have no curvature: it is a meaningless concept when applied to them, and so it is nonsensical to estimate force at a polygon vertex from the curvature at that point. In effect, Taylor's rational mechanics of the string used two

---

[39] *Ex particulis rigidis* (Taylor 1713, 27). That they are "equal" is ambiguous. Taylor likely means them to be of equal *mass*, but he draws them equal in *length*. The two equalities are not coextensive unless he tacitly assumes that his rigid particles are of equal mass density.

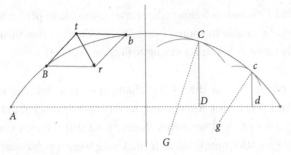

Fig. 8.6. Taylor's two theories of matter. Left: *Bt* and *bt* are rigid particles; tension forces at *t* pull them away from each other, toward *B* and *b*, respectively; *tr* is their resultant, and also the restoring force on the spring at that location. Right: Mass distribution in the string is along *ABbC*, which is a true curve. *CG* and *cg* are the radii of curvature at *C* and *c*, respectively; they're also the direction and strength of the restoring force at those locations. *CD* and *cd* are the respective distances from the rest axis. Because they differ from *CG* and *cg* by a second-order infinitesimal amount, Taylor takes them to be proportional to the restoring force. Adapted from Taylor 1713, where he assumes a rigid-particle picture of matter (left); and from Taylor 1717, where the official picture is a true continuum (right).

distinct matter theories—the rigid body and the deformable continuum—but he never paused to ask, which is the true one?

Things get little better in *Methodus*. There he rehearses his earlier derivation (of the restoring force) but now all talk of rigid particles has been expunged. Instead, Taylor says just that his string consists of "very narrow matter, uniformly dense," and that forces act on its "points."[40] A note upstream from his treatment gives more details, in a lemma to Prop. 18:

> The next four propositions will treat the shape of cords, of cloth fabrics filled with water, and of vaults supporting some given weight. Insofar as they are made of physical matter, these shapes have some true density [*crassities*], are little fit to bending, and give some way to forces that stretch or compress them. Thus, he who wishes to describe these shapes precisely ought to take these facts into account. However, these factors are difficult to handle mathematically, and greatly complicate our calculations, which are already too long. For that reason, I disregard entirely these factors' effect,

---

[40] Taylor 1717, 89.

and instead I assume as follows: the shapes are made of perfectly flexible matter, wholly unable to contract or dilate, and so thin that their density practically vanishes in respect to length. (Taylor 1717, 74)

Hence, his official matter theory has changed—from rigids to deformable continua. And yet, Taylor still chooses to rely on premises about the string regarded as a set of rigid segments. Namely, he still inferred the restoring force by the Parallelogram Rule, as if the string were a polygonal set of tiny rigid masses, not a true continuum.

Just as with Taylor, Johann Bernoulli's underlying theory of matter is likewise opaque and equivocal. He sets out to treat a discretely loaded string, by induction. Then he switches to a string modeled as a physical continuum, "uniformly dense" and deformable into a true curve, with an osculating circle at every point.[41] But, he gives no thought to which of these two pictures may be *true*, and on what grounds.

Clairaut's substitutive maneuver above—his replacing an extended body with a representative point—has a companion in his strategy for theory building. In this respect, his most significant choice was to *curtail the explanatory ambitions* of his theory. In effect, Clairaut inaugurated the explanation-bracketing approach at home in modern analytic mechanics. Namely, he bypassed—or rather, refused to answer—all matter-theoretic questions about the physical constitution and specific causal powers of the materials that might instantiate the constraints.

Plainly put, he tackled MCON2 by removing it entirely from the *physics of body*. Thereby, in effect he closed off his work from solving constrained motion qua problem in *philosophical* mechanics. What properties of body do external constraints instantiate? What corporeal powers underpin their action as they impede the free motions of other bodies? That is, how do Nature and Action link up explanatorily with his rational mechanics of unfree motions? Clairaut would not even raise these questions—and with good reason, because they were intractable *then*. We will soon see his bracketing maneuver turn rampant in much 18th century mechanics. However, we must not lose sight of these questions just because Clairaut and others did not ask

---

[41] Bernoulli 1742b, 207, 209. He proceeds by induction, as follows. He supposes the string loaded with one weight, and derives its frequency. Next, he considers the same string, loaded with two weights, and infers *its* frequency. Then he reiterates for three, four, and then seven equally spaced weights. He seeks to extrapolate to a general formula for the frequency N of a string with n discrete equal weights.

them. They go to the very heart of philosophical mechanics as a project that straddled philosophy, physics, and the mathematical science of motion.

(iv) Epistemology: sources of evidence

There is a baffling aspect to Taylor's and Bernoulli's treatments of the vibrating string: they derive their key results *twice*, by two different routes. Why? We think they did it for evidential reasons. Bear in mind, there was *no* empirical evidence about the shape and frequency of a taut vibrating string. So, Taylor and Bernoulli could not justify their conclusions by induction from data. Not even more weakly, by showing the conclusions to be at least compatible with data. There *were* no empirical facts to verify their rational-mechanical models. Absent that, they decided that *multiple derivation* could supply the evidence otherwise unavailable.

More precisely, they each infer their key results from dynamical principles: Taylor in *Motus nervi* starts "from the principles of mechanics," viz. the Parallelogram of Forces; and Bernoulli in *Meditationes* starts from "the principle of live forces." Then Taylor in *Methodus* says the harmonic condition likewise follows "from the principles of *Statics*." His antagonist re-obtains them as well "from statical principles," in the second part of *Meditationes*.[42] Recall that at the time statics was the much older science, born in the work of Archimedes. Its principles thus counted as fully secure, not up for debate, and immune to skeptical doubts. And so, the ability to derive a novel result from the ancient and accepted laws of statics counted as *evidence* for it; and it served as a cross-check on its derivability from properly dynamical principles. The latter were much more recent, far from universally accepted, and their scope unclear to the community of research.

This epistemological verdict gains support from our study above of Jakob Bernoulli's paper on the compound pendulum. He first had devised this novel source of evidential support, specifically for the Huygens Formula. Recall, Huygens before him had obtained it from a dynamical law—an energy principle in disguise. But then some doubted that the principle was legitimately applied; they could not see that dissolving the constraints in a system leaves the motion of its mass center unchanged. To assuage their doubts, Jakob Bernoulli started from a principle of statics: the law of equilibrium for a bent

---

[42] See, respectively, Taylor 1717, 89, and Bernoulli 1742b ("*Meditationes*"), 203; our emphasis.

lever. He thought his starting point was epistemologically impeccable, and so the Huygens Formula, if validly derived from it, must be true. The same logic of evidence is at work in the Taylor-Bernoulli papers on the vibrating string: to preempt doubts about the legitimacy of their dynamical premises, they show that the intended results follow from laws of statics too, and *those* are beyond dispute. That must count as evidence for the result, in the absence of empirical facts to confirm it.

However, their evidential appeal to statics is not without problems. For one, it is cognitively obscure: they rarely declare overtly on *which* principles of statics they rely for confirmation. Taylor and Johann Bernoulli tend to invoke it as a shibboleth. *Constat ex Staticis*—it is plain from the principles of statics, they say, without further detail. But their evasive language does not inspire confidence.[43] For another, it is explanatorily opaque: these authors do not explain how a law from statics (the theory of masses at mutual *rest*) is supposed to confirm a result about masses in *motion*, or points in a vibrating string. Lastly, some of these appeals are not quite legitimate, because they credit statics with principles it did not *have* before then. That is the case of Jakob Bernoulli's principle above. It is a statical claim, to be sure; qua statement about the equilibrium of a bent lever, it does belong in statics. However, before 1703 no theory of statics had contained it as a principle. *He* first stated it, but gave no evidence for it: no account of what justifies his law of the bent lever. So it seems illicit to treat it as an ancient and accepted truth of statics, able to transfer truth to conclusions inferred from it.[44]

His own brother rejected it, and set out to re-derive the Huygens Formula from yet another, supposedly safer, premise. As he did so, Johann felt the need to justify his move: Huygens had "assumed a principle . . . that some found too rash and not fit to be accepted without demonstration."[45] Later he tried another method, again driven by evidential concerns about previous solutions to the center-of-oscillation problem:

I leave it to expert readers to see that I established the Formula [giving the center of oscillation] from a *foundation much safer* than it has been done so far. For, not only did I not need to use Mr Huygens' gratuitous hypothesis [on the descent and ascent of the gravity center]; but also, I did not need to

---

[43] See Bernoulli 1746. Commentators do no better. None of them has tried to elucidate the statical principles at work in the treatments above; cf. Truesdell 1960, Feigenbaum 1985, and Villaggio 2008.

[44] Recall, the canonical Law of the Lever—which stems from Archimedes' treatise, *Equilibrium of Planes*—is about two weights (not more) on a straight lever, not bent.

[45] Bernoulli 1715, 242.

suppose—which my Brother did—that the center of oscillation lies on the line of center. (Bernoulli 1717, §29; our emphasis)

These complications further illustrate the evidential predicament that beset mechanics after 1700. For their chosen setups, Taylor, the Bernoullis, and their posterity had neither the clean experimental results that collision theorists had received from Huygens and Mariotte; nor the rich body of observational data that Newton had drawn on for his gravitation theory. As a task in philosophical mechanics, Evidence would dog them throughout the century.

(v) Epistemology: principles

Examined through PCON as our analytic lens, the half-century of researches above reveals commonalities that have eluded all historiography of mechanics so far. In particular, a great amount of cutting-edge theory struggled with the overarching question, *How does an extended body with constraints move over time?* Most figures then believed that success on this front required the discovery of a new principle; the term "principle" becomes a leitmotif in the rational mechanics of constrained motion. Two aspects of it are philosophically significant.

Ambivalence
Regarded as an agent's category, the term "principle" denoted two distinct ideas. For some, a principle was a *common premise*. Namely, a proposition used as a premise in many derivations of effective parameters, orbits, or equations of motion for a broad spectrum of mechanical systems. In this role, the premise had an evidential role: its truth served as evidence for the truth of the conclusions of the derivations. Two favorite principles in this sense were the Parallelogram Rule and Conservation of *Vis Viva*.

For others, a principle was a *fruitful heuristic*. Namely, a generic recipe, policy, or guideline for solving a great number of different mechanical problems, viz. for treating systems importantly different (e.g., a sphere rolling down curved inclines; many particles strung together and given arbitrary impulses; etc.). Some offered their heuristic principle as a discursive recipe stated in a general, non-specific way; an example is d'Alembert and his *General Principle* (see Chapter 9). Others stated it as epitomes, or illustrative examples, applying the heuristic to a few systems while hoping that the reader would extract a general lesson from their epitome. That is the case of Clairaut: "I give here a good number of problems, and try to explain them so that readers can infer the general *Method* for solving any other similar topic."[46]

To sharpen the insight, we give distinct names to these two senses, as follows.

E-principle: a premise common to many deductive derivations of specific equations of motion.

H-principle: a heuristic that works across a broad range of setups in rational mechanics.

These are different things, so we ought to keep them distinct. One is a knowledge claim; it is truth-apt, and it requires evidence if we are to accept it. The other is a pragmatic injunction, and we judge it from success in action, or instrumental adequacy for its declared purpose. More to our point, the distinction will prove an efficient analytic tool for the next four chapters, where talk of principles proliferates.

Generality
Another aspect of principles in these years is a rising push for generality. Specifically, some *wish* that rational mechanics had truly general principles:

> One would be wrong to regard this sort of problem [on constrained motions] as just an isolated research whose sole fruit is the pleasure to have solved it. . . . Such problems never fail to enlighten us: often they yield *new principles*, and teach us *new general laws* that nature follows in all her operations. . . . The wonderful System of the world, whose important discovery was reserved for the great Newton, is simply a consequence of Mechanics. And, what is left to discover in this System depends solely on achieving a greater perfection—in my opinion, not so much by improving the calculus as by getting closer to the *general laws* of motion, which are *still hidden* to us. (D. Bernoulli 1746, 55; our emphasis)

Others claim to have them already: "A Principle General and Direct for solving all Problems of determining the Motion of several Bodies acting on each other, whether by means of strings, levers or any other way."[47] As the

---

[46] Clairaut 1739, 3.
[47] Clairaut 1745, 21.

century progresses, these claims will only get louder and bolder, as we will see in Chapters 9–11.

Since their notion of principle was ambiguous, it turns out that calls for generality were just as equivocal. Some asked to see a premise that entails all the equations of motion—for every mechanical system, treated or in need of treatment. Others claimed to have a recipe (or at least the proof-of-concept for one) that will guide successful research into mechanical systems not yet treated (e.g., the motion of air currents). A modern way to put this duality is: some sought a *dynamical law* for a comprehensive theory, others proposed a *uniform heuristic* for a forward-looking research program. Examples of the former are the Euler–Cauchy laws of continuum mechanics; or the Principle of Virtual Work, in analytic dynamics. For the latter, an example is the method of Lagrange multipliers.[48] Any heuristic that works analogously counts as uniform. That is what some 18th century principles of rational mechanics aimed to be.

(vi) Epistemology: fumbling for insight

Though Jakob Bernoulli's Lever Principle was evidentially opaque, in the century after him it played a major role as a guideline to solving further problems. In particular, path-breaking figures in rational mechanics learned two things from his terse paper.

First, they learned that if a system has internal constraints, they are *workless*. That is, the net virtual work they do on the system is null. Jakob Bernoulli put it as the idea that, in the compound pendulum, the internal actions form a generalized bent lever in equilibrium. Two later figures reworked his insight into two related heuristics. D'Alembert transformed it into the idea that, *if* motion-changes due to internal constraints *were* caused by applied forces, these *fictitious* extra forces would be jointly balanced. In the next chapter, we explain his idea and the innovative uses to which he put it. Lagrange took from Jakob Bernoulli the insight that, if a system of rigid

---

[48] Other examples are the Euler-Cut approach, in continuum mechanics, where we let an arbitrary plane cut through the body, then we identify the balance of forces and torques on either side of that plane; or the approach where we treat a system by identifying the relevant potentials (monogenic or dissipative), subtract the kinetic energies, and thereby obtain a function L that we can plug into the Euler–Lagrange equation, which then—once we have described the system in terms of its generalized coordinates—yields the equation of motion for it.

masses is in equilibrium, the net virtual work of the applied forces is zero. He too generalized this idea in his own, equally novel way; see Chapter 11 for explanation.

Second, they learned that, if the constraints are *external* to the moving system and *rigid*, their action is *normal* to the constraint. From Bernoulli's analysis above, recall that in the compound pendulum the only motions compatible with the constraints (viz. the rod swinging rigidly) are circular, around the suspension point. Hence, he infers, the constraint's action destroys (so to say) the component of the external force normal to the circular path that every mass in the bob is allowed to have; see the discussion around Fig. 8.3. The lesson that Clairaut and d'Alembert took from Bernoulli's insight was that, if the external constraints can be modeled as rigid surfaces, their effect on the system is perpendicular to the constraint surface at that point.

Important as Clairaut's paper was, however, it did not quite do enough. It fell short of teaching a principle that applies to *all* the problems then studied. That kind of general law still eluded mechanics; hence the hope that continued work on constrained systems might just yield that general principle one day. Though valuable, their principles and associated insights were just partial, thus ultimately insufficient. Jakob Bernoulli saw, correctly, that internal constraints do zero net work—provided they are *rigid*, however, not in general. Clairaut saw, correctly, that the constraint combines vectorially with the external force, thereby determining the body's effective, actual acceleration. However, his insight is inert without help from diagrams: applying his principle requires *indispensably* that we read the direction and strength of constraint actions from a geometric picture—the diagrammatic representation of the body and the obstacles to its freedom.[49]

It will be left to d'Alembert and Lagrange to overcome these limitations. The former would find an ingenious way to combine Bernoulli's and Clairaut's respective insights into a novel heuristic, or H-principle; see Chapter 9. The latter, building on his three predecessors, would at last devise a general way to solve MCON: by combining the Principle of Virtual Work, d'Alembert's insight, and his radically new recipe, viz. the method of Lagrange multipliers; see Chapter 11.

---

[49] In his paper, every solved problem comes with at least one diagram, for a total of thirty-two figures; see Clairaut 1745, 52ff.

8.7 Conclusions

We have examined five decades of work in rational mechanics relevant to the problem of constraints, PCON, in its two variants: the motion of a body constrained to move as one extended volume of matter (MCON1); and the motion of bodies under external constraints, or impediments to their kinematic freedom (MCON2). As exemplary setups for this duality, we chose the vibrating string and the compound pendulum, respectively.

Despite strenuous efforts, the tangible results were modest: for each setup above, they were able to find just one effective parameter, or pragmatic quantity. Specifically, the frequency of the principal mode in a vibrating string, and the location of the center of oscillation in a pendulum bob. These results fall short of the whole-body descriptions of motion they pursued in studying the two objects above. And so, in the late 1730s rational mechanics remained far from any promising solution to PCON—hence a philosophical mechanics of extended matter still eluded natural philosophers. Again, that illustrates how difficult BODY was, and how little help Newton's laws of motion had been for its solution.

Still, those efforts were not wasted; slowly, they bore fruit. For one, they focused everyone's attention on vibrating systems—the need to mathematize the string's motion at every point, not just one; and then to extend the approach to higher-dimensional objects, such as elastic membranes. Another was a lasting benefit: Jakob Bernoulli's insight of 1703 (his principle of the bent lever) taught the 18th century an invaluable lesson, namely, to regard *internal* constraints as workless, or jointly balanced. Yet another benefit was a growing awareness of the multitude of principles in rational mechanics— and a new sense of urgency about their anarchic proliferation.

Research on the compound pendulum in these decades matters greatly for philosophical mechanics, not least because it yielded a seminal heuristic: the virtual-work approach, a novel way of studying constrained systems that proved greatly successful, in retrospect. This breakthrough was Janus-faced: while useful, the heuristic gave figures in rational mechanics permission to postpone the hard task of asking what material forces and actions underpin the work of constraints in a system.[50] The reason is that Jakob

---

[50] At this juncture we use the phrase "virtual work," which is slightly anachronistic; the term is coined and becomes entrenched about a century after the fact. Figures in the early 1700s did not have a stable vocabulary for it; some (such as Johann Bernoulli) called it "energy," others (including Lagrange later) called it "virtual velocity."

Bernoulli's heuristic insight lets us disregard the *physics* of constraints: it says that their combined virtual work vanishes, and so their underlying makeup and causal import may be circumvented as we seek to infer the motion of the body, or system, that they constrain. We elaborate on this fact in Chapters 9 and 11. Put in our terms, the heuristic came at the price of disregarding Nature and Action entirely. Moreover, their struggles required them to pay renewed attention to Principle and Evidence: to seek out new principles beyond those offered by Newton and the 17th century, and to examine sources of confirmation for them.

The person who made the most of these chances was d'Alembert. Soon after 1740, he would make great strides on the three counts above. He first produced a proper rational mechanics of the vibrating string. He made real progress with both MCON1 and MCON2, thanks to a powerful heuristic that came to bear d'Alembert's name. And, he did much to unify rational mechanics, to broaden its descriptive scope, and to breathe new life into philosophical mechanics. Accordingly, we turn to him next.[51]

---

[51] For developments in 18th century rational mechanics outside the scope of problems treated in this book, see Hepburn and Biener 2022.

9

Constructive and principle approaches
in d'Almbert's *Treatise*

9.1 Introduction

During the first half of the 18th century, constructive approaches to the problem of bodies (*BODY*) were dominant. Beginning from material commitments such as the qualities and powers of matter, philosophers sought to construct a body concept, one demonstrably consistent with the demands of mechanics, specifically the rules of collision (see Chapters 2–5). This approach persisted into the second half of the century (see Chapter 6), but by mid-century developments in rational mechanics had changed the philosophical problem space in important ways (see Chapter 7). Enter Jean Le Rond d'Alembert who, in 1743, sought to set rational mechanics on a new foundation in his *Traité de dynamique* (or *Treatise on Dynamics*, henceforth his *Treatise*), with lasting consequences.[1]

For two reasons, we think the *Treatise* is pivotal for the development of philosophical mechanics in the latter half of the 18th century. First, the structure of his book suggests taking a "principle approach," in which the theory's principles—such as the laws of motion—are the primary resource for solving *BODY*. We explain what we mean in detail below. Though contrasting with the constructive approach, the principle and constructive approaches are complementary: the interplay between the two is productive for theorizing in physics, as later chapters will show. However, the principle approach comes with a risk: there may be no single body concept, or indeed any single ontology, uniquely consistent with the principles. Thus the second consequence for philosophical mechanics: a shift from ontology to principles as the source of theoretical unity. Whereas the first half of the century is replete

---

[1] In what follows, page references are to the 1743 French edition; and translations are our own. D'Alembert's *Traité de l'équilibre et du mouvement des fluides* (1744) is also important for our story, but in this chapter we restrict our attention primarily to the *Traité de dynamique*.

Philosophical Mechanics in the Age of Reason. Katherine Brading and Marius Stan, Oxford University Press.
© Oxford University Press 2023. DOI: 10.1093/oso/9780197678954.003.0009

with theories exhibiting *ontic* unity, in the second half of the century we see theories unified axiomatically, by shared principles. We call this "nomic unity," and it is an explicit goal of d'Alembert's *Treatise*.[2] Nomic unity is consistent with a proliferation in ontologies associated with a single theory—or no articulated ontology at all. An alarming outcome for philosophers concerned with Body.

In this chapter, we argue for an interpretation of d'Alembert's *Treatise* that foregrounds: the challenges facing a constructive approach; the availability of a principle approach; and the search for nomic unity. In later chapters, these features of his *Treatise* will be important for understanding developments in philosophical mechanics through the turn of the century.[3]

We proceed as follows. We begin by explaining the constructive/principle distinction (section 9.2). We then outline the structure and content of d'Alembert's *Treatise*, by way of background (section 9.3). We evaluate the theory first for its contribution to rational mechanics (section 9.4), and then as a work in philosophical mechanics. For the latter, we argue that the *Treatise* can be evaluated from the perspective of either a constructive (section 9.5) or a principle (section 9.6) approach. We then explain how this is connected to diverging conceptions of theoretical unity, ontic and nomic (section 9.7), before returning to Body to draw out the implications for Nature, Action, Evidence, and Principle in the context of d'Alembert's theory (section 9.8). We summarize our conclusions at the end (section 9.9). One important lesson is that philosophers interested in Body have no choice but to engage with the complexities of constrained-motion mechanics. This is true quite generally from mid-century onward; and is especially the case with d'Alembert where, in order to address Body, we cannot rely on his *definition* of body, but must investigate the details of his rational mechanics.

Philosophers have given d'Alembert's *Treatise* very little attention.[4] Yet, as we hope to show, it is enormously important—both in and of itself, and for its role in the transition from early 18th century conceptions of physics to a new notion of physics as an independent discipline with its own goals, methods, and norms.

---

[2] See Chapters 1 and 7 for our explication of ontic versus nomic unity.
[3] See especially Chapter 12.
[4] This is changing. See the special issue of *Centaurus* (2017, 59:4) commemorating the tercentenary of d'Alembert's birth for the articles and references therein, including Gilbaud & Schmit 2017, and Schmit 2017. See also Le Ru 2015 and 2021.

9.2 Constructive and principle approaches

We take the "constructive" and "principle" terminology from Einstein.[5] We use it to label two approaches to the development and interpretation of theories, and—in line with the narrower task of this book—to tackling BODY.

The *constructive* approach to theorizing begins from the material constitution of the objects or systems of interest, and from there develops an account of the laws that such objects or systems satisfy. In the case where our objects are bodies, we first develop a material account of bodies, including their properties and powers. Then we move on to the rules describing the behaviors of those bodies (such as rules of collision). We may seek to develop, derive, or explain the rules from the account of body, and must at least demonstrate consistency.

The constructive approach to BODY may be stated as follows:

Constructive approach (bodies): The qualities and properties of matter are the primary resource for solving BODY.

We begin with material commitments and from them we construct concepts of body (Nature) and bodily action (Action) consistent with Principle and Evidence (see Chapter 1) to arrive at a theory of the behaviors and motions of such bodies. We have seen examples of this throughout Chapters 2–6 and, though the label is ours, we have seen that the associated approach dominated physics during the early 18th century.

The constructive approach to BODY comes in two varieties, a stronger and a weaker. The stronger begins from an explicit theory of matter, and from there constructs bodies. Only with this in hand do we develop a theory of the behavior of such bodies. For example, Du Châtelet *first* provides her account of the nature of matter, *then* using this offers her account of the nature of bodies, and *finally* derives the laws from the nature of bodies thus understood.[6] The weaker eschews a foundation in matter theory, and works directly

---

[5] In 1919 Einstein distinguished between constructive and principle theories (see Einstein 1982). The distinction has been much discussed in the philosophy of physics literature. See especially Brown & Pooley 2006 and Janssen 2009. For applications in quantum theory, see Bub 2000. See also Flores 1999. This discussion has been carried out in terms of constructive versus principle *theories*. However, we prefer the categories of constructive and principle *approaches* to the development and interpretation of a theory. Hence, from our vantage point, a given theory is not a constructive or principle theory in itself. Rather, we may *interpret* it constructively or principledly, and each approach yields different insights.

[6] See Du Châtelet 1740, chapter 7 for matter; 8 for body; and 11 for laws.

with bodies. The qualities and properties of bodies are first enumerated, then invoked to justify the behavior of these bodies (e.g., the Newtonians' efforts to demonstrate the collision rules for different kinds of body; see Chapter 3). All the attempts to address BODY discussed in Chapters 2–6 are most naturally interpreted and assessed constructively.

More generally, the constructive approach finds a natural home in *philosophical physics*: the subdiscipline of philosophy charged with the account of body in general, and the dominant conception of physics in the early 18th century. Philosophical physics offered a qualitative approach to the theory of bodies, frequently rooted in matter theory,[7] and the task was to integrate the quantitative rules of collision—taken from rational mechanics—into this physics to arrive at a philosophical mechanics.

With this in mind, we can see how the constructive approach provides us with a useful tool of analysis. A theory will be amenable to constructive interpretation if it offers an account of its objects (such as bodies) *prior* to the statement of principles that pertain to the behavior of those objects. To assess a theory constructively requires us to determine whether, and to what extent, that theory offers a material account of its objects compatible with and adequate for its needs.

In contrast, when adopting a *principle* approach, we begin our theorizing from some principles that we hold to be very general, perhaps universal. Importantly, these principles are independent of the particular material constitution of the target systems. For instance, in special relativity (SR), Einstein adopted the light postulate and the relativity principle, on the grounds that these principles were consistent with all known empirical evidence. We then derive the consequences. These are conditions that any material system, no matter its constitution, must satisfy. In the case of SR, Einstein used the light postulate and the relativity principle to derive the Lorentz transformations. SR tells us that all material systems, regardless of their individual material constitution, must satisfy the Lorentz transformations: for example, a rod put into motion must contract.[8]

---

[7] By "matter theory" we here mean the philosophical enterprise familiar from the 17th and early 18th centuries of seeking the nature of matter through largely a priori reasoning. The term undergoes a subtle but important change of meaning later, around the turn of the 18th century, as seen most especially in French matter theory and Laplacian physics (see Chapter 12).

[8] Here is another example: Einstein contrasts thermodynamics (principle) with the kinetic theory of gases (constructive). In the latter case, we build up to the principles of thermodynamics from the behaviors of the particles constituting the gases; in the former, we specify the principles of thermodynamics to which all gases, whatever their material constitution might be, must conform.

The principle approach to theorizing is powerful because it allows us to set aside some aspects of the dynamics of material systems: we do not need to know the details of the dynamical behavior of a material rod in order to know that it must contract. In this sense the Lorentz transformations are a *kinematic condition*: no information about the forces acting on or within a particular body is required in order to know that it will satisfy them. However, the Lorentz transformations encode *dynamical* commitments about the forces pertaining to bodies (e.g., the forces that yield the stability of measuring rods). These commitments are placed into the kinematic condition, bracketed out of explicit consideration by the assumption that all systems satisfy the principles of the theory, and therefore the Lorentz transformations.

These are the three most important features of the principle approach: the adoption of principles that are independent of the particular material constitution of the target systems as the starting point for theorizing; the generality of the principles; and the bracketing of dynamical considerations by means of a kinematic condition.

Assessing a theory through the lens of the principle approach sets three tasks for the philosopher. First, to clarify the principles, including their scope. In the case of SR, the scope of the principles is maximal—it is universal—and we think of the resulting kinematic condition (the Lorentz transformations) as expressing the structure of space and time themselves. But the technique of bracketing dynamics into kinematics does not require this maximal generality, as we will learn from a principle reading of d'Alembert's *Treatise*. Second, to investigate the epistemic status of the principles. Einstein's justification for adopting the light postulate and the relativity principle was inductive: they were two principles consistent with all known empirical evidence. However, justification need not be inductive, and with d'Alembert it seems that it is not (more on this below). Finally, the third task is to make explicit the dynamics that has been bracketed into the kinematics. As noted above, to adopt the relativity principle is to make *dynamical* commitments concerning the behavior of material systems, and we can seek to illuminate what these are.[9]

If we take a principle approach to theorizing, the question remains: what are the objects or systems to which this theory applies? To what extent does the theory itself specify its own objects? On this approach, rather than specifying the objects prior to theorizing, we rely on the principles themselves. For

[9] See Brown 2005 for extensive philosophical discussion of this aspect.

example, in the case of bodies and the laws of motion, adopting the principle approach means drawing our concept of body, and of bodily action, from the laws themselves, without appeal to any prior theory of the material constitution of bodies.

> **Principle approach (bodies):** Theoretical principles, such as the laws of motion, are the primary resource for solving BODY.

The question is then whether, and to what extent, the principles contain adequate resources for this task. If the principles prove insufficient for determining a unique ontology, we will be left with ontic pluralism, or even with ontic under determination. More on this below (see especially section 9.7).

Though anachronistic, our use of the constructive/principle terminology in an 18th century context is well-motivated. Einstein did not create his distinction from nowhere. We suggest that it originates in philosophical mechanics: in the amalgam of physics (as it was once understood, with its constructive approach) and rational mechanics (with its search for principles that may serve as axioms) that went on to become the physics of the 19th century.[10] If this is right, then using the distinction to analyze d'Alembert's *Treatise* may be productive of new insights. We argue that it is.

9.3 D'Almbert's *Treatise on Dynamics*: its structure and contents

The *Treatise* consists of a Preface; Definitions and Preliminary Notions; Part I containing his three principles or axioms; and Part II containing his *General Principle*,[11] fourteen problems demonstrating the utility of the *General Principle*, and a discussion of *vis viva*.[12] In what follows, we outline the content of each of these elements, focusing on the aspects most relevant to BODY.

---

[10] See Chapter 12. For the 19th century background to Einstein's distinction, see Howard 2005.

[11] The *General Principle* of his *Treatise on Dynamics* is often referred to as "d'Alembert's Principle." However, this term is thrice ambiguous. First, there is the 1743 *General Principle*, discussed here. Then, there is d'Alembert's attempt to generalize it in his 1744 *Treatise on the Equilibrium and Motion of Fluids*. This latter was the primary inspiration for Lagrange, who in 1788 coined the term "the principle of Mr d'Alembert" when he attributed his own version of the principle to him. It is important we keep the three versions distinct, since they each contain new insights. For this reason, we restrict the term "d'Alembert's Principle" to d'Alembert 1744. We use "Lagrange's Principle" for Lagrange's *Mechanique* (1788), discussed in Chapter 11.

[12] We will not discuss *vis viva* here. See Iltis 1970.

This provides us with the background necessary for evaluating d'Alembert's *Treatise* from the perspective of philosophical mechanics.

(i) Preface

D'Alembert opens his Preface with the issue he plans to tackle. Recent decades had seen an expansion in the range of mechanical problems treated, with a proliferation of solutions to particular problems, derived from a variety of different principles.[13] Many of them, he claims, are obscure. His stated goal is to provide mechanics with a minimum number of clear principles that are maximally general in terms of the problems they can solve. Specifically, he claims that all the various principles of mechanics hitherto relied upon can be reduced to just three: force of inertia, composition of motion, and equilibrium. We discuss each in turn below.

D'Alembert's aim in the *Treatise* is to show that his three principles suffice for a mechanics of the motions of bodies acting on one another in any way through contact (such as collision, being tied together by threads, and so forth).[14] His strategy consists of four steps. First, he introduces his "Definitions and Preliminary Notions." Second, he presents his three principles. The justification for their truth is *clarity*: each principle should be sufficiently clear so as to command assent as certain.[15] Third, he uses his three principles to formulate his *General Principle*, which he takes to be the general solution of all problems in mechanics. The value of his *General Principle* lies precisely in this generality. For, insofar as it is successfully formulated using only the resources of the three principles, and insofar as it is indeed a general solution for all problems in mechanics, this justifies the claim that not only are the three principles true, but they are sufficient to serve as the axioms of mechanics: they are the foundation from which all of mechanics can be developed.[16] Hence the fourth and final step: d'Alembert offers solutions to fourteen problems as evidence for his claim of generality.

The Preface is important for setting out d'Alembert's epistemology, and for his summary of the contents of the *Treatise* that is to follow. It also contains

[13] See Chapter 8.
[14] This is what we mean by "mechanics" for the remainder of this chapter.
[15] For the relationship between truth, clarity, and certainty, see d'Alembert 1743, i.
[16] In this *Treatise*, he is concerned with solid bodies. He seeks to extend his principle to fluids in his 1744 work.

his arguments against incorporating any notion of force into mechanics (1743, xxv), and we return to this topic below. For our present purposes, however, we leave our exposition of the Preface here.[17]

(ii) Definitions and Preliminary Notions

This section is divided into four: body and space; rest and motion in space; motion in time; and a collection of issues including the infinite divisibility of space and time, uniform motion, and accelerated and retarded motion in a straight line. The section deserves a full treatment, but for our purposes we focus our attention on body and motion.

D'Alembert introduces his notion of body by appeal to extension and impenetrability: if two portions of extension, equal in size and shape to one another, cannot be imagined unified and combined so as to make one portion of extension whose total size is less than that of the sum of the original two, then these two portions of extension are impenetrable to one another, and each is what we call a body.[18]

In these remarks, he presupposes that bodies are extended, taking this to be unproblematic, at least for the purposes of mechanics. Like Descartes, he takes the idea of extension to be clear. However, unlike Descartes, he states that what distinguishes bodies from space is their mutual impenetrability. He gives a geometric conception of impenetrability for which the criterion is imagination: bodies are entities for which we cannot imagine their being combined in such a way as to result in a decrease in the total volume.

That bodies are mobile requires that bodies be distinguished (at least conceptually) from the space in which they move. The impenetrability of bodies plays this role: it is that by which we distinguish, in our minds at least (i.e., conceptually), body from space.[19]

Following Newton, d'Alembert says that the place of a body "is the part of space that a body occupies, that is to say, the part of space with which the extension of the body coincides." This definition of place enables d'Alembert to define the rest and motion of a body: "A body is at rest when it remains in the same place; it is in motion when it passes from one place to another." For the

---

[17] For further discussions of d'Alembert's *Treatise*, including his epistemology, see Grimsley 1963; Hankins 1970; Firode 2001; Schmit 2017; and references therein.

[18] d'Alembert 1743, 1.

[19] See d'Alembert 1743, v. See also d'Alembert 1754 and 1759.

case of motion, he further elaborates that in going from one place to another, the body must pass through all the parts of space immediately contiguous to one another, successively and without interruption. Next, d'Alembert explicitly introduces time, writing that since a body cannot occupy several places at once, it cannot go from one place to another instantaneously, but must take time to do so: motion requires a duration of time.[20] Finally, d'Alembert claims that the space traversed by a body and the time in which it does so are each divisible to infinity. This is d'Alembert's concept of motion, and his epistemology requires that we have a clear idea of it.

Two questions arise. First, is d'Alembert right that we have a clear idea of body and motion, thus conceived? Second, even if so, can his mechanics be developed on these ideas of body and motion? For example, no force concepts have been introduced, in striking contrast to the definitions found in Newton's *Principia*; we return to this issue below.

(iii) Part I. General laws of motion and equilibrium of bodies

Part I of d'Alembert's *Treatise* sets out the three principles from which he develops his mechanics. From the perspective of rational mechanics, we may think of these as his *axioms* (see section 9.6, below).

1. Force of inertia
D'Alembert asserts that he follows Newton in using "force of inertia" for "the property that bodies have for remaining in the same state."[21] By itself, this would be uninformative without details about what properties or attributes of a body are part of its "state," and d'Alembert writes that "a body is necessarily in a state of rest or of motion."[22] Given this, two laws follow, he claims:[23]

---

[20] But see Le Ru 2015.

[21] d'Alembert 1743, 1.

[22] d'Alembert 1743, 3–4. The strength of the necessity here is as follows. "Body" was defined with mechanics, as the science of motion, in mind: we required a definition of body such that bodies are mobile (see his section "Definitions and preliminary notions"). Given this definition, it follows necessarily that such a body is determinately at rest or in motion. See his later *Essais sur les éléments de philosophie* (1759) for more on this topic of necessity, as well as Hankins 1970.

[23] These two laws together correspond to Newton's first law of motion. Notice that d'Alembert splits the law into two, rests both on his principle of the "force of inertia," and seeks to argue for them from this principle. For more on his principle in relation to Newton's laws, see Hankins 1970, chapter 8. Hankins points out (176) that whereas Newton offered three laws of motion, many of those who were primarily responsible for the development of rational mechanics in the 18th century, including the Bernoullis, Clairaut, Euler, and Lagrange, did not formulate their mechanics in terms of laws.

Law 1. A body at rest remains at rest, unless a foreign cause removes it. For a body cannot set itself in motion. (d'Alembert 1743, 3)

Law 2. A body once put into motion by some cause, must remain in uniform, straight-line motion, so long as a new cause, different from that which put it in motion, does not act on it. That is to say, unless a foreign cause that differs from the motive cause acts on the body, it will move perpetually in a straight line, and traverse equal spaces in equal times. (d'Alembert 1743, 4)

Law 1 seems to follow unproblematically from "force of inertia." D'Alembert draws an important corollary: regardless of the cause whereby a body receives motion, it cannot then further accelerate itself; a cause that sets a body in motion cannot also give the body the power of self-acceleration. As we will see presently, he means that a body cannot change its own speed.[24]

Law 2 is a little more complicated. Why should a body put into motion remain at a constant speed and move in a straight line? D'Alembert discusses this issue at length.[25] For our purposes, two points are most important. First, for d'Alembert constant speed follows from the first corollary to Law 1, whereas straight-line motion requires a further argument. He writes that, after the first instant, there is no reason for a body to turn right rather than left, so it will move in a straight line. Thus, it seems that he supplements force of inertia with a principle of sufficient reason so as to infer that a body, when not acted upon by an external cause, continues to move not just at a constant speed but also in a straight line.

Second, d'Alembert believes that Law 2 can be demonstrated whether or not bodies are assumed to have either resisting force or force of motion. The result is independent of one's views on force, he claims. Still, he faces the same difficulties as Descartes when trying to remove any notion of force from his discussion, for d'Alembert ends his argument for Law 2 with an apparently throw-away comment that bodies have a *tendency* to move in a straight line: nothing in his arguments has said anything explicit about any such tendency, or about how we should think of it. In short, though his arguments demand further attention, we are already in a position to conclude that his first principle, "force of inertia," introduces additional conceptual resources beyond those in his "Definitions and Preliminary Notions."

---

[24] d'Alembert 1743, §4.
[25] d'Alembert 1743, §§6–7. This deserves more treatment than we can give it here. See Hankins 1970, 179–85.

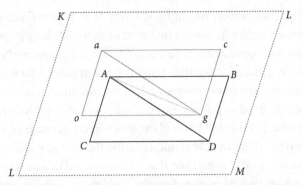

Fig. 9.1. D'Alembert's proof of the Parallelogram Rule. Throughout the *Treatise*, d'Alembert uses a combination of geometric and algebraic reasoning. In his statement of the composition of motions, *ABCD* is the parallelogram and *AD* the actual motion. The term "power" is used neutrally; what matters is the consequent motion. For example, two "powers" give rise to the quantities of motion *AB* and *AC*, and all theorizing is then done in using these quantities of motion, with no further reference to "powers." Cf. d'Alembert 1743, fig. 6.

2. Composition of motions

By this phrase, d'Alembert means the Parallelogram Rule for the addition of velocities, which he states as follows:

> If any two powers act at the same time on a body or point A (fig. 6 [Fig. 9.1])
> to move it, one from A to B uniformly in a certain time, and the other from
> A to C uniformly during the same time, and we complete the parallelogram
> ABCD, I say that the body A will traverse the diagonal AD uniformly, in
> the same time that it would have traversed AB or AC. (d'Alembert 1743, 22)

He is very deliberate in this formulation of the Parallelogram Rule and in his ensuing demonstration, as he explains in the remark that follows. He wants his demonstration to be neutral between whether the two powers on the body act only at the initial instant, or whether they both act continuously throughout. Puzzlingly, he also thinks the addition of motions along the *same* line is a *corollary* of the Parallelogram Rule.[26]

His demonstration of the Parallelogram Rule is as follows (see Fig. 9.1).[27] Suppose that body *M*, moving freely, traverses some unknown straight line

[26] d'Alembert 1743, §§22–23.
[27] d'Alembert 1743, §21–22.

Ag, resulting from powers *AC* and *AB*. Once *M* arrives at point *g*, suppose that two powers act on it, powers that tend to move *M* along *gc* and *go* with speeds equal and opposite to those along *AC* and *AB*, respectively. It is clear, d'Alembert says, that *M* would then remain at rest at point *g*, for it is animated by equal and opposite powers along each of the two directions.

Now keep *M* fixed at point *g*, but consider its path relative to a moving reference plane. Let this reference plane move freely in two directions: first, in the direction *AC* so that *M* would describe the line *gc* parallel and equal in length to *AC*, in the same time that *M* traversed *AC*; and second, in the direction *AB* so that *M* would describe the line *go* parallel and equal in length to *AB*.

It is evident, d'Alembert asserts, that the path of *M* relative to the moving plane is the line *ga*, equal and parallel to the diagonal *AD* of the parallelogram *ABDC*. But this simply reverses the original motions, and *M* can only arrive back at *A* if *AD* = *Ag*. Therefore, the unknown path *Ag* must be the diagonal of the parallelogram formed by *AC* and *AB*.

There are many puzzling things about this attempted demonstration, both implicit and explicit. For one, the demonstration presumes the intended result: in considering the motion of the reference plane d'Alembert treats the two motions independently and assumes that they add as vectors. This is explicit, but hidden from view, because the body *M* is supposed stationary in this case, and the motions belong to the reference plane—not to the body with whose motions we are concerned. But the addition rule is in use nonetheless. Moreover, in treating the trajectory of *M* first from a resting reference plane and then from a moving one, d'Alembert implicitly assumes a transformation rule between the two. And, as we know with the benefit of hindsight following SR, such transformations hide assumptions about the behaviors of physical systems. Lurking in the background, and never explicitly addressed by d'Alembert, is the lack of a relativity principle in his mechanics.

3. Equilibrium

D'Alembert's axiom of equilibrium states: "bodies that have equal and opposite quantities of motion are in equilibrium," where "quantity of motion" is "the product of the mass and the speed."[28]

---

[28] See d'Alembert 1743, §39, for the statement of equilibrium, and §49 for the quantity of motion. Notice that no notion of mass has been explicitly introduced by d'Alembert prior to this, raising questions about the place of mass in his mechanics. We will have more to say about this in section 9.6.

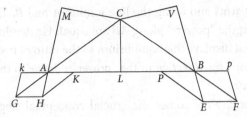

Fig. 9.2. D'Alembert's equilibrium condition for the bent lever. The proof proceeds as follows. Let AM be the extension of the line AH through A, all the way to point M, such that CM is perpendicular to AM. The length of the line CM then varies with the angle that AH makes with AB; that is, with the angle at which the power acts through A when the system is in equilibrium. Let CV be the analogous line constructed for the power acting at B. Now decompose AH and BE (i) parallel to the respective arms of the lever (AG and BF, respectively), and (ii) along the line joining A to B (AK and PB, respectively). Then our equilibrium condition is that $AK = PB$. By similar triangles, $AK/AH = CM/CL$ and $BP/BE = CV/CL$. Therefore, $AH{:}BE$ is as $CV{:}CM$; that is, the ratio of the "powers" yielding equilibrium of the bent lever is equal to that of the lengths CV to CM. The standard lever (one with a straight arm) is then recovered as a special case (1743, §47). Cf. d'Alembert 1743, fig. 16.

In earlier chapters we saw proposals for arriving at new principles adequate for mechanics by generalizing accepted principles of statics. Here, d'Alembert is attempting to take the *familiar* notion of equilibrium from statics (of bodies at mutual rest) and to *re-theorize* it so that it can be used for systems of bodies in motion. He explains the relationship between "quantities of motion" and "equilibrium" thus: given a system of bodies in equilibrium, we assume each body has a certain quantity of motion, and then by our axiom we conclude that the total quantity of motion in any given direction must be zero.[29]

D'Alembert's version of the Lever Principle illustrates his axiom of equilibrium as follows.[30] Take a bent lever, ACB, fixed at point C (see Fig. 9.2). Let the lever be acted upon at A and B by two "powers," and let it be in equilibrium. We represent the direction and magnitude of these "powers" by the lines AH and BE. D'Alembert decomposes the "powers" in two directions,

---

[29] See the first two corollaries that follow §39.
[30] See Chapter 8 for the importance of Bernoulli's Lever Principle in early 18th century attempts to treat new problems in mechanics.

along the lever arms and along the line joining A and B. The equilibrium condition is that the "powers" along AB are equal. He develops a geometric construction that displays this equilibrium as the ratio of two lengths sensitive to two factors: the *direction* of the "power" relative to the balance arm, and its *magnitude*.

So far, so good. Now comes the crucial conceptual step: according to d'Alembert the "powers" in the above scenario are to be understood as nothing but "quantities of motion"; this is how statics is to be incorporated into a general theory of mechanics.[31] Lines AH and BE in the geometric construction represent the magnitudes and directions of quantities of motion, and when two such quantities are equal, equilibrium obtains.

Prima facie, this analysis of the lever in terms of quantities of motion is puzzling because when a system is in equilibrium there are no *actual* motions. But this is exactly d'Alembert's new theorization of "equilibrium": bodies that have equal and opposite quantities of motion are in equilibrium. To make this work, he will need a concept of *counterfactual* (or *virtual*) motions, and we return to this later when we discuss his *General Principle*.

D'Alembert's goal is more general than statics: he aims to treat bodies in motion. He arrives at his principle in three steps. First, he considers the motions of bodies that encounter obstacles. Next, he introduces his notion of equilibrium. And then finally, moving through a series of four cases, he arrives at his generalized principle of equilibrium. There is much in these discussions that demands closer scrutiny. However, we will focus our attention on those elements most relevant for our own purposes here.

In the first step, where d'Alembert is considering obstructed motions, there is one argument that stands out as especially important. This is his argument for the claim that when a body without elasticity collides perpendicularly with an immobile and impenetrable plane, it must come to a stop after the collision and remain at rest.[32] What could justify this claim?

In Chapters 2–4, we examined several attempts to justify the kinematic behavior of elastic and inelastic bodies on the basis of their geometric behavior, and of their material composition. D'Alembert offers no such argument. Instead, he states simply, "it is obvious that if the body moves after the encounter with the plane, this can only be backwards and in the perpendicular

---

[31] See d'Alembert 1743, §43, where he argues that his framework is sufficiently general to contain the theorems of statics, and that his proofs are more precise due to his identification of power (an unclear notion) with quantity of motion.

[32] d'Alembert 1743, §29.

direction." And since "there is no more reason why m should be one number rather than another"—where m is a coefficient such that if u is the speed before impact then mu is the exit speed—m must be zero. That is, the speed after impact will be zero.[33] But an equally good argument would set m equal to 1 (viz. no speed is lost). Indeed, nothing in d'Alembert's arguments keeps m from being greater than 1, such that motion is *created* in the impact. To rule this out, d'Alembert would need to appeal to both the principle of inertia and the claim that interactions do not generate new quantities of motion, both of which are substantive claims about the behavior of bodies and forces. Finally, notice that m being *less* than 0 is ruled out by the *stipulation* that the obstacle is immobile, rigid, and impenetrable; this is posited without regard for whether *any physical body* could satisfy it. So, d'Alembert's claim—that a body without elasticity colliding perpendicularly with an immobile and impenetrable plane must come to a stop and remain at rest—is both a theoretical stipulation and a substantive physical claim about bodies and their motions. It is one for which he offers no sound justification.

So much for the speed after impact. As for direction, d'Alembert claims that, were there any motion after impact, it would have to be perpendicular to the impenetrable plane, and this is taken to be "obvious," as we saw. Perhaps he relies on Descartes' proof that angle of incidence equals angle of reflection, but this proof itself relies on the assumption we are interested in: that the perpendicular component of speed in a collision with a rigid, flat surface is not deflected into a new direction.[34] This is a substantive claim about the interaction that occurs when one physical body collides with another, and about the conservation laws that hold for such interactions. In sum, there is a great deal of physics in just this first step of d'Alembert's argument for his principle of equilibrium.

In the second step of his argument, d'Alembert introduces his notion of equilibrium:

If the obstacles that the body encounters in its motion have precisely the resistance necessary to prevent the body from moving, we say that there is equilibrium between the body and these obstacles. (d'Alembert 1743, §38)

[33] d'Alembert 1743, §29.
[34] See the essay on dioptrics in Descartes 1637.

However we seek to understand this claim, one thing to notice is that he introduces the idea of a body's "resistance" to another, but without any discussion of the justification for introducing it as a concept of rational mechanics (let alone its physical basis). Indeed, recall that in arguing for Law II, he maintained that this law holds regardless of whether one believes that bodies have a force of resisting associated with them. So his principle of inertia cannot give him the "resistance" of bodies that he appeals to here.

In the final step of his argument, d'Alembert moves to demonstrate the principle of equilibrium itself. Up to this point, he has considered a body encountering an obstacle. Now, he replaces the obstacle with a second body, so that we have a pair of bodies, each of which is an obstacle for the other. Then he states his principle of equilibrium as follows:

> If two bodies whose speeds are in inverse ratio to their masses have opposite directions, in such a manner that neither can move without displacing the other, there will be equilibrium between these two bodies. (d'Alembert 1743, §39)

He demonstrates this via a series of four cases, beginning with two equal bodies of equal and opposite speeds. Of this case he says: "it is evident" that such bodies will remain at mutual rest.[35] He uses his earlier argument to support this conclusion, and in light of our discussion we know it is far from obvious that "equal" bodies with equal and opposite speeds will be at mutual rest.[36] Nevertheless, d'Alembert takes this to be the only scenario "where equilibrium manifests itself clearly and distinctly."[37] It seems that underlying his claim must be the thought that, when two bodies act on one another in an equal and opposite way, they are in equilibrium. The head-on collision of "equal" bodies with equal and opposite speeds realizes this idea, and the concept can be made precise by introducing "quantity of motion": such bodies

---

[35] See d'Alembert 1743, §39. The four cases raise interesting questions concerning the concept of mass in d'Alembert's rational mechanics. It might be thought that he is illegitimately helping himself to a dynamical notion of mass as primitive, but thus far in the *Treatise* mass seems to be a derivative concept. That is, speeds and equilibrium are empirically accessible, but "mass" and "quantity of motion" are theoretical constructs accessible via the behaviors of bodies. We will have more to say about this in section 9.6.

[36] See our discussion of d'Alembert 1743, §29. This situation is problematic because of d'Alembert's epistemology: he requires that his mechanics rests on clear ideas. The mutual rest of such "equal" bodies might be something on which widespread agreement could be reached (as, for example, with Newton's first law of motion), but that is not the same as its being a principle that is sufficiently clear that it is thereby certain.

[37] d'Alembert 1743, xiv.

have equal quantities of motion that are oppositely directed, and so when the two bodies come into contact they find themselves in equilibrium.

For the purposes of exposition, let us grant this. D'Alembert then varies the masses and speeds, and arrives at his general conclusion, restating his principle of equilibrium thus: *bodies having equal and opposite quantities of motion are in equilibrium*.[38]

We have indicated some of the difficulties with d'Alembert's attempt to justify his principle of equilibrium, without dwelling on too many details. He hoped that his principles would be mutually independent, yet his demonstration implicitly relies on his principle of inertia. He appeals to a priori reasoning in arguing that a body without elasticity will come to rest on impact with an immovable obstacle, but either this argument fails (for the reasons given above), or the claim should be taken as stipulative (viz. definitional of what it is for a body to be without elasticity)—in which case it would not require the kind of extended argument that d'Alembert provides. And, his conception of quantity of motion applies to bodies not actually moving, which seems puzzlingly disconnected from the concept of motion introduced in the "Definitions and Preliminary Notions," where he states that a body is in motion "when it passes from one place to another."[39] Finally, his principle of equilibrium invokes some notion of mass (via "quantity of motion"), yet he nowhere introduces an explicit mass concept.

Inertia, composition of motion, and equilibrium are the three principles at the foundation of d'Alembert's mechanics. His stated goal was to rest mechanics on clear principles. In this section, we offered an initial exposition of the principles, raising some questions about their content and clarity. These will be important when we turn to the interpretation of d'Alembert's *Treatise* as a book in both rational and philosophical mechanics (sections 9.4–9.6).

(iv) Part II. General principle for determining the motions of several bodies acting on one another in any manner, with several applications of this principle

D'Alembert claims that his three principles of inertia, composition of motion, and equilibrium are sufficient for a rational mechanics of bodies in motion.

---

[38] d'Alembert 1743, §39. He refers to this principle as an "axiom" (§39, 39–40).
[39] d'Alembert 1743, II.

His evidence is: from them he obtains a result—his *General Principle*—that is a general form for solutions to all such problems. Or so he believed. He hoped that his *General Principle* would enable the solution of "the most difficult problems" of mechanics.[40] This is the sense in which he calls it a "method": it is a generic solution, a general schema for solving problems in mechanics. To support his claim, he works through fourteen problems. We begin with the *General Principle* and then turn our attention to the problems.

General Principle

The *General Principle* concerns a system of any number of bodies arranged in any relations to one another whatsoever. Given the initial motions a, b, c... of bodies A, B, C... we are charged with finding the subsequent motions of each of the bodies. He states his principle thus:

> Decompose the motions a, b, c etc. impressed on each body, each into two others: a', α; b', β; c', γ; etc., and let these be such that, had we impressed on the bodies only the motions a', b', c' etc., the bodies would have been able to conserve these motions without mutual impediment; and such that, had we impressed on them the motions α, β, γ etc., the system would have remained at rest. It is clear that a', b', c' will be the motions that these bodies take in virtue of their action. (d'Alembert 1743, §50)

In other words, we separate out the components of the motions a, b, c... that yield a system in equilibrium (α, β, γ...), and then the remaining motions are those that the bodies in the system in fact undergo (a', b', c'...).[41]

Note that the *General Principle* appeals to motions that, *had* they been impressed, the bodies *would* have conserved them, or (for other motions) *would* have remained at rest. D'Alembert is comfortable with non-actual motions—motions that bodies *would have had*, had they not been impeded. These "virtual" motions are at the heart of his *General Principle*, and they are the technique by which statics falls under his general approach: the "powers" holding the lever in static equilibrium are to be understood as equal and opposite virtual quantities of motion (see earlier).

---

[40] d'Alembert 1743, 49–50.

[41] To better understand the principle, see d'Alembert's fourteen problems where he shows the principle in action (section 9.5). For detailed discussion of the *General Principle* and several of these problems, see Fraser 1985.

What is the scope of the *General Principle*? How general is it? The first step toward addressing this is to get a sense of how the *General Principle* functions as a general method for solving problems in mechanics. Then, we can consider the scope of this generality, as well as its limitations.

Fourteen problems

D'Alembert breaks his fourteen problems into four groups. Section I (Problems I–VI) contains problems of constrained motion in which mass points are variously connected (by rods and threads) and constrained by rigid surfaces (such as horizontal or curved planes). The rods and threads are massless, mobile, and perfectly inextensible, but differ in the rod being rigid while the thread is flexible. The rigid surfaces may be either fixed or mobile. D'Alembert's goal is generality: he aims to bring prior solutions to particular problems, including center-of-oscillation problems and the compound pendulum, under one set of principles, as well as solving new problems.[42] We can get a sense of how the *General Principle* works in these cases by looking at the simplest, Problem I (see below).

The remainder of Section I concerns several complex cases involving multiple point masses connected either rigidly via massless inflexible rods or by means of massless non-stretching threads that can bend around obstacles (such as fixed points, pulleys, and so forth), including compound pendulums constructed with threads rather than rigid rods.[43]

Section II, Problem VII concerns an oval shape touching a horizontal plane such that the vertical through the shape's gravity center does not pass through the contact point. The problem is to determine how the shape will move. This is a constrained-motion problem for an extended body, in which both the shape of the body, and the position of its center of gravity within that shape, affect its motion along a constraining surface. The body is assumed to be rigid, and so is the constraining surface. Section III, Problem 8 concerns two mass points connected to a thread, one at a fixed point and the other free to move along the thread. Once again, we have a constrained motion, but this time the constraining surface (the thread) is not fixed in place—it changes position as the constrained bodies move.[44]

---

[42] He refers to previous solutions of particular problems by J. Bernoulli, Clairaut, Hermann, D. Bernoulli, and Euler.

[43] For detailed treatment of Problems II, V, and X, see Fraser 1985.

[44] See also Chapter 7, where we see Clairaut treating the same class of problems. For a comparison of d'Alembert's *Treatise* with contemporaneous work by Clairaut, see Schmit 2017.

Finally, Section IV (Problems IX–XIV) is entitled "Of bodies that push or collide." D'Alembert begins with the familiar case of head-on impact of two spherical bodies, in which the motions of the bodies are treated by means of representative points located at the center of each, while the extended shapes of the bodies provide the constraints. He then goes on to treat collisions of bodies constrained in various additional ways: by rods, threads, and surfaces, where the impact is oblique, and where more than two bodies collide at once. In this way, d'Alembert locates the problem of collision (PCOL) under the general umbrella of the problem of constrained motion (PCON), to be treated by the same principles and techniques.

We can get a sense of how the *General Principle* works in these cases by considering the most straightforward (see below). What matters for our purposes is not the complexity of the scenarios but the simplifying assumptions, present in even the simplest case. D'Alembert supposes spherical bodies with uniform mass distribution. They hang from inextensible strings, so that they swing into one another rather than roll, removing the need to consider the complexities of any rotation of the sphere as it impacts another sphere, and the role of friction for a body rolling on a surface. Jointly, these simplifications ensure that dynamically each body can be treated as a point, with the mass located at the center of the sphere, while the size and shape of the body play a role as purely geometrical constraints, restricting the spatial paths available to the bodies and thereby impeding their motions.

In both the cases we analyze (Problems I and IX), we will see d'Alembert's general strategy at work. He shows how a simple case (such as the linear compound pendulum or head-on collisions of equal bodies) can be treated by his *General Principle*, and then shows how progressively more complex cases can be treated by means of the same method. D'Alembert claims to have incorporated and extended a disparate variety of problems previously treated by others, all by means of the *General Principle*.[45] He is explicit that his goal is generality. For example, in the scholium following Problem III d'Alembert emphasizes his aim of showing that all these problems can be treated by his *Principle*, even if in some cases such a treatment is longer than one available by means of another principle (see §90).

---

[45] D'Alembert's treatment of many of these problems is challenging to follow: the notation is often awkward; the methods are unfamiliar; and the problems are often extremely difficult (becoming more readily tractable only in light of later developments due primarily to Lagrange). Problem II has been analyzed in detail by Fraser 1985. See also Schmit 2017 (and references therein) for his analysis of the problems.

The problems he discusses incorporate all known results and, importantly, show a way forward for a unified treatment of more complex results of the same kind. An open question is whether this exhausts all the kinds of problems we might hope to treat within mechanics. We take this up in section 9.4.

Section I, Problem I: the linear compound pendulum
This problem concerns a rigid, compound pendulum: multiple weights are attached to a straight rigid rod at different points along its length; the rod itself is assumed to be massless and suspended from one end, and it is assumed to swing freely about its fixed point of suspension. The problem is to find how the motion of the rod is affected by having various weights placed along its length. The technique works for any number of weights, $m_1, m_2, \ldots m_N$, and for simplicity we will work with three: m_A, m_B, and m_R located at positions A, B, and R as the rod hangs vertically from point C (see Fig. 9.3). Throughout, displacement from the vertical is small.

The method is as follows. We consider a given unit of time, so that our task is to find the displacement of the rod in that unit time, and for this it is sufficient to find the displacement of m_R from R in that time, say from R to S (see Fig. 9.3). That is, we are to determine the distance RS.

First, consider a simple pendulum associated with each weight. Let the impressed speeds be such that, in the given time interval, m_A would displace

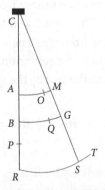

Fig. 9.3. D'Alembert's treatment of the compound pendulum. C is the point of suspension, with weights located at A, B, and R along the rod CR. The problem is to find the displacement RS, different from the displacement RT that the weight R would have had, were it not constrained by the rod CR to which the weights A and B are also attached. Cf. d'Alembert 1743, fig. 22.

from A to O; m_B from B to Q; and m_R from R to T. That is, since we are considering a unit time interval, the impressed speeds are represented by the distances RT, BQ, and AO.

Now build the compound pendulum: m_A, m_B, and m_R are constrained by the rod to move together so that, in the same given interval of time, m_A displaces from A to M; m_B from B to G; and m_R from R to S. The positions M, G, and T are, of course, collinear: they are the positions of the weights along the rod, as they swing together about the fulcrum C. Now we can write the impressed speeds as consisting of two contributions:

$$RT = RS + ST$$

$$BQ = BG - GQ$$

$$AO = AM - MO$$

This decomposes each of the impressed motions RT, BQ, and AO into two: those that the bodies *would* have been able to conserve without impediment, despite their being attached to the rod (RS, BG, and AM), and the remainder (ST, $-GQ$, and $-MO$).

Next, we apply the *General Principle*. This states that impressed motions can be decomposed into two contributions: those that balance one another out and those under which the bodies move freely. Hence, the former in this case are ST, $-GQ$, and $-MO$. The "lever," as d'Alembert calls it—in this case just a single rigid rod swinging from C as a pendulum—would remain at rest if the bodies m_A, m_B, and m_R were in equilibrium. Thus, our equilibrium condition is given by the masses multiplied by the speeds:

$$m_A \cdot MO \cdot AC + m_B \cdot GQ \cdot BC = m_R \cdot ST \cdot CR$$

and from here we solve algebraically for RS.

The givens of our problem are: the masses m_A, m_B, and m_R; the lengths CA, CB, and CR; and the displacements RT, BQ, and AO. The displacement RS (which d'Alembert re-labels x) is straightforwardly obtained in terms of these givens. In other words, from our knowledge of the behavior of the simple pendulums associated with the masses, the *General Principle* allows us to determine the behavior of the compound linear pendulum.

Section IV, Problem IX: two bodies in head-on collision

We consider a body of mass m and initial speed u, moving along the same line as another body of mass M and initial speed $-U$. The problem is to find the speeds of the bodies after the collision.

Applying the *General Principle*, we decompose each of the initial motions mu and MU into two parts: the part that produces equilibrium (say mc and MC) and the part that the bodies are able to preserve notwithstanding the impediment (the final motions mv and MV, say). Since mc and MC produce equilibrium, $mc + MC = 0$; that is, $m(u - v) + M(U - V) = 0$, as he says.[46]

To proceed from here to knowledge of the final speeds v and V, more information is required. D'Alembert states that, by his principle, it must be the case that the bodies move together with the same speed after collision.[47] Given this, $v = V$, and the final speed v (or V) is given by $(mu + MU)/(M + m)$.

However, as we noted above, the claim that the bodies move together with the same speed after collision is a substantive, dynamical claim, and not one that follows from the *General Principle*. It is a stipulation about the behavior of bodies that requires justification.

9.4 D'Alembert's *Treatise* as a contribution to rational mechanics

From the perspective of rational mechanics as it stood at the time, the *Treatise* offers a remarkable unification under a single set of principles.[48] That said, there are important limitations in the scope of the problems covered.

First, insofar as d'Alembert considers extended bodies, all the target bodies treated in his fourteen problems are "perfectly hard": they are impenetrable and rigid. Moreover, so are many of his constraining bodies, including his surfaces and rods. The only exceptions are the threads that he

---

[46] d'Alembert 1743, 138.
[47] d'Alembert 1743, 138.
[48] For comparison of the principles adopted by d'Alembert with those of his contemporaries, and of the problems covered with those by Clairaut and others, see Schmit 2017.

uses to constrain the motions of bodies: these are extended and impenetrable but they are flexible (they bend but do not change length).

Second, d'Alembert's bodies are non-rebounding. As we know from previous chapters, this kinematic assumption is independent of the geometric assumption about rigidity (lack of shape change).[49]

Third, the primary means for treating extended bodies is to treat them as composite: constructed from point masses connected to one another by massless rods or threads. D'Alembert succeeds in giving highly general treatments for arbitrary numbers of such masses. But this generality is rather deceptive, for it is a generality in principle only: the initial motions of the point masses are presumed known, and while this might allow for a quantitative solution of particular cases when the numbers of point masses are small, it seems less illuminating once the numbers increase. Indeed, when it comes to his treatise on fluids published the following year, d'Alembert is quick to point out that additional principles are needed due to our lack of access to the mutual actions of the particles composing the fluid.[50]

Fourth, even in the case of collisions, the extension of bodies appears primarily as a kinematic constraint, limiting the trajectories available to bodies. In most cases, the trajectories of the target bodies are determined within these constraints via representative points. For example, in Problem VI d'Alembert claims to treat a compound pendulum with an extended bob of arbitrary shape, but it seems that by restricting consideration to infinitesimally small oscillations he is able to determine the motion of the whole via a known center-of-gravity, that is, via a representative point. There is only one case where the target body is non-composite and extended, and where the shape of the body plays a dynamical role: Problem VII. In this case, the shape is placed in a non-equilibrium position on a surface, and the challenge is to determine how it will move. This is a particularly interesting problem, combining rotational and linear motion on a constraining surface. However, note that even in this case he again assumes the target body to be rigid, and, importantly, that mass is uniformly distributed across the shape.

---

[49] While hard, non-rebounding bodies are the central case for d'Alembert, he does offer some remarks at the end of the fourteen problems on how to extend his analysis to treat elastic and soft bodies. However, this involves stipulation as to how the rebound behavior differs from the hard body case, using the analysis of hard-body collision as the basis from which to arrive at results for elastic and soft bodies (see Fraser 1985, 151–5). Hence, he added more dynamical information, beyond the content of the axioms and General Principle, and it is unclear what justification d'Alembert takes this additional content to have.

[50] d'Alembert 1744, viii–ix.

Fifth, all of d'Alembert's fourteen problems treat time-independent (i.e., scleronomic) constraints. Extension to constraints with an explicit time dependence lies in the future. More generally, though d'Alembert offers generalized treatments of existing problems, it seems to us that no new *kinds* of problems fall within the fourteen problems of his *Treatise*.

Alongside these limitations in scope, there are conceptual limitations. It is far from clear that the concepts articulated at the outset are adequate for expressing the *General Principle* and solving the fourteen problems, let alone that they meet d'Alembert's epistemological requirement of being so clear as to be certain (section 9.2). All of these considerations pertain to the assessment of d'Alembert's rational mechanics.

Notwithstanding these considerations, the overall *structure* of d'Alembert's *Treatise* is important for our purposes, for he is attempting to *unify* rational mechanics by means of a single set of principles, and to *derive* all of rational mechanics from these principles. We might call this d'Alembert's axiomatization project in the foundations of rational mechanics.[51] This will be important in what follows.

9.5 D'Alembert's *Treatise* as a contribution to philosophical mechanics: a constructive approach

According to d'Alembert, mechanics is "the science of the motions of bodies that act on one another in any manner whatsoever."[52] What, then, are the bodies of d'Alembert's mechanics, and how do they act on one another?

To answer this question would be to give a philosophical mechanics, one that integrates d'Alembert's rational mechanics with an account of the nature and properties of the bodies that his theory is about. Though d'Alembert's

---

[51] The unification of mechanics through axiomatization is an early element of a broader interest in unification that d'Alembert later develops in his *Preliminary Discourse* of 1751 (d'Alembert 1995) and *Essai* (1759). In the *Preliminary Discourse* d'Alembert says (1995, 29): "The universe, if we may be permitted to say so, would only be one fact and one great truth for whoever knew how to embrace it from a single point of view." This can be seen in the context of discussions over systems and systematization in philosophy (see, for example, Condillac 1749). One of us (KB) is indebted to Matias Navarro Crespo for extended discussions of d'Alembert's views on the unity of the sciences.

[52] See d'Alembert 1743, xxiii. He adds that dynamics properly signifies the science of powers or motive causes, whereas he conceives of mechanics as primarily a science of effects rather than of causes. But he chooses the word "dynamics," since he intends mechanics to include the actions of bodies on one another. From our perspective, he intends to treat such actions from the perspective of rational mechanics (as in Book I of Newton's *Principia*) in preference to physics, understood as a branch of philosophy concerned with causes (see Chapters 1–5).

Treatise primarily addresses rational mechanics, we may also attempt to read it as contributing to philosophical mechanics. That is, we may examine whether the *Treatise* itself offers an account of its own objects. We may ask: to what extent does d'Alembert's *Treatise* provide us with a solution to BODY?

Viewed from this perspective, it is natural to read the *Treatise* constructively, as beginning with a material account of body—its qualities and properties—and moving from there to a theory of the motions of such bodies. For, the very first definition that d'Alembert offers is of "body." And by the end of the definitions we know that bodies are extended, impenetrable, and mobile. Perhaps these are the properties of body, and this is d'Alembert's account of Nature.[53]

Such a reading leads to problems, however. For example, when introducing his first principle, "force of inertia," d'Alembert writes that this is the *property* bodies have for remaining in the same state of rest or of motion in a straight line. That is, he seems to introduce a further property of bodies not included in his definition, and one by which he attributes a "tendency" to bodies (see the discussion above). When discussing his third principle, "equilibrium," d'Alembert invokes a notion of mass, and speaks of bodies having "resistance." Moreover, in arriving at this principle, d'Alembert makes commitments concerning the kinematic behavior of bodies when encountering obstacles, behaviors that do not follow from the properties already attributed to bodies, as we saw above. So his principles seem to attribute additional properties to bodies beyond those given in the definitions.

Therefore, to pursue a constructive approach would require us to work through his principles, his *General Principle*, and his fourteen problems, to enumerate the properties that he in fact attributes to bodies in the course of his *Treatise*. This is a task that remains to be completed.

For our purposes, the important points are these. On the one hand, attempting a constructive approach to the bodies of d'Alembert's *Treatise* is not unreasonable. It is consistent with the book's structure, and is encouraged by the conception of physics at the time. Indeed, we suggest that Boscovich undertook just this in his *Theoria* (see Chapter 6). On the other hand, even the preliminary examination we have offered here is sufficient to show that it faces significant obstacles.

---

[53] See Chapter 1 for the four elements of BODY: Nature, Action, Evidence, and Principle.

9.6 D'Alembert's *Treatise* as a contribution to philosophical mechanics: a principle approach

Sometimes, there are moments in theorizing when a principle approach is more appropriate than a constructive one. In the latter, we build up to the objects of our theory (such as bodies) from prior material commitments; and then we show that these objects satisfy relevant principles (such as laws of motion). In contrast, the former *begins* with the adoption of general principles and develops the theory from there. Pursuing this, we can then ask whether the objects of the theory are to be presumed from outside (such as when philosophical physics provides bodies as the objects of mechanics), or whether the theory itself, in its principles, has sufficient resources to specify its own objects. In the second option, the objects of the theory are whatever satisfy the principles.[54] In what follows, we suggest that adopting such a reading of the *Treatise* is philosophically fruitful.

To be clear: our question is *not* "Is d'Alembert's theory a constructive or a principle theory?" The categories of "constructive" and "principle" approaches are ours, drawn from Einstein's distinction explained in section 9.2. D'Alembert did not have these categories, and his theory is not best understood as being of one kind or the other. Rather, we believe that bringing these categories to bear on d'Alembert's theory yields philosophical insights into the structure and content of that theory. Some we obtain from reading the theory constructively (see section 9.5); some we obtain from reading the theory as taking a principle approach; some we obtain from seeing the interplay between the two in the context of the one theory.

So now let us turn to reading d'Alembert's *Treatise* through the lens of a principle approach. Recall that when adopting a principle approach we seek principles that are *independent* of the material constitution of the target systems; are *general* in scope; and enable *bracketing* of (hitherto) dynamical considerations by means of a "kinematic condition" (see section 9.2 for more details). Consider first d'Alembert's *General Principle*. It has the three central features associated with a principle approach. First, the formulation of the *General Principle* is independent of the particular material constitution of the systems that it encompasses. Second, it is intended to be general, covering any system of bodies interacting in any way whatsoever. Third, it reduces dynamics to kinematics by restricting the space of available trajectories for the

---

[54] See Brading 2012 for her "law-constitutive" approach to the bodies of Newton's *Principia*.

system: no matter the dynamical details of any particular system, its trajectory must satisfy the *General Principle*. In other words, the *General Principle* is appropriately understood as a kinematic condition.[55] Adopting the principle approach allows us to see these features of the *General Principle*, and encourages further exploration of d'Alembert's theory from that perspective. We begin this here.

(i) Assessment

Assessing a theory through the lens of the principle approach involves three tasks.

Task 1

Clarification of the principles. This includes assessing whether the principles satisfy the axiomatization conditions of being mutually independent as well as jointly necessary and sufficient for deriving the kinematic condition.[56] When introducing d'Alembert's three principles in section 9.3 we raised some queries about their content, mutual independence, and scope. Adopting the principle approach requires us to pursue these issues with vigor (a task beyond our scope here). Moreover, since the kinematic condition is the *General Principle*, we are required to analyze its relationship to his three principles. D'Alembert claims to have derived the former from the latter, but are the three axioms necessary and sufficient? It seems there are no concepts being used to state the *General Principle* that have not already been introduced in the three principles. But this is not the same as its being strictly derivable. More work is needed.[57]

Task 2

Investigate the epistemic status of the principles. In the case of Einstein's SR, the principles are justified inductively, as we noted earlier. This is not

---

[55] For d'Alembert, rational mechanics picks out the allowed spatiotemporal trajectories from among all the curves that may be treated by means of geometry (1743, vii).

[56] For example, recall from section 9.2 that the "kinematic condition" in the case of Einstein's SR is the Lorentz transformations. Much work has been done on the relationship between Einstein's postulates and the Lorentz transformations (see Brown 2005 and references therein). We are grateful to Oliver Pooley for a discussion of the analogy between Einstein's special theory of relativity and d'Alembert's theory.

[57] For example, although Hankins (1970, 190) states that d'Alembert derived his principle from the last two of his laws, he does not show that this derivation can be carried through.

the case for d'Alembert's *Treatise*. His claimed justification for the truth of each of his principles is its clarity: each principle should be sufficiently clear as to command our assent as certain (see our discussion of the Preface, above). Interestingly, the clarity of the principles of a given science arises from the simplicity of the object of that science, he avers; algebra and geometry have the simplest objects, and from there we move to the object of mechanics: bodies. So, the justification demanded by d'Alembert for his principles takes us back to the objects of his theorizing, and to the constructive approach. As we noted at the outset, d'Alembert's *Treatise* is not best interpreted along either constructive or principle lines; rather, there are insights to be gained from each.[58]

What about the *General Principle*? If it is not derivable from the three axioms, what are its status and justification? Insight into that comes from d'Alembert's remark that the *General Principle* provides a "general method" for the solution of problems in rational mechanics (indeed *all* such problems, he claims).[59] From this perspective, what we have called a kinematic condition—a condition that the trajectories of all systems must satisfy—may be viewed as a heuristic by which to analyze systems and arrive at solutions. The justification for the *General Principle* (and thereby for the claim that the three principles "axiomatize" all of mechanics) lies not in its strict derivation from the axioms, but in the extent to which the generality claim is true. The role of his fourteen problems is to support the claim of generality, and thereby justify the *General Principle*. In this way, the question of justification is intimately tied to that of scope: the greater the scope of the problems falling under the *General Principle*, the stronger its justification.

Task 3
Make explicit the dynamics encoded into the kinematic condition. Newton's theoretical architecture of masses subject to impressed forces is utterly absent. We have already noted that d'Alembert uses the term "mass" without explicitly introducing his intended concept. And, famously, he sought to

---

[58] Not all sciences can proceed in the same way as mechanics, according to d'Alembert, for not all have a simplicity of object sufficient for us to formulate clear principles without appeal to the details of experience or experiments. He is explicit about this in his treatise on fluids, where he argues that, unlike for the mechanics of solid bodies, "the theory of fluids, on the contrary, must necessarily be based on experience/experiments." Our goal is then to arrive at principles that maximize the combination of "simplicity and certitude." The reason for this is that we don't have access to the shapes and motions of the particles of a fluid, and so in order to have a successful theory of fluids we need new principles beyond those presented in the *Treatise*. See d'Alembert 1744, vi–vii.

[59] d'Alembert 1743, 50.

eliminate any notion of force from mechanics. He is explicit that whenever he uses the term "force" this is merely for convenience: the term has no
meaning beyond that which results from the three principles.[60] So our first
research question must be: exactly what dynamical information is present in
the three axioms?

In our earlier discussion of these axioms we made some observations about
the dynamical commitments they bring. In light of the principle approach,
these issues become pressing. To pursue them, the most important point to
note about d'Alembert's mechanics is that "quantity of motion" is his central
notion. He argues that we can treat the effects of force (i.e., motions) without
treating the forces themselves (the causes of the motions).[61] His equilibrium
principle requires that each body, at any given time, has a quantity of motion
associated with it, where this quantity incorporates speed, direction, and also
mass. Two conceptual limitations are apparent. First, d'Alembert presumes
that quantities of motion obey his principle for the composition of motion.
But this principle was developed for velocities only, and quantity of motion
involves not just speed and direction but also mass. The claim that quantities
of motion compose in accordance with the vector addition of their associated
velocities is a further assumption that d'Alembert does not justify. We may
obtain it either by direct stipulation or by assuming that mass is a velocity-
independent scalar, but the important point is that it does not follow from
d'Alembert's rule for the composition of motions. In the *Treatise*, d'Alembert
never offers an explicit account of what he means by the term "mass."

Elusive mass

The place of mass in d'Alembert's mechanics is a subtle issue. He first appeals
to mass when he argues for his equilibrium principle, and there it looks as
though he needs such a concept, but this conclusion would be too hasty. We
can think of his appeal to mass at that point in his argument as being tem-
porary: d'Alembert allows his readers to help themselves to a notion of mass
because it aids our intuitions about equilibrium, but once we arrive at the
principle of equilibrium a notion of mass no longer plays any role. Rather, it

---

[60] d'Alembert 1743, xxv.
[61] He argues that the language of "force" is entirely a matter of convenience by which increments
of speed and time are set in relation to one another, and it can be eliminated in the following
manner: "motive force" is the product of the mass of a moving body with the element of its speed, or
(what is the same thing) by the small space that it traverses in a given instant in virtue of the cause that
accelerates or retards its movement; "accelerative force" is simply the element of speed (d'Alembert
1743, 18–9).

is the product of mass and speed ("quantity of motion") that he is interested in, not the apportioning of that quantity between mass and speed. Thus, unlike for Newton, for whom mass is an invariant quantity associated with a body, measuring the quantity of matter, for d'Alembert mass by itself would then have no meaning or signification. The crucial quantity—"quantity of motion"—can be determined from the speeds of bodies and their equilibrium states, for both of which we have a clear idea (unlike mass) and (not unrelatedly) empirical access, at least in principle.

However, two passages challenge this interpretation of mass significantly. The first lies in d'Alembert's treatment in chapter II of center-of-gravity, where his definition seems to rely on mass as a separable parameter. The second is more important for our purposes, for it is in the treatment of collisions. There, the desired outcome is to find the post-collision speeds of the bodies, and the masses of these bodies are among the givens of the problem. As we saw above, the very first collision problem asks us to consider a body of mass m and speed u, moving along the same line as another body of mass M and speed U, and to find the speed of these bodies after the collision.[62] This is solved using the *General Principle* plus the assumption that the bodies move off together at the same speed after impact. The solution explicitly includes the masses.

These "masses" could be understood as theoretical constructs, derived via the consideration of other equilibrium problems. However, this is not the justification d'Alembert gives, nor is it what he says about mass. Moreover, the fact that mass seems to be a constant associated with a body, independent of any particular scenario or circumstance, whereas "quantity of motion" varies, might seem to indicate that mass is a primary property of bodies, with quantity of motion being derivative. That said, we see no evidence d'Alembert took mass to be primary: he was explicit in allowing only clear concepts into the basis of his mechanics, and he would surely have commented on mass had he believed it to be a primitive notion. So we find ourselves with an interpretative puzzle.

Where does this leave BODY? Once we have worked through the above tasks, we will have the conditions that d'Alembert's principles place on adequate objects for his theory. Are these sufficient to specify a body concept? If so, this will be d'Alembert's solution to BODY, arrived at via a principle approach.

---

[62] d'Alembert 1743, Problem IX, §125.

Boscovich and d'Alembert

A comparison with Boscovich's *Theory* is informative. Like d'Alembert, Boscovich begins from a force of inertia, composition of motion, and considerations of equilibrium, and as for d'Alembert the place of a mass concept in his theory is a tricky issue (see Chapter 6). Unlike d'Alembert, however, Boscovich offers a theory that is explicitly and thoroughly constructive. This has the advantage of making visible just how difficult such a constructive theory is going to be, given the state of mechanics at the time.[63]

However, it also has disadvantages, and the most telling is that we lose sight of the power enjoyed by the newly emerging rational mechanics of constrained motions. For, in Boscovich's theory, all elements of a system are treated dynamically by means of Boscovichian points and forces. By contrast, in d'Alembert's theory the *General Principle* allows constraints to be treated kinematically. From the perspective of 18th century physics, this leaves d'Alembert's theory incomplete, pending a full causal account of the system in question, which is what Boscovich's theory seeks to provide. Looking back from our present-day perspective, with the constructive/principle distinction in hand, the issue is not so one-sided. First, the fact that the constraints can be treated kinematically provides important dynamical information about our system. The principle approach encourages us to elucidate that information in ways that do not require a constructive account of our system. Second, that we can successfully treat a wide range of motion behaviors *without* having a constructive account in hand turns out to be a powerful achievement. The significance of this last point becomes increasingly apparent as the century progresses, and is seen most especially in the work of Lagrange (see Chapter 11).

In the end, as we will see in Chapter 12, it is the *interplay* between constructive and principle approaches to theorizing that turns out to be most fruitful.

---

[63] Indeed, we think Boscovich's *Theory* is helpfully read in light of d'Alembert's *Treatise*, revealing his philosophical mechanics to be engaged with recent developments in rational mechanics. Nevertheless, as of the early 19th century, it remained the case that no single constructive theory was adequate for the needs of mechanics. See Chapter 12, and our discussion of ontic unity below.

9.7 Ontic and nomic unity

D'Alembert's explicit goal in his *Treatise* is to identify a minimum number of clear principles from which all problems in mechanics can be solved. It is an axiomatization project in *rational* mechanics, one that seeks to unify mechanics through a set of principles.

Read as a *philosophical* mechanics, the result is a theory that exhibits *nomic* unity. Notice that this theory may or may not be consistent with a unique ontology. In attempting to specify the objects to which the theory applies, we may adopt either a constructive or a principle approach, and if we take a principle approach there is no guarantee that the theory successfully specifies a unique object. Given the expectations of early 18th century physics, this is a weakness, perhaps even a catastrophe.

However, we argue that it must also be viewed as a strength, for it respects the state of enquiry as of mid-century. At that time, it was utterly unclear *what* the appropriate underlying ontology should be—point particles, finite rigid masses, deformable continua, etc.—or how to relate the candidate ontologies to one another. There was not sufficient evidence to decide the issue. Nomic unity is consistent with a proliferation in ontologies associated with a single theory—or with no articulated ontology at all—and allows theorizing to progress in the face of this epistemic challenge. D'Alembert explicitly criticized Daniel Bernoulli for justifying his choice of axiom (conservation of *vis viva*) by appeal to a specific matter theory (an underlying ontology of elastic bodies).[64] He argued instead that axioms must be justified either by their clarity or inductively, as we have seen. The epistemic situation was that *we did not know* what ontology to posit. In those circumstances, pursuing nomic unity and a principle approach to the associated ontology of a theory is, we maintain, exactly the appropriate philosophical move.

The interplay between constructive and principle approaches to a given theory, and the desire for ontic unity alongside the search for nomic unity, are engines of theoretical development during the remainder of the century, as we will see (Chapters 10–12). The resulting changes bring us closer to a conception of theoretical physics recognizable from the perspective of the 21st century, in which physics is no longer a sub-discipline of philosophy but an enterprise with its own goals, methods, and standards of acceptance.

---

[64] d'Alembert 1744, xvi.

9.8 Nature, Action, Evidence, and Principle

What, then, of BODY? We have offered two different ways of addressing Nature using the resources of d'Alembert's *Treatise*—the constructive and the principle approaches—and both leave questions unanswered. Viewed from a constructive perspective, d'Alembert's definition of body (the very first definition he offers) takes on particular importance. But the properties he attributes there to bodies are inadequate for his mechanics, and it is only by working through his axioms, *General Principle*, and his problems, that we learn the additional properties he includes, and the restrictions in scope on the kinds of bodies he explicitly treats. This moves us toward the principle approach (see section 9.6). Either way, a philosopher seeking to develop a philosophical mechanics from, or at least consistent with, d'Alembert's theory cannot work simply with his definition of body: they must work through the details of his rational mechanics.

The two approaches differ in the role ascribed to the principles (Principle) and in their response to Evidence. Constructively, the evidence for the properties of body lies in the clarity of our ideas of body qua object of mechanics. From a principle perspective, the evidence rests entirely on that accruing to the principles themselves. This is of two kinds. First, d'Alembert takes his principles themselves to exhibit clarity, and thereby command assent. But there is a further justification: the *generality* of the principles, displayed in the *General Principle* and demonstrated by our ability to solve a wide variety of problems by their means. Notice that neither of these is concerned with *empirical* support for the principles or for the attendant body concept.

In addition to addressing Nature, a response to BODY also requires a treatment of Action: of how bodies act on one another, if at all. D'Alembert intended his *General Principle* to cover bodies acting on one another in any manner whatsoever, but in his *Treatise* he restricts his attention to non-gravitational contact action.[65] Unfortunately, he did not explicate his notion of "contact," and we have already seen the emerging recognition that it is a problematic notion in need of explication (Chapter 7). What about "action"? Here, one can turn to d'Alembert's discussion of causes.[66] In the end, however, his methodological goal seems to be to treat motions (effects) without

---

[65] d'Alembert 1743, 49.

[66] d'Alembert 1743, x. For him, impenetrability is an epistemically privileged kind of cause, with contact action having a correspondingly privileged epistemic status.

appeal to forces, causes, or any analysis of the means by which one body acts on another.[67] He stipulates a category of bodies (hard), and offers an unsatisfactory attempt to justify their behavior on contact (their rebound behavior), but ultimately these dynamical considerations are swept up into the kinematic condition that is his *General Principle*. D'Alembert would rather not address Action at all.

9.9 Conclusions

We have argued that d'Alembert's *Treatise on Dynamics* is a pivotal work for the development of philosophical mechanics in the 1700s. By mid-century, the primary locus of research in rational mechanics pertinent to BODY had shifted from PCOL to PCON, and d'Alembert offered the first systematic attempt at a general treatment of the mechanics of constrained motions. His book exemplifies the enormous difficulties involved in PCON, and the ensuing consequences for philosophical mechanics.

The first difficulty is that a general theory of the motions of extended bodies is extremely hard. We outlined d'Alembert's strategy of axiomatizing rational mechanics under a set of three principles, justified by their clarity and by the generality of the problems whose solution they enable (via his *General Principle*). Though he successfully captured a wide array of problems this way, their scope remained somewhat limited. These limitations transfer to any associated philosophical mechanics.

The second difficulty is that of making explicit the properties of bodies treated in such a theory. We suggest that it is helpful to analyze this through the lens of two distinct approaches, constructive and principle, each offering different insights (though see section 9.8, above). The principle approach requires us to clarify the principles as axioms, determine the scope of the systems falling under these principles, and make explicit the dynamics being bracketed into a general kinematic condition by means of the principles. This last point, about the relationship between dynamics and kinematics as probed by a principle approach, is particularly important for understanding the insights into physical systems coming out of the study of constrained motions. It is an issue to which we will return in the chapters that follow. However, here our goal has been more modest. We hope to have done enough

---

[67] d'Alembert 1743, especially 18–9.

to indicate the benefits of reading d'Alembert's *Treatise* through the lens of a principle approach, regarding both the structure and the content of his theory. We highlighted one consequence. Whereas a constructive approach seeks to build a single ontology as the object of theorizing, the principle approach allows for a proliferation of ontologies, and indeed leaves open the possibility that the theory fails to specify a coherent object at all. This latter outcome would clearly be disastrous for those committed to solving BODY. In the place of the ontic unity that is explicit in constructive approaches, the principle approach to d'Alembert's axiomatization project offers us nomic unity: a theory unified by its principles, but uncommitted as to the unity or otherwise of its material objects.

Whether one adopts a constructive or a principle approach when interpreting d'Alembert's theory (and we suggest that both are needed to adequately explore the theory), the contribution the theory makes to BODY is both necessary (due to PCON) and demanding: it requires us to work through the details of the theory so as to uncover its object. The philosopher interested in BODY has no choice but to pursue the complexities of the mechanics of constrained motions. We continue this task in Chapters 10 and 11.

10

Building bodies: Euler and impressed force mechanics

10.1 Introduction

The main reason for the problem of constrained motions (PCON) pressing on philosophy to look beyond impact for a solution to the *Problem of Bodies* (*Body*) came from rational mechanics. Mathematicians in those years moved into a new domain untrodden by their 17th century predecessors, viz. the study of bodies that are constrained: certain regions of space and certain possible motions are forbidden to them.[1] This new domain—constraints—was vast and not yet mathematized. Theorists had the enormous task of building a *rational mechanics* of constrained motions. We call that task MCON, for brevity. Note that it is a problem in rational mechanics, not philosophical mechanics. It requires that constrained extended bodies be mathematized as effectively as Newton and Huygens had done with free particles.

Insightful theorists soon came to understand that constraints come in two broad classes, internal and external. The former are types of conditions on the motion of body parts relative to each other; for instance, rigidity, incompressibility, and flexibility. The latter are limits on the parts of space (regions) that a body is allowed to be in; for instance, motion outward in a rotating massless tube. And so, the rational mechanics of constrained motion came to unfold along two tracks of theory-building. MCON splits into two problems, and so we have given them custom names, as follows:

MCON1: *Given an extended body subject to* internal *constraints, how does it move?* Namely, given an extended body whose parts are mutually

---

[1] A good example is a ball rolling down an inclined plane. Gravity pulls it vertically downward, but it cannot move that way—because the rigid plane obstructs its downward motion. So, the ball is *constrained* to roll along the plane.

Philosophical Mechanics in the Age of Reason. Katherine Brading and Marius Stan, Oxford University Press.
© Oxford University Press 2023. DOI: 10.1093/oso/9780197678954.003.0010

constrained among themselves (i.e., held together to form *one* body), what is the motion of each of the parts?

MCON2: *Given an extended body subject to* external *constraints, how does it move?* Namely, when the motion of a body is impeded by an obstacle, what is its resulting motion?

In this chapter, we reconstruct the milestones for work on MCON1 in the 1700s, and their implications for philosophical mechanics. We begin by explaining clearly what a solution to MCON requires (section 10.2), and why theorists lacked the resources for it (section 10.3). Next, we survey the milestones on their way to solving MCON1 (section 10.4). Then we subject their results to a critical evaluation, and we take stock of the state of the problem as the century turned (sections 10.5 and 10.6).

Before we proceed, a word on method. This chapter is not a history of mechanics after 1735. In line with our declared interest, we focus only on results and developments conceptually relevant to the *philosophical* problem of constrained motion. Accordingly, it is a narrative of the key results on the way to a generic object, viz. the equation of motion for a material point in an extended body. From history, we take just what made a difference to the problem we study. But, of course, mechanics in those decades made fateful progress on many fronts, not all related to our topic.

10.2 Solving MCON

We mentioned above the need to mathematize the motion of extended bodies, but that idea is vague. Spelled out exactly, it amounts to three distinct steps—a full solution requires three distinct formulas:

Step I a *local* expression for the instantaneous kinematic change at any material point in the body. That is, an equation of motion.

Step II a *global* expression that quantifies how the whole body changes in an instant dt. That is either a volume integral or a finite sum over discrete particles.

Step III an *evolution* formula describing the body's motion over time: after some finite interval, not just an instant.

These steps differ significantly in representational content and in regard to what they require for success. Step I demands *physical insight*: a grasp of the agencies at that point, viz. internal and external forces, body forces and stresses, tractions and shear, etc.; and of the kind of actions that internal constraints can do. It also requires a grasp of the various response factors present there: mass, moment of inertia, and mass flow; translation, spin, and strain; effective acceleration and motion lost to constraints, etc. Step II requires an *architecture of matter*: a picture of mass distribution at basic scales, and its kinematic behaviors, or possible ways to move. Without that picture, the task of representing a whole-body motion is indeterminate (see section 10.5). Lastly, Step III requires *mathematical power*: that certain branches of pure mathematics have developed enough to yield exact answers to that task; for example, closed-form solutions to the relevant differential equations (or at least some approximation techniques to supplant them, such as methods of numeric approximation, finite-element approaches, and the like).[2]

The figures at issue in this chapter were keenly aware of these distinct needs for solving MCON, and also of how much they lacked as they set out to solve it. Facing up to rigid rotation around a variable axis, Euler commented: "if the axis of spin changes orientation, the principles of Mechanics we have so far do not suffice to determine the motion. So, we must find and prove *new principles* fit for the task."[3] Likewise, on the shortcomings of mathematics for many Step III tasks, required *after* the equation of motion has been found:

> if we are unable to reach a complete knowledge of fluid motion, it is *not the fault of Mechanics and its known principles* of motion. Rather, it is *Analysis itself that lets us down* in this matter. . . . Seeing as a general solution must appear impossible, on account of *shortcomings in Analysis*, we must content ourselves with knowledge of a few particular cases. (Euler 1757, 298; our emphasis)

> However, the integration of these equations [of perfect-fluid motion] exceeds the powers of the calculus as we know it. (Lagrange 1788, 456)

---

[2] In the cases we study below, the salient branches of mathematics were integral calculus, the theory of differential equations, and (after 1800) the theory of analytic functions.

[3] Euler 1752a, 189; our emphasis.

Plainly put, as they set out to quantify extended bodies moving under a net force, they lacked both the physical insight and the mathematical strength needed for it.

In Chapters 10–12 our focus is on Step I. One reason: that sort of task (deriving the equation for a specific body type or mechanical system) has long been regarded as the key problem in rational mechanics. Here is Max Planck, in 1919:

> The motion of a material body is fully determined just in case we know the motions of *every single material point* we may suppose to make up that body, i.e., when we have given the position of every such point as a function of time.
>
> To describe a definite point in it, we must represent the body's position at some time $t = 0$. Then to every body-point we assign its three coordinates a, b, c at that instant.... These three quantities should be used to describe that material point also at any later time t, from position a, b, c to x, y, z. ... The *whole* motion of the *body* is determined in every last detail if, for *every* material point a, b, c that the body consists in, its coordinates x, y, z are given as functions of t. ... We take these functions to be continuous, because we presuppose that, as the body moves, it *does not disintegrate* into disjointed parts. (Planck 1919, 3f.; our emphasis)

Another reason: a focus on Step I tasks is the most powerful interpretive lens for rational mechanics in the Long Enlightenment. Before 1740, theorists did not *have* any equations of motion. All they had or hoped to obtain were specific instances of Step III results. Only with Johann Bernoulli and d'Alembert did equations of motion take center stage in rational mechanics. By 1742 they had enough resources for that task to no longer seem hopeless.[4] Once they succeeded, the quest for *further* Step I results became the central agenda for mechanics through the 1840s, as Chapter 12 makes clear.

Throughout this book we forgo any discussion of Step III. It received scant attention in the Enlightenment; only the 19th century began to give it sustained study. And, as the collective focus switched to it, the aspect migrated into pure mathematics, with just a few, faint connections to then-ongoing work in philosophical mechanics.[5] Next, we explain how rational mechanics

---

[4] For an account of those new resources and their impact on mechanics, see Stan 2022.

[5] On the mathematics side, Step III becomes a key focus of the theory of elliptic integrals, in the 1830s and beyond; and then of the theory of differential equations, as part of the general problem of the existence, uniqueness, and stability of exact, analytic solutions. On the rational-mechanics side,

after 1735 made the first significant progress toward solving MCON1 in terms of Planck's explanation above.

10.3 Newton's *Lex Secunda*, Euler's principles, Cauchy's laws of motion

The struggles we survey in this chapter yielded a physical insight that was hard won and deep. It is the principle that the balance of force equals the net kinematic change. Later in the chapter we call it "Euler's First Law," following Truesdell.[6] From today's vantage point, it is easy to miss or trivialize how novel and deep it was, because it *looks* like an older idea coming from Newton, which he called *Lex Secunda*: "A change in motion is proportional to the motive force impressed," and is along the force's line of action.[7]

The insight was the threefold realization that (i) "the net force equals the total kinematic change" is necessary for mathematizing a very broad class of motions; (ii) it suffices to represent mechanical change at every point in an extended body; and (iii) it comes with a heuristic, or recipe for applying it. The mistake is to think that, to gain that insight, Enlightenment theorists just took Newton's idea and applied it to new cases; or at least that his idea can yield the insight above—no new *physics* being required, just mathematical prowess:

Newton's principles suffice for solving *every* mechanical problem we encounter in practice, whether in statics or dynamics. We need not appeal to any *new principle* for that. If we run into obstacles, they are always just mathematical. Not difficulties with the *principles*. (Mach 1883, 239; our emphasis)

Step III mattered just to the question of systems with anholonomic constraints (which are generally *not* integrable) and, later in the 19th century, to the long-term evolution of many-body gravitational systems.

[6] Truesdell 1991, 64f. Here and in every allied context below, we follow 20th century commentators in their somewhat unfortunate use of the term "balance." Note that, in these contexts, the term denotes the net amount, or sum, rather than a case of mutual rest, or static equilibrium: it is a concept from accounting, not statics.

[7] Newton 1999, 416. For discussion of the meaning of Newton's phrase, see Pourciau 2006. We employ the term *Lex Secunda* to distinguish it from the ambiguous phrase "Newton's Second Law." The latter covers a variety of inequivalent expressions and formulations; see below.

The mistake is as old as it is pernicious (see also Chapter 7). Our case for that is the remainder of this book; for now, we give the reader a capsule account. The thought that Euler's and Cauchy's respective laws and rational mechanics merely extend Newton's *Lex Secunda* to further systems rests on distorted history and conceptual confusion.

(i) Illusions of hindsight

The thought above supposes that after Newton it was clear to everyone that any further theorizing ought to start with his *Lex Secunda*. Specifically, that it was the preferred law, and that it clearly applies to objects other than a kinematically free particle (which the *Principia* had treated admirably). These assumptions are false; they are illusions of hindsight—looking back at history with weak eyes, from the wrong angle.

It was *not* clear that *Lex Secunda* was the natural candidate law. When Euler and his age set out to solve MCON1, they had many other laws available, and used them: the Torricelli-Huygens principle; Conservation of *Vis Viva* (really, one name for two distinct laws); the principle of least action; and d'Alembert's *General Principle*. *Nothing* about Newton's law singled it out for primacy.[8] And, it was not clear *at all* that his law applies to the motion of an extended body. If anything, they knew early on that it does not:

> In a vacuum, a material point in projectile motion describes a parabola. From that, we can understand why a body too will cross a parabola if we throw it. But, that point motion alone will *not* teach us the laws governing the motion of *individual parts* in a finite body. . . . What Newton has demonstrated about motion under centripetal forces applies *just to a single point*. (Euler 1736, v–vi; our emphasis)

Accordingly, they set out to find new, better laws that do suffice for MCON. If some of these laws ended up looking like Newton's *Lex Secunda*, this was not a preordained, expectable outcome.

---

[8] In the 18th century, conservation of *vis viva* denoted two ideas: (i) when several particles interact, a quantity K (viz. twice their total kinetic energy) remains constant through the interaction; (ii) when an extended body moves, its "actual descent" equals its "potential descent" (i.e., the change in kinetic energy equals the negative of its change in potential energy); see section 10.4; and Calero 2008, chapter 7. For d'Alembert's *General Principle*, see Chapter 9; for the principle of least action and its application to mechanics, see Veldman (n.d.), Pulte 1989, and Fraser 1983.

(ii) Conceptual confusions

We misunderstand Newton's *Lex Secunda* if we think that Euler's and Cauchy's respective principles are equivalent to it. These principles assert content that *Lex Secunda* cannot even express, and they apply to cases where his law is impotent.

First, if we express *Lex Secunda* as $F = ma$, here is what it says: On a *free* (unconstrained) *point* endowed with constant mass m, the net force F has a fixed ratio (viz. m) to the point's linear *acceleration* a. But that is not enough content to support a mechanics of extended bodies. It leaves out, or even disallows, concepts and distinctions absolutely crucial for a mathematical theory of deformable extendeds in motion. Here are some. In fluids and elastics, the relevant factor is mass density, not just mass; but a point has no density (it's a meaningless notion for it). In solids and fluids, there are body forces and contact forces (stresses), but they are dimensionally different; only the former kind count as an instance of F as in *Lex Secunda*. For extended bodies, rational mechanics must distinguish between internal and external forces; but for a mass point, that distinction is meaningless: *all* forces on it count as external, because it has no interior.

Second, Euler's and Cauchy's principles are *dynamical laws* in the modern sense: they entail equations of motion for every point in an extended body. *Lex Secunda* does not do that; that fact has long been known, but its lesson remains unlearned.[9]

Third, Euler-Cauchy principles come with an associated recipe: an iterative scheme that yields the particular equation of motion for the target system. We call it the Euler Heuristic (see section 10.5). Newton's *Lex Secunda* has no such scheme. It does not require it, for it applies directly to its intended objects.

Fourth, it is a mistake to think the phrase "the Second Law" denotes a single, univocal expression. When we state the vague ideas behind that phrase exactly, we obtain three expressions mathematically and physically distinct, not one; see section 10.5. Newton gets credit for just one expression, the easiest; the other two are due to Euler and Cauchy. But Euler's and Cauchy's respective laws were hard won, with no help from Newton. To discover them, it took radically new *physical* insights—about the ways that

---

[9] Truesdell first drew attention to that fact (1968b, 92ff.). For the modern concept of dynamical law, see Stan 2022.

internal constraints work—plus a picture of matter and a program to tackle extended bodies; not merely more mathematics, as Mach wrongly believed. Here and in Chapter 12 we will show precisely what it took for Euler and Cauchy to establish their much broader expressions.

To sum up, some are tempted to think that *Lex Secunda* was the obvious and sufficient frontrunner for solving MCON. That temptation is wrong twice. Historically, Bernoulli, Euler, and Cauchy did not set out assuming it is the preferred law and was sure to apply beyond Newton's chosen objects. Conceptually, it gets things backward. It is a mistake to see Euler's and Cauchy's mechanics of extended bodies as a further application of *Lex Secunda*. Rather, once Cauchy created the concepts of stress and strain, Newton's and Euler's respective theories turn out to be just inchoate precursors, valid for a few degenerate cases, or types of object (see Chapter 12). That realization is old, by the way. Lamé had it not long after Cauchy's breakthrough:

> In the past, geometers thought that, before we can study elastic bodies, we ought to tackle thin strings and membranes first, viz. lines and surfaces before solids. This approach seemed natural and reasonable—but it *failed completely*, because the true theory of elasticity has borrowed *nothing* from those first attempts. In fact, it was born wholly outside them.
>
> . . . the theory of elasticity applies in full generality to solid bodies of comparable dimensions. In exceptional cases, it applies to thin membranes, and to strings even more rarely. Which is the exact opposite of the order suggested by the abstractions of Geometry. A failure to heed this anomaly (which was hard to foresee), combined with an abuse of the *laws and methods of celestial mechanics*, ended up hampering real progress in elasticity. (Lamé 1852, 93, 107; our emphasis)

By the laws and methods of celestial mechanics, Lamé meant reliance on mass points and *Lex Secunda* (the joint basis for Laplace's vast work in celestial mechanics). Thus Mach's verdict above starts at the wrong end. Newton's principles are not a true basis for a realistic account of extended bodies; they are an easy corollary of that basis, once Cauchy's laws of motion are in place.

10.4 Solving MCON1

Most of the breakthroughs below come from Euler, who worked on them for many decades. He first grasped the problem of constraints in its enormity at St. Petersburg in the mid-1730s, when he also drafted a research program for solving it. Then he moved to Berlin, where from 1740 to 1764 he obtained his most significant results, on the motion of rigid bodies, inviscid fluids, and harmonic oscillators. In the twilight of his career, back in St. Petersburg, he worked on extending his approaches to the motion of elastic bodies, a difficult terrain for anyone.

The other figures that matter here are d'Alembert, along with Johann Bernoulli and his son, Daniel. Their work served mostly to inspire Euler, by giving him a glimpse into the way to a solution (e.g., with fluid motion, where the Bernoullis had made crucial breakthroughs earlier). Or, it challenged him, by solving a key problem before he could—which spurred Euler to find his own solution, from first principles he trusted (e.g., with rigid bodies, fluid motion, and vibration, where d'Alembert at first bested him). Still, Euler towered over them all. In scope, generality of treatment, and internal unity, his rational mechanics of extended bodies far exceeds whatever these figures had obtained in that area.

For two reasons, in the 1730s the path forward (to mathematizing whole bodies) was neither easy nor clear. First, they had inherited *many* laws from the 1600s. Recall, those laws apply merely to the motion of a point in the body (a representative centroid). No one then knew a priori which laws can handle the motion of extended bodies; whether they can handle it at all; and how one might apply those laws to a volume of moving matter. Second, while the matter volume moves as one body, its parts *act* on each other. They must act, so that the thing remains a connected whole; their workings secure the one-body constraint that drove the very rational mechanics at issue in this chapter. And, the body parts act in *particular* ways. The strength and content of their actions depend as well on material specifics: properties that distinguish one class of bodies from another. However, these facts—which properties are relevant, and what effects they induce—is not known a priori, at the outset of a rational mechanics that must quantify them.[10] Hence, to

[10] Put modernly, no one then knew whether those properties were forces, energies, charges, fields, temperature gradients, enthalpies, or their higher derivatives. In fact, this question remains an open problem in our time. It is the main topic in the theory of constitutive relations, within the modern science of materials.

18th century figures the task of mathematizing extended bodies seemed an intractable riddle. With characteristic insight, Euler described their conundrum sharply:

> These principles [inherited from the 17th century] are of no use in the study of motion, unless the bodies are infinitely small, hence *the size of a point*. Or at least we can *regard* them as such without much error: which happens when the direction of the net force passes through the center of gravity.... But if it does not pass through that center, we *cannot* determine the entire effect of the force. That is all the more true if the body to be moved is *not free*, viz. is constrained by some obstacle, depending on its *structure*. (Euler 1745, §17; our emphasis)

In one breath, Euler expresses why mathematizing bodies in motion had eluded his age: the available laws were too weak; and constraints were a problem in general—both as external obstacles restricting the whole body's motion, and also as internal actions of the body parts among themselves, mutually affecting their motions.

Euler had glimpsed their predicament well before he voiced it in print. Around 1734, he began work on a comprehensive study of mechanics "treated analytically" by a uniform method. He aimed to treat extended bodies too, not just free particles. At work on this task, however, he soon realized that it was not yet solvable; in particular, the motion of rigid bodies seemed an opaque mystery (see below). In response, Euler greatly restricted his ambitions, and reconceived the work as a treatise on the dynamics of a particle; this became his *Mechanica*. Still, in the preface he took care to motivate and outline a long-term research program well beyond the book's narrow scope:

> The parts of my work come from the very division of bodies whose motion I treat; and of their state into free and constrained. My criterion for dividing them is their makeup [*indoles*]. And so, in the present book I investigate the motion of infinitely small bodies, or point-sized, so to say. In later works, I will move on to finite-sized bodies: which I divide into rigid, flexible, and bodies whose parts are completely loose [viz. fluids]. (Euler 1736, v)

Through the analytic lens of PCON, we can see that Euler charted the way to a very broad rational mechanics of internal constraints. Accordingly, our synopsis will follow his roadmap. We begin with his work on rigid dynamics, then we move to vibration theory, and we end with a survey of results on fluid motion.

(i) Rigid bodies

Impulse for work in this area came from two external factors, viz. the French state's interest in engineering applications; and theoretical challenges in celestial mechanics. Common to these factors was a single task: to *give the equation of motion* for an inflexible body of arbitrary shape when a force acts on it.[11] Euler confidently set out in 1734 to tackle it, but the question defeated him quickly. All he gained was the shocked realization that mechanics was not yet strong enough to answer that question (see above). His early struggles with rigid bodies did pay off in a key respect, however. They helped him grasp a lack in the foundations: *For the general task of mathematizing an extended body in motion, the laws of 17th century mechanics are insufficient. They do not have the expressive power to fully describe its motion as extended. And, they do not have enough physical content to predict its full motion over time.*

Much of Euler's subsequent work in this area amounts to overcoming these two shortcomings. In effect, he set rational mechanics on a new basis, representational and explanatory.

Sobered by the difficulty of the problem, Euler resolved to conquer the rigid body stepwise. For many years, he was able to grasp only what happens when a force turns the body around an axis fixed in space. In that case, he discovered, the result is as follows:[12]

$$Px = 2Mk^2 \frac{d^2\omega}{dt^2}. \qquad (1)$$

---

[11] A complete study of Euler's work in rigid-body dynamics is Verdun (forthcoming), on which we rely in this section. In a historiographic first, his study draws on much unpublished work by Euler. However, we must forgo those sources; our case relies just on results *publicly* available in the 18th century. Relatedly, Verdun 2015 is an extended study of Euler's work on rigid-body motion in the context of celestial mechanics.

[12] See Euler 1746a, 66 (§48). P is the x-component of the force (expressed by a weight), M the mass, x the arm of that force, ω the angular speed, and k the particle's distance to the axis of rotation.

The insight behind this formula is: the balance of torque on a particle equals its change of angular momentum.[13] A step forward, to be sure, but far from enough. Euler later made more progress by learning that an exact account of rigid motion requires novel concepts: angular velocity and acceleration; instantaneous axis of rotation; and moment of inertia.

Still, he could not see how these concepts must combine to yield the formula for the arbitrary motion of a rigid body. His result (1) holds when the body must turn around an axis fixed in space. But suppose the axis of rotation is *not* stationary. Then external forces will change the speed of every body-point, and also the direction of that axis. To infer those changes, one might think it would be sufficient to reapply the insight above (about the balance of torque). However, that is insufficient.[14] If the applied force tilts that axis, the effect seems intractable—no one at that time had *any* principles describing how a force moves a line passing through a fixed point. Euler struggled with this question for some ten years; then he spent ten more making it mathematically tractable. He struggled to find the equation of motion (Step I), and then to find workable ways to integrate it over time (Step III).

D'Alembert was able to find an answer to the former question before Euler. Spurred to action by a newly-discovered astronomical phenomenon, by 1749 d'Alembert had discovered the equation of motion for a rigid sphere rotating under gravity as its axis of spin changes direction.[15] However, his derivation was meandering, obscured by unhelpful notation, and short on elucidations at critical junctures; he was notorious for his opaque style. Euler grappled with it at some length, frustrated by its proof structure. Eventually, he glimpsed in d'Alembert's paper some things that showed him a way to his own proof, from different premises. Those things were two kinematic insights: (i) when a rigid body moves, there exists an instantaneous axis of

---

[13] A word on notation: 18th century mechanics was born before the coordinate-free language of modern vector algebra and analysis. The results we examine in this book were all written in component form. Generally, for illustration or discussion we choose the component (of the equation or principle) in the x-direction.

[14] The missing knowledge is of two kinds. Kinematic: the body's points are in circular motion, and the rotation axis moves in space. It was not clear what quantities are relevant to mathematize these motions. Dynamical: the axis of rotation moves on account of the force (external to the body) and of the body parts' mutual actions. It was unclear how—and to what extent—a force moves a line; and how much the parts act on each other.

[15] The phenomenon was a new type of terrestrial motion, viz. nutation, a wobbling of the earth's axis of spin as it precesses across the northern sky. James Bradley, Astronomer Royal, had discovered nutation in the 1740s, and announced it around 1748.

rotation; (ii) the linear velocity of a body-point can be described as the cross-product of two quantities.[16]

Euler's way to the equation of motion goes through a few key stages. He begins with a terse expression that connects the balance of forces and the linear accelerations; in the x-direction, it reads:[17]

$$2P = M \frac{d^2x}{dt^2} \qquad (2)$$

It is a Step II formula, or integral expression for the balance of force over an extended body. Euler says two important things about it. He calls it the "general and fundamental principle of mechanics." That is his way of saying that (2) counts as what we called an E-principle (see Chapter 8) and modern science calls a dynamical law.[18] And, he explains that the balance P above comprises

> both the external forces (whereby a body is urged from outside) and also the *internal forces whereby its parts are mutually connected*, such that they cannot change their relative distances. (Euler 1752a, §42; emphasis added)

This latter thought carries an immensely valuable insight. Euler at this point has grasped that, when an extended body moves, the change in motion at every point comes from two sources: external and internal forces. To find the equation of motion, both kinds must be taken into account. Crucially, the

[16] The two quantities are the radius vector **r** (to the origin of a reference frame) and an angular speed ω around the axis of rotation. Of course, the term "cross product" is anachronistic; Euler expressed it geometrically, as a difference of two segment products, or areas. And, (i) above says that at any time t there exists a line around which every point in a rigid body sweeps the same infinitesimal angle θ. There is some debate about what insights Euler got from d'Alembert. Wilson said it was the knowledge of how to apply his *General Principle* to a rigid body (1987). Verdun argues, more persuasively, that it was the two insights above (cf. Verdun forthcoming, chapter 2). We follow him on this.

[17] P is the external force, M the mass, and x the coordinate function in the x-direction. The factor 2 is because Euler represents mass by an equivalent weight, with the value of **g** (the acceleration of gravity) set by convention at 1/2. For details and context, see Ravetz 1961.

[18] Euler calls it the general and fundamental principle in his 1752a, §§26, 11. A dynamical law is a differential expression that has been proven to entail a broad range of equations of motion for specific mechanical systems. By the time Euler wrote, the formula (2) had been used to solve the vibrating string (by d'Alembert), systems of interacting mass points (by Euler in *Mechanica*), and one-dimensional fluid motion in a tube of variable width (by Johann Bernoulli in 1742). For a discussion in the context of 18th century work, see Stan 2022.

latter depend—for their strength and manner of action—on the *internal constraint* specific to the body type at issue.[19]

Now let us return to the rigid body. To obtain the equation of motion, Euler applies the "general principle" (2) above to Z, which is an "arbitrary element" in the rigid body, and "let *d*M be its mass."[20] That requires him to do two things. First, to express the *linear* velocity of Z, which in fact *rotates*. Second, to take the time derivative of that velocity, so as to get the linear acceleration that (2) governs. For the first task, Euler found a clue in d'Alembert's paper on precession. That clue amounted to a geometric way to infer the result of $\mathbf{r} \times \omega$, the cross-product that gives the linear velocity of rotation. He found the *x*-component of that velocity to be[21]

$$dx = (\lambda y - \mu z)dt \tag{3a}$$

For the second task, he took the derivative *ddx* of this linear velocity, which gives the linear acceleration relative to an external frame. Then he let that acceleration equal the "elementary force" $Z\gamma$, viz. the balance of force on the mass element at Z:[22]

$$Z\gamma = 2\frac{dM}{dt}(yd\lambda - zd\mu) + 2dM\ [\lambda vz + \mu vy + (\lambda^2 + \mu^2)x] \tag{3b}$$

Behind this complicated expression lies a deceptively simple insight: the total force on a body-point (the dynamical term on the left) equals its total change of motion (the kinematic terms on the right). The latter depend on two factors. First, the internal constraint specific to that body type—in particular, the condition that a rigid body remains self-congruent as it moves. Second, the way the instantaneous axis of spin changes orientation over time; the terms λ, μ, and ν above encode that knowledge.

---

[19] For example, in rigid bodies internal forces "mutually destroy each other" such that no internal displacements occur (Euler 1752a, §42), whereas in flexible bodies they are tensions analogous with the structure of Hooke's Law actions, and in fluids they are internal pressures.

[20] Euler 1752a, 190. The term "element" is ambiguous; we discuss it in section 10.5.

[21] See the next footnote for an explanation of the variables he employed.

[22] Euler 1752a, 207. $Z\gamma$ is the net force, but represented by a *length*, viz. the length of the force's projection on the γ-axis of a frame $Z\alpha\beta\gamma$. Further, *x*, *y*, and *z* are coordinates of the element Z, and λ, μ, and ν the components of its angular velocity vector (relative to the instantaneous axis of rotation, which passes through the body and changes orientation over time).

Lastly, Euler then derived an analogous expression for the net *torque* and its effect, viz. the rate of change in the body's angular momentum.[23] He implied that his two differential expressions (for the balance of force and torque, respectively) jointly suffice to describe the motion of a rigid body.

Euler's equation (3b) was an enormous advance, but just a partial success. In the form above, the equation was true, but useless for the aim of predicting how the body moves over time (i.e., for Step III as we defined it).[24] By 1758, Euler had found an ingenious way around this difficulty—a way to turn the intractable quantities from variables into constants, so as to make a rigid body's behavior over time epistemically determinate. Specifically, his workaround was to obtain *two* expressions for the instantaneous motion of a rigid body. One expression describes how a body-point (acted on by external and internal forces) moves in an instant relative to a frame fixed to the body. That expression is Euler's *dynamical* equation. The other describes how the co-moving, body frame changes place relative to another, stationary frame fixed in space. That is Euler's *kinematic* equation. Together, his two formulas make the motion both determinate and tractable. They solve MCON1 for the class of rigid bodies; Euler had them in place by 1758.[25]

By the end of his career Euler had come to grasp all the ingredients that a complete rational mechanics of rigids demands. Synoptically stated, they are as follows. Any finite motion of a rigid body consists in two displacements, viz. a pure translation in space and a rigid rotation around an axis through the body's mass center. These two are distinct: they are dynamically independent (the quantity of one has no influence on the other), and can be studied separately, on their own:

> An inflexible body can engage in two motions independent of each other. One is progressive, whereby the body's center of gravity moves uniformly ahead. The other is a rotating motion whereby the body turns uniformly around an axis passing through its gravity center. . . .

[23] For details, see Verdun (forthcoming, chapter 2.1).

[24] The reason is that, in the expression (3b), the quantities λ, μ, and ν are variable: they are functions of time, so they change from one instant to the next—because they are referred to a spin axis that changes position over time. For the equation of motion to be integrable, they ought to be constant quantities.

[25] Admittedly, he kept on doing research in this area of rational mechanics. That work, however, falls outside the purview of PCON as a philosophical problem; it bears just on the mathematical aspects of rigid motion and its applications. For complete details, see Verdun (forthcoming, chapter 4).

Now let some forces act on the body, whether it be at rest or in motion. Likewise, those forces will produce two effects in the body. One regards the progressive motion, which they will increase or change. The other regards its rotation around the said axis, which they will increase or change. These two effects are such that there is no connection between them; neither influences the other. (Euler 1745, 43)

In general, however, a full representation of the latter motion (rotation) requires two descriptions: the kinematic change at every material point relative to a frame fixed in the body, co-moving with it; and that frame's change of orientation relative to a second, stationary frame (or to absolute space).

(ii) Flexible bodies

This term meant things that vibrate (i.e., oscillate about an equilibrium position); and things that revert to a reference shape, following deformation. A comprehensive theory of the latter type of object (elastic solids) was too difficult; theorists had been able to offer only a few special solutions for lower-dimensional bodies.

Here too, d'Alembert first broke new ground, by solving MCON1 for a vibrating string.[26] He supposed it to be *continuous*, uniformly dense, and fixed at both ends, but not necessarily at rest. And, he made a few restrictive assumptions: all oscillations are small, all forces on it are coplanar, the tension in the string is nearly equal everywhere, and the accelerations are parallel.

Against this backdrop, d'Alembert approached the problem as follows (see Fig. 10.1). For each point in the string, he supposes: "Let t be the time elapsed after the string has begun to vibrate. Then it is clear that the ordinate PM [of a point P can only be expressed as a function of time t and of its abscissa."[27] In our terms: the string's motion is a transverse wave. To represent any particle P in the string, he uses a function $y(x, t)$, with y the ordinate, x the abscissa, and t a continuous variable ranging over instants. Some cryptic reasoning follows. Buried under it are two key premises that d'Alembert imported without argument from Brook Taylor, which we express as follows:

---

[26] See Chapter 8 for earlier attempts to tackle the vibrating string.
[27] d'Alembert 1747, 215.

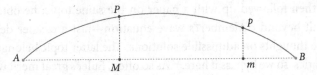

Fig. 10.1. D'Alembert's derivation of the wave equation for a vibrating string. As the string moves, the location of any point P in it depends just on the time t and the point's location on the string, viz. its x-coordinate. Based on d'Alembert 1747.

a = F/m: for any string particle, its effective acceleration equals the restoring force on it divided by its mass;

F ∝ T ∂²y/∂x²: the restoring force at every point is proportional to the tension T (in the string at that point) and the local curvature.[28]

To describe the string's motion, d'Alembert aims to solve this problem: at any given time t, find the value of y for every material point in the string (just as Planck above required in general, for any possible body).[29] At this juncture, the solution is in sight; a few more algebraic manipulations are all it takes. D'Alembert expresses the particle's acceleration above as the second derivative of y with respect to time. Substituting into the two premises above yields[30]

$$\frac{\partial^2 y}{\partial t^2} = \left(\frac{T}{m}\right)\frac{\partial^2 y}{\partial x^2} \qquad (4)$$

That is the Step I formula for a string vibrating in a plane. "Thus, after so many near misses, the wave equation finally appears."[31]

[28] Taylor, following Newton, had used curvature as a measure of the restoring force (cf. Chapter 7). Equipped with a concept of function as he is, d'Alembert expresses that Newtonian idea as the second partial derivative of the ordinate y with respect to the abscissa, x.

[29] d'Alembert 1747, 216f. He thinks no string particle moves sideways, just up and down from the rest axis AB. So, its x-coordinate never changes. In modern terms, he assumes the string's motion is a transversal wave. Taylor (1713) had tried to justify that assumption: he granted that each point (as it moves up and down) also shifts sideways, so its x-location changes as well. But, he explained, those changes count as second-order infinitesimals, so they can be ignored. (By assumption, the string's vibration was first-order infinitesimally small, not finite.) So, he inferred, a point's x-coordinate remains practically unchanged; see Chapter 8.

[30] See d'Alembert 1747, 216.

[31] Truesdell 1960, 238.

Euler then followed up with a paper on the same topic; he obtained no new result beyond d'Alembert's wave equation—just a cleaner derivation, and some thoughts on admissible solutions. The latter topic belongs in pure mathematics, so we bypass it here.[32] As so often, Euler's great merit was to see how a predecessor's breakthrough can be generalized. In this case, he subsequently broadened d'Alembert's result to two new types of whole-body motion. Neither was remotely easy.

First, Euler derived the equation for a string vibrating *in space*, viz. under forces that drive it to oscillate out of a fixed plane (the only motion that theorists from Brook Taylor to d'Alembert had treated). Euler called this novel behavior "whirling motion," *motus turbinatorius*.[33] Here too he takes the direct method, or balance-of-force approach. He regards the string as a continuum of uniform density q, and picks an arbitrary line element Zz of length ds. The element has a tiny mass (*massula*) equal to qds. Euler next expresses its acceleration (in component form), and infers the strength of the "elementary force," viz. the net impressed force on the element Zz. Then he reasons that the elementary force equals the tiny mass times the "accelerative force" of Zz. That is one principle of mechanics he relies on. However (because the string vibrates in three dimensions, not one) Euler realizes that he needs *another*, second principle for his solution. Namely: the moment of the elementary force equals the moment of the "motive force."[34] Together—and only together—these two principles yield his desired equations of motion. "And thereby the entire theory of the equilibrium and motion of strings, both flexible and elastic, has been brought to the greatest degree of perfection we desire for it."[35]

Second, Euler derived the equation for a body of *two* dimensions. Specifically, for a very thin membrane uniformly dense.

---

[32] See Euler 1748. In terms of sections 10.2 and 10.3, it concerns the type of formulas that emerge at Step III. The exchanges between Euler and d'Alembert on this matter turned polemic, and dragged on for some two decades, with little benefit for mechanics. Cf. the critical verdict by Truesdell 1960, 237–44.

[33] Suppose a string at rest between two points A, B fixed in space. Pinch it at several locations, and stretch it from rest in different directions, then release all pinched locations simultaneously. Euler seeks to find how every point of that string will then move.

[34] In modern terms: the torque on ds (about the three axes of an inertial frame) equals the rate of change in angular momentum. Even at that late stage, Euler still used the rapidly obsolescent terminology of "accelerative force" and "motive force" to denote the acceleration and momentum, respectively.

[35] Euler 1771, 347–8.

All that Geometers have studied so far about vibrating motion applied just to
bodies of one dimension, or which we can regard as such: a stretched string
and an elastic lamina.... But, if I am not mistaken, no one has determined how
a surface made to vibrate will move, e.g. a tight canvas or a stretched membrane. . . . I have resolved to set down here the first foundations of this doctrine. (Euler 1766, 243)

He starts with a rectangular membrane supposed *continuous*, uniformly dense,
of very small thickness k, and held fixed at the edges. Suppose that a perpendicular force depresses the membrane downward by a small amount, and the
resulting tension force is nearly equal everywhere. He mentally decomposes the
membrane into extremely thin threads criss-crossing at right angles. The result
is a grid with each cell being a squarelet (*quartulum*) of side w and thickness k.
He lets w become arbitrarily small, and picks the centerpoint Y of any such cell,
with the justification: "we can conceive each element, viz. the matter that fills a
squarelet, to be concentrated in this point." His aim is to find an "equation that
expresses the membrane's vibrating motion."[36]

To that end, he represents every material point Y in the membrane by a
function z of three variables: two space coordinates x, y orthogonal in the
equilibrium plane; and time t.[37] He reasons that Y is subject to the tension
force acting in the x- and y-directions, respectively. The restoring force, which
"presses the element Y downward" to its rest position, is the resultant of the
tensions on Y. Euler expresses it accordingly, then claims that the sought motion of Y "arises from the principles of mechanics."[38] One principle, in particular, is important—unstated but visible just below the linguistic surface:

$$\text{mass of } Y \cdot \text{acceleration of } Y = \text{restoring force on } Y$$

Two of these three quantities are given or already inferred above, and the acceleration is the second derivative of the z function. That yields for Euler the
equation of motion he required:[39]

---

[36] Euler 1766, 244–5.
[37] z is the height of that point above or below the plane. The sought equation gives that height for
any point at any instant.
[38] Euler 1766, 245.
[39] h and f are coefficients proportional to the tension force (in the x- and y-directions, respectively),
and g is the acceleration of gravity, which Euler took to be proportional to the membrane's weight.
There was a mass term on either side of the equation, but it canceled out.

$$\frac{\partial^2 z}{\partial t^2} = gh\left(\frac{\partial^2 z}{\partial x^2}\right) + gf\left(\frac{\partial^2 z}{\partial y^2}\right)$$

In essence, that solves Stage I of the one-body problem of constraints for a thin membrane. Euler then proceeds to Steps II and III: he infers the integral form of the equation above (over the whole membrane), then eliminates time from it to make clear that, at any instant, the shape of any infinitesimal thread in the membrane is a sinusoidal curve.[40] With these results, the 18th century rational mechanics of flexible bodies reaches its zenith.

(iii) Fluid bodies

Progress in this area was slow and uncertain; from Torricelli to Newton and Varignon, researchers managed to obtain just two meager results—two effective parameters of whole-body motion. One was the efflux speed of a "perfect" fluid, the other was the resistance force on the head of a solid submerged in a moving fluid.[41]

Two decades of separate efforts by the Bernoullis and d'Alembert then yielded a few key results that Euler soon combined into the basis for a rational mechanics of frictionless fluids. In modern terms, they were the concepts of internal pressure and "convective" acceleration; the use of partial derivatives to represent motion changes; and the balance-of-force approach to the equation of motion. Next, we survey the birth of these results; then we explain what Euler made of them.

In a 1738 treatise, *Hydrodynamica*, Daniel Bernoulli inferred inter alia the speed of water flowing out of a small hole in a large tank. A trained physician, he had an eye for practical consequences of his research. One implication is knowledge of the pressure that a fluid moving in a tube exerts on its walls.[42] He reasoned that it must be proportional to a certain change in speed: "Clearly, the water's pressure on the container sides will be proportional to the acceleration," or the velocity increment it would acquire if the obstacle to its motion (the tube wall with the hole in it) disappeared at that instant, and the water could gush free. He found a way to compute

---

[40] Euler 1766, 247.
[41] For work on the efflux problem then, see Calero 2008, chapter 6.
[42] This gives a tool for studying blood pressure (on arteries and veins) quantitatively. For Bernoulli's experimental work on "dynamic" fluid pressure, see Calero 2008, 318–20.

that increment, and thus infer the "dynamic" pressure, or fluid force *on the walls*.[43] To get some of his main results, Daniel Bernoulli used an approach then called *methodus indirecta*: to obtain a parameter value at a point, he would start by imposing a condition on an *extended* mass of fluid (usually a thin slice) and then infer to *local* values (at a point) from that supposition.

An odd impulse to compete then led his father, Johann, to publish his own tract, *Hydraulica*. Though it had no new major results, it came with new insights into how fluids move; three, in particular. First, he realized that contiguous parts in a fluid also press—and so exert a force—on *each other*, not just on container walls. Second, that in a tube of variable width a fluid particle undergoes a certain change of motion he obscurely called *gurges*.[44] And third, Johann Bernoulli found a way to handle instantaneous change at every point in a fluid by an approach then called *methodus directa*. Specifically, he inferred the resulting acceleration (of a fluid particle) from the net balance of all the forces acting on it.

D'Alembert studied fluids at length, but two of his works are especially important here. In 1744, a year after his influential treatise on solid mechanics (see Chapter 9), he published a sequel on fluid motion. In it, he reobtained the Bernoullis' key results above but from a new premise that he first devised: a powerful version of "d'Alembert's Principle." Among those key results was the integral form (our Step II expression) of the equation of motion for one-dimensional flow in a variable tube.

Then he made a real breakthrough in a 1747 paper on the supposed cause of winds. Euler studied that paper closely, and drew much inspiration from it, in two regards. D'Alembert showed that the Bernoullis' results on one-dimensional flow can be extended to the much more difficult case of fluid motion in *two* dimensions. And, he showed how the calculus of partial

[43] See Bernoulli 1738, 259. Prior to *Hydrodynamica*, the notion of pressure then current was hydrostatic, viz. as the force on the *container* walls by a fluid *at rest*. Daniel Bernoulli first investigated the "dynamical" pressure, viz. by a fluid in *motion*, and showed that it is less (than hydrostatic pressure). His father's subsequent researches led to Euler's discovery of the modern concept of pressure qua *internal* force, exerted on a fluid particle—not on container walls, as they had thought until then—by the surrounding fluid. For the hydrostatic notion of pressure, see Chalmers 2017.

[44] "And so, for that infinitesimal width [of the fluid at a point] . . . a sort of throat [*gurges*] gets formed . . . that goes from wide to narrow . . . through which the fluid must pass while it accelerates, by continuous degrees" (Bernoulli 1742a, 398). A *gurges* is literally a throat (cf. "to regurgitate," etymologically "to go backward through the throat"). Truesdell translated *gurges* as "eddy" (1954, 32f.). However, that seems excessive; Bernoulli's notion does not denote a gain in vorticity (as Darrigol 2005, 10, correctly notes). Rather, it indicates the rate of change of position at that point (experienced by a fluid particle that happens to be there). The nearest idea is our modern notion of "convective" acceleration, which (combined with the "local" acceleration $\partial v/\partial t$) gives the particle's total acceleration at that instant. See the perceptive Darrigol & Frisch 2008.

derivatives (then a nascent branch of pure mathematics) fruitfully applies to fluids, especially to their acceleration at a point.[45]

Maturation

Euler read *Hydraulica* in installments, before it came out in 1742. After the first, he wrote back in excitement: "You have given me great light on this topic, which until now had seemed covered in thick fog."[46] The great light he got from Bernoulli was really three lights. One, that *methodus directa*, or the balance-of-force approach, is the path to the equation of motion for that element. Instead of taking Daniel Bernoulli's integral approach, one must start with an infinitesimal volume element dV, and identify the causes that change its motion at an instant. Two, that pressure is an action—on a volume element dV, caused by the surrounding fluid—and so it contributes to the net force on it. Namely, internal pressure is a force-like cause of changes in fluid motion at every point: "for the first time, the *local and internal* dynamics enters the scene."[47] Three, that *gurges* is part of the kinematic change induced in that element, in *addition* to the change in speed that forces cause. Finally, here too d'Alembert's mathematics taught Euler to represent Johann Bernoulli's vague notion of *gurges* exactly: as a sum of the partial derivatives that we now call convective acceleration; see below.[48]

With these lessons absorbed, Euler in 1755 presented the rational mechanics of inviscid fluids in a lucid, strikingly modern paper on the "general principles of the motion of fluids." There he obtained the equation of motion for a volume element—the key result of his century, really.[49]

He sets out with a clear statement of the "general problem" of fluid dynamics: to discover the "principle whereby we can determine the motion of a fluid, in whatever state it may be, acted on by arbitrary forces." For that task,

---

[45] D'Alembert there showed how a fluid particle moves ahead while getting deflected sideways (specifically, how an air particle gets deflected by centrifugal force as it moves around the earth). For quantitative details, see Darrigol & Frisch 2008, 1862f. For background, context, and reception, see Grimberg 1998.

[46] Euler to Bernoulli, May 5, 1739, in Eneström 1905, 25.

[47] Truesdell 1954, 36; his emphasis.

[48] These results hold for the "Eulerian" description of fluid motion, which describes changes in the motion of any fluid particle that happens to pass through a location X fixed in space (the equation of motion is about changes in velocity and pressure at that location). In that description, the change of motion can be written as a sum of four partial derivatives: one with respect to time ($\partial u/\partial t$), and the others with respect to each of the three directions in space. The latter type of change is sometimes called the "convective" acceleration, and the former is the "local" acceleration. In the alternative, "Lagrangian" description of fluid motion, we track an individual particle as it moves through space. In that description, the split above (between convective and local acceleration) does not obtain.

[49] Euler 1757.

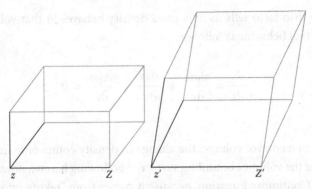

Fig. 10.2. Change of volume. Left: A volume element dV in the reference configuration. Right: The same element, but in the stressed configuration. Euler showed how to express the change in volume using the partial derivatives $\partial x_i /\partial x_j$ of the changes in length (of the sides) in the three directions.

certain quantities are given, or known: the positions and velocities of every particle in the fluid at some instant; that is the fluid's "primitive state." Also given, or known, are the "external forces" that act on the fluid: "I call them external to distinguish them from the internal forces (whereby the fluid's particles act on each other). These latter forces are the topic of my researches below." Other quantities are unknown, and must be inferred: "to know the fluid's motion, we must determine, at every instant and every point, the velocity and pressure" on the fluid there.[50] An equation of motion does just that, and so his primary task is to discover it.

To that end, first Euler derived two kinematic results: an expression (A) for the size-change in an infinitesimal volume element (see Fig. 10.2); and an expression (B) for the components of the acceleration vector du/dt at a point in a deformable continuum—the four partial derivatives that sum up to it:

$$\frac{du}{dt} = \frac{\partial u}{\partial x} + \frac{\partial u}{\partial y} + \frac{\partial u}{\partial z} + \frac{\partial u}{\partial t} \qquad (6a)$$

Here is why he needs expression (A). Density q is proportional to volume, V.[51] And, in a fluid the total mass is constant. Then knowing a volume's rate

[50] Euler 1757, 276.
[51] In fact, Euler assumes the density q (in the fluid's reference state) to be 1 by convention.

of change ipso facto tells us how *mass* density behaves in that volume. He describes that behavior as follows:

$$\frac{dq}{dt} = \frac{\partial qu}{\partial x} + \frac{\partial qv}{\partial y} + \frac{\partial qw}{\partial z} = 0 \qquad (6b)$$

In words: in a control volume, the change in density comes *only* from mass flow across the volume's bounding surface.[52] Following his cue, nowadays we call it the Continuity Equation, because it comes from "taking into account the continuity of the fluid."[53]

This result meets a critical need for him. Recall the main topic of this chapter: motion under *internal constraints*. In Euler's framework, a fluid in motion experiences internal actions that cause accelerations at every point: density changes, and internal pressure. His key to taming the latter was to treat it as a force—that was Johann Bernoulli's clue—and to infer its action from the internal constraint, viz. the condition that the fluid is incompressible. To grapple with the *former*, however, he needed the Continuity Equation above.

At this juncture, the sought result is in sight. To find out how a fluid moves, Euler chooses an arbitrary infinitesimal volume, and applies to it his "general and fundamental principle of mechanics," namely:

the balance of force = the increment of linear momentum

In the particular case of a fluid, on the right side go the changes in speed and in relative position.[54] On the left side go external forces (such as gravity) and internal forces (viz. pressure).

---

[52] Recall, his symbol q stands for mass density, hence $\partial qu/\partial x$ is the rate of mass flow in the x-direction, and so on.

[53] Euler 1757, 284.

[54] There are two ways to track the motion of deformable bodies, "Eulerian" and "Lagrangian." In the former, the equation of motion refers to a fixed point of *space*, and it quantifies the flow of matter there: the change of density, velocity, temperature, etc. at that location. The Lagrangian approach picks out matter points—constituents of the body—and tracks *their* motion *through* space, over an instant of time. Choosing between these approaches is based on pragmatic criteria (e.g., ease of experimental access). In fluid dynamics, the Eulerian approach is more convenient (in lab settings, we place pressure gauges at fixed locations, and we monitor their readings over time). In elasticity theory, the Lagrangian approach works better (we use strain gauges to measure the relative-distance changes between neighboring particles).

The other factors [that determine the acceleration] come from taking account of the forces that affect each fluid particle. Now, beside the components P, Q, R of the [external] force acting on the fluid at location z, there is also the [internal] pressure acting from all sides on the volume element situated at z. (Euler 1757, 284)

That yields the equation of motion for a perfect (inviscid) incompressible fluid. In the x-direction it reads:[55]

$$P + \left(-\frac{1}{q}\frac{\partial p}{\partial x} \right) = \frac{\partial u}{\partial t} + \left(u\frac{\partial u}{\partial x} + v\frac{\partial u}{\partial y} + w\frac{\partial u}{\partial z} \right) \tag{7}$$

In words: the net force (on the left) equals the net kinematic change (on the right). Specifically, because the fluid is a deformable continuum, the total force (from outside the fluid) plus the internal pressure—at a point, caused by the surrounding fluid—equals its speed change plus its rate of position change.[56]

In effect, by 1755 Euler had obtained a key result—his formula (7), known as Euler's Equation for a perfect fluid—by a route that has become canonical: the balance-of-force approach. That solved MCON1 for inviscid fluids, at least in the key respect: "at this stage, the entire theory of fluid motion reduces to solving analytic formulae," namely, to finding Step III, closed-form solutions to the equations of motion.[57]

And so, five decades of collective work slowly produced a vast rational mechanics of extended bodies. After much hesitant fumbling, Euler succeeded in developing a rational mechanics of impressive scope, including rigid, flexible, and fluid bodies. We turn now to assess its strengths and weaknesses, with an eye to the demands of philosophical mechanics.

---

[55] Euler 1757, 286. P is the net force external to the fluid, q is the mass density, p is the internal pressure, and u, v, and w are the components of the velocity vector at that point in the fluid.

[56] In our terms: the body force plus the pressure gradient equals the "local" acceleration plus the "convective" acceleration. That is, $\Sigma f + (-\nabla p) = m\mathrm{D}v/dt$. Here, D is the so-called material derivative, taken with respect to a moving frame, that is, $\mathrm{D}v/dt = \partial v/\partial t + (v \cdot \nabla v)$.

[57] Euler 1757, 298. Good accounts of Euler's derivation are Truesdell 1954, 84–90, and Darrigol 2005, 23ff.

10.5 Assessment

In 1700, mathematicians knew just enough to treat two interacting particles, and one mass point under elementary constraints. By 1780, they had quantified the motion of broad classes of extended bodies. Several figures drove these enormous advances, and Euler built on their work by giving wide-ranging solutions to MCON1. The keys to his success were two: "Euler's First Law," and the "Euler Heuristic." Still, his treatment had lacunae, as we will see.

(i) Euler's First Law and the Euler Heuristic

Overall, the basis of Euler's success was his decision to rely on balance laws—of mass, force, and torque. One law in particular proved indispensable for *any* solution to the problem of internal constraints. That makes it an E-principle, in the language of Chapter 8. Following a recent tradition, we call it "Euler's First Law":[58]

> the balance of force, internal and external
> = the local increment of linear momentum

To be sure, he stated it in quantitative terms—but that statement varies with the context of application. A significant fact; see below.

Though their theory fell short of full generality—as of 1780 there were important types of extended body that rational mechanics could not treat—Euler put in place an approach that, based on his track record, seemed the most promising approach to solving MCON1. To credit him, we call it the "Euler Heuristic," and summarize it as a series of seven steps:

1. For every "element" in an extended body, refer its position to a frame external to the body, via three coordinate functions x, y, z.
2. Identify all the forces, internal *and* external, on that element at an instant t_0. Resolve each force into components relative to the three axes.

---

[58] Truesdell (1991, 64ff.) first called it so, with good reason.

3. Apply Euler's First Law to the element; write it for each of three orthogonal components (e.g., $\Sigma P + \Sigma N = md^2x/dt^2$, etc.).
4. The external forces P are known, or given by experiment.[59]
5. The internal forces N are generally unknown, viz. not given as above. Infer their effects on the element from the *internal* constraint *specific to that particular type* of extended body.
6. Plug the found formulas for P and N into the force law above. The result is the equation of motion for that species of body. In essence, that solves MCON1 for that kind of object.
7. Mutatis mutandis for the torques on the body, if there are any; use Euler's Second Law.

For this chapter, step 5 is especially noteworthy. Euler first grasped it, and even he took decades to do so. To determine the motion of a fluid, he explained, we must "know the motion of every element" in it. But that seems a daunting task: for a rigid body, "if we know the motion of three non-collinear points in it, we have thereby determined how every other point" of that body moves; whereas in a fluid its elements "can move independently of each other," and so its overall motion would "remain undetermined even if we knew the motion of a thousand particles" in that fluid. He goes on:

> Still, we should not think that the individual elements' motions do not depend *at all* on one another. For, if the fluid's density cannot vary, then plainly its individual particles cannot flow such that they may disperse into a greater volume or squeeze into a smaller one. Hence there follows a certain *condition on the motion of every individual* particle. And, even if the fluid is allowed to condense or rarefy, that change cannot happen without regard to pressure [in the fluid]. Hence, that pressure *imposes a certain law on the motions* of every particle. (Euler 1761, 276; our emphasis)

> We must assume the state of the fluid at an instant to be known; I call it the primitive state. It is nearly arbitrary, so we must be given the position of the particles making up the fluid, and their impressed motion. . . . However, this initial motion is not wholly arbitrary: both the *continuity* and the

---

[59] For instance: "among the givens [of the problem] we must count the external forces; the fluid is subject to their solicitation. Here, I call these forces external to distinguish them from internal ones [*forces intestines*] whereby the fluid's particles act on each other. These latter forces are the topic of my study below. In consequence, we may suppose the fluid to be solicited by no external force at all; or just by natural gravity, which we regard as equal and parallel everywhere" (Euler 1757, 275).

impenetrability of the fluid *limit that motion in certain ways* that I will study in the following. (Euler 1757, 275; our emphasis)

In other words, the *internal* constraints specific to that body type—for instance, rigidity in solids or incompressibility in fluids—allow us to infer the actions of its parts on each other; see also below.

With the above synopsis as a lens for historiography, we may regard the subsequent discovery of the Navier-Stokes equation as a further and formidable vindication of the Euler Heuristic. That equation differs from his formulas for fluid motion because it quantifies an internal force (viscosity) that his age could not tame. The same is true of the equation for elastic bodies in three dimensions, another victory of French mechanics after 1820.[60]

And so we see the extraordinary power of the Euler Heuristic: it yielded solutions to MCON1 for a wide range of bodies, and promised a solution for all. The path to reliable success turned out to be the "direct method": start with a mass element in the extended body, and infer its motion from the sum of the forces on it, external and internal. The latter had to be derived from the internal constraint, viz. the restriction (on the relative motion of parts) specific to that body type: rigid, flexible, incompressible, elastic, etc.

And herein lies a crucial point for the purposes of philosophical mechanics. The unity of Euler's rational mechanics lies in its principles—laws of force and torque—that were shown to apply to a broad variety of kinds of bodies: rigid, fluid, flexible, and elastic. This allowed Euler and his predecessors to build a rational mechanics of broad descriptive scope from a small set of "first principles," as we say nowadays. But this nomic unity belies an ontic *dis*unity. The body kinds are independently specified—independently of the principles, that is—and *irreducible* to one another. This latter fact points to a telling lacuna.

(ii) Lacunae

Two limitations are most important for our purposes. The first is an instability in the classification of body types, whose root cause leads to a failure to treat the most important kind of body of all: deformable bodies (such

---

[60] A complete account of the discovery of the Navier-Stokes equation is Darrigol 2002; for Cauchy's equation for elastics, see Chapter 12.

as the human body). The second stems from a lack of micro theory for the matter out of which the body kinds are constructed. We call them "Kind" and "Architecture," respectively.

Kind

In some key respects, the forced motion of extended bodies—its full geometry and causal structure—defeated even Euler and his emulators. In particular, the kinematics and dynamics of *deformable* bodies remained beyond their grasp. Importantly for our purposes, this includes not only *our* bodies (in particular), but *all* the bodies of common, terrestrial experience: wind blowing sand, seas lapping at the shoreline, blood coursing through veins, trees bending in the breeze, and so on.

With hindsight, 18th century mechanics fell short of discovering two concepts that are fundamental and indispensable for the motion of three-dimensional bodies that deform. The two concepts are *strain* and *stress*: one to describe mathematically any change of body shape, and the other to quantify the mechanical agency causing those changes. Put differently, six decades of work by d'Alembert, Euler, and Lagrange did not yield enough concepts to answer the broad question: how does a continuous deformable body move if forces act on it? Not until Navier, Cauchy, and their posterity was rational mechanics able to answer it at last. And so, in effect, around 1800 the problem MCON1 remained unsolved, Euler's great advances notwithstanding.

Their lack of insight into the true basis for representing extended matter (the concepts of stress and strain) kept them from seeing another foundational distinction, namely, between "mechanical" and "material" properties. The former are features that all bodies have simply qua objects of mechanics in general (e.g., inertial mass and space filling, or resistance to deformation). The latter denotes properties that only a class of bodies has, and serves to distinguish it mechanically from other classes. Examples: fluidity, linear elasticity, plasticity, and hysteresis (or mechanical memory). In modern, post-1900 formulations of classical theory, mechanical properties are encoded in the *dynamical laws* (such as Euler's First Law, or the principle of linear momentum) valid for all extended bodies whatsoever. Material properties, on the other hand, are encoded in *constitutive relations*, or equations that govern the kinematic response (to forces) specific to that body type. Examples are the general form of Hooke's Law, which contains the stiffness tensor k qua material property common to linearly-elastic bodies; the drag equation, which contains the dynamic viscosity (μ) as the property specific to "Newtonian" fluids; the Kelvin-Voigt equation, which contains the dynamical modulus (σ), a property of viscoelastic materials; and so on.

The crucial aspect of the above distinction is: a constitutive relation (defining a body type) is an equation that relates stress and strain. Absent that pair of concepts, the very notion of constitutive relations—the gateway to our modern taxonomy of bodies—is meaningless. And, without *that* notion, classifying bodies into general kinds is not principled, or theory-driven: it is not based on first principles of rational mechanics, but on extraneous considerations, such as engineering practice, rules of thumb, or empirical regularities found in laboratory settings.

These lacunae matter for MCON1. Only the two insights above allow us to give a well-grounded account of body and its main species: fluid, elastic, plastic, solid, etc. That account was a desideratum of philosophical mechanics from the earliest days of the problem of collision (PCOL). Recall, a central part of that problem was to explain why elastic, soft, and hard bodies behave as they do in impact. To speak of elastic, soft, and solid in *general*, we need four concepts: stress, strain, material property, and constitutive relation. The 18th century did not see that. And so, in an important sense their answers remain incomplete.[61]

Architecture

There is a recalcitrant dark corner in Euler's philosophical mechanics: the architecture of matter at microscales. For decades, he took extended bodies to be composed of "elements" of matter—his equations of motion apply to single elements; and for internal forces (between body parts) he coined the term "elementary forces," because they are the mutual actions of intra-body elements. However, the makeup of an Euler element has three obscure aspects: the geometry of its mass distribution; its kinematic possibilities; and the kinetic actions it can exert.

The first aspect amounts to whether an element is a mass point, a very small but finite rigid volume filled with mass, or a deformable, continuous infinitesimal volume dV. The second regards the range of motions it can perform. A mass point's trajectory is always a (curved) line. A rigid particle both moves in curves and spins around its gravity center—whereas no mass point can spin. Finally, a volume element dV can translate, spin, and also

---

[61] From our vantage point, we can readily see that they collectively left out certain broad species (e.g., brittle solids and plastic bodies) that were not yet tractable. So, their partition of extended bodyhood into classes is certainly incomplete.

deform (change shape), which the other two entities cannot do. The third aspect concerns the types of force an element can apply. Mass points exert only actions at a distance. Rigid particles do too, but they also apply contact forces on each other's surfaces—which cause translations and rigid rotations in the element as a whole. Volume elements can apply contact forces of a different kind (namely, stresses), which deform such volumes.[62] In sum, an Euler element can be one of three possible, but conceptually distinct, types of object.

Unsurprisingly, Euler saw deeper than most into how different these ultimate objects are. And yet, he never squarely faced up to the question, *What is the geometry, kinematics, and dynamics of the element common to all bodies?* The difficulty of describing an "element" exactly contaminates philosophical mechanics downstream. Any attempt to state his balance laws exactly—in sharp, univocal mathematical terms—results in three *distinct* expressions for any single law, and none is general, or sufficient by itself to describe the motion of *all* extended bodies.

Take conservation of mass. If the Euler element is a mass point (hence an extended body K is a lattice of i equal elements of individual mass m_i), the principle becomes:

$$\frac{dM_K}{dt} = \frac{d(\Sigma m_i)}{dt} = 0$$

But if the body is a physical continuum, then conservation of mass is given by the Continuity Equation:

$$\frac{\partial \rho}{\partial t} + \nabla[\rho \mathbf{v}] = 0$$

Namely, for any volume element in a continuous body, the change in mass density at any instant equals the rate of mass flow into or out of the element.

Now take his second principle, viz. Euler's First Law. If a body is a lattice of mass points, the principle must be stated as:

$$\Sigma \mathbf{F}_i = \Sigma m_i \mathbf{a}_i$$

---

[62] Mass points and rigid particles do not deform, so they cannot possibly exert stresses—nor can they respond by undergoing strain.

That is, the total force on the body equals the net mass times acceleration of its individual elements. But if we regard the body as continuous, we must state the principle as:[63]

$$\sum \mathbf{b} + (-\nabla \mathbf{T}) = \frac{d\rho\mathbf{v}}{dt^2}$$

Our initial point now stands out sharply: none of Euler's principles admits of a single, univocal expression in quantitative terms.

This outcome points again to an ontic disunity, this time at the level of underlying matter theory. There is no single theory of matter from which to construct the distinct body kinds to which Euler's principles apply.

In Chapter 9 we distinguished between principle and constructive approaches to theories. Euler's successful quest to base rational mechanics on exact *principles* yields a tension with attempts to produce a *constructive* account of the bodies to which his principles apply. This tension—between general principles and the three competing notions of matter element described above—persists into the early 19th century. We return to it in Chapter 12, when we examine how the Laplacians sought to anchor their rational mechanics in a single, specific theory of matter.

Solving for internal constraints

Though he built on real breakthroughs by his predecessors, in his century Euler went the farthest in the direction of general treatments, viz. mathematical descriptions that hold for very broad classes of extended body. In one respect, however, even Euler fell short: a truly general solution to MCON1 eluded him. That is, he could not find a way to state exactly a *uniform* recipe for tackling internal constraints. Granted, this charge is somewhat difficult to grasp clearly at this point. But it will become more transparent by the end of the next chapter, after we survey the method of Lagrange multipliers. That method yields, as a very valuable corollary, a genuinely uniform treatment of internal constraints. But, it must be paired with a different dynamical basis— a principle significantly unlike Euler's balance laws of force.

---

[63] b is the body force, T the net stress (the surface tractions and the internal stresses), and ρ the mass density. Cauchy established this result; see Chapter 12.

10.6 Conclusions

While philosophical mechanics pondered its way out of the morass of PCOL, *rational* mechanics around 1730 turned to the task of mathematizing extended bodies. These objects—the bodies around us on earth, not the free particles in the sky that Newton had treated—generally come with constraints, internal or external.

The constraints' geometry and physics affect the motion of extended bodies, and so constraints have to be taken into account. In this chapter, we have reconstructed the 18th century struggle to tame internal constraints: the restrictions that make a body count as rigid, flexible, elastic, fluid, etc. For this particular struggle, we coined the term "MCON1." Euler made the greatest advances on this front, often by building on breakthroughs and insights by d'Alembert and the Bernoullis. His genius for generalizing led him to an iterative recipe that we have called the Euler Heuristic. It is a powerful mix of three ingredients: the balance-of-force approach to the equations of motion; a broad dynamical principle, viz. Euler's First Law of Motion; and the deep insight that internal *constraints* induce effects (on the moving body) on a par with the external *forces* acting on it—and so they can be treated by generic laws of force.

The central problem we have surveyed in this chapter may now be stated more precisely: what is the mathematical description of a body moving (under exogenous forces) as a connected whole while its parts constrain each other?

Euler gave the most comprehensive answer, but his approach to a solution was *piecemeal*. Namely, first he identified the broadest species of extended body that he could tackle. For that task, his speciation criteria were geometric, viz. based on shape behavior: he classified bodies as either rigid or elastic or fluid-compressible or incompressible. Then from their *kinematic* possibilities—the changes of shape admissible for that species of body—he sought to infer the type of *internal forces* that would yield those kinematic behaviors. For instance, the forces required to maintain rigidity, or sameness of shape over time; tension qua the internal force responsible for shape restoration in elastic solids; or pressure as the internal force that drives the shape behavior of compressible fluids.

Euler's recipe for handling internal constraints deserves a further note. In the next chapter we will witness efforts to tame *external* constraints, and they culminate in *another* recipe, which we call the Lagrange Heuristic, built around a different set of ingredients. That fact will be momentous. In effect, thanks to Euler and Lagrange, two powerful heuristics—two recipes for expanding the descriptive scope of rational mechanics—arose by 1800, but they differed, and they pulled mechanics in different directions. Their divergence threatened to break it apart, or at least seriously subvert all attempts at a unified rational mechanics. That ominous prospect (its growth and real force) will be a focus of enquiry in Chapter 12.

11

External obstacles: Lagrange and the mechanics of constraints

11.1 Introduction

The zenith of theory-building in the century after the *Principia* was Lagrange's treatise on "analytic" mechanics. In this chapter, we address two themes: his results in the rational mechanics of constrained motions (MCON), and their relevance for philosophical mechanics. We focus on edition A (1788) of his work, which we abbreviate as *Mechanique*, in line with its peculiar spelling. Edition B (1811–5) has significant changes, some of them philosophically important, which we defer to Chapter 12.[1]

Philosophically, our route into constrained motion has been from the *Problem of Bodies* (BODY). As we have seen in earlier chapters, the early to mid-18th century quest for an account of whole-body motion led to a shift in the locus of research pertinent to BODY, from the *philosophical physics* of the early 18th century to *rational mechanics*. The challenge facing rational mechanics can be broken into two: extended bodies subject to *internal* constraints (MCON1) and bodies impeded by *external* constraints (MCON2).[2] In Chapter 10, we saw one approach to MCON1, led by Euler. Here, we turn our attention MCON2, and to Lagrange.

> **MCON2:** Given an extended body subject to external constraints, how does it move?

Formulated in this way, MCON2 concerns *bodies*. And yet, as the rational mechanics of external constraints developed, its connection to BODY

---

[1] *Mechanique* was Lagrange's second theory of rational mechanics. Previously he had devised and used a different, narrower theory based on a principle of least action. For lucid accounts of the earlier theory, see Fraser 1983. Chapters 11 and 12 are based on wider, ongoing research by MS.
[2] See Chapter 7 for the transition from philosophical physics to rational mechanics, and for the specification of MCON and its division into MCON1 and MCON2.

Philosophical Mechanics in the Age of Reason. Katherine Brading and Marius Stan, Oxford University Press.
© Oxford University Press 2023. DOI: 10.1093/oso/9780197678954.003.0011

became ever more tenuous, with momentous consequences for philosophical mechanics.

In this chapter, we analyze *Mechanique* as a contribution to rational mechanics, and then evaluate it from the perspective of philosophical mechanics. To that end, first we lay out the logical structure of Lagrange's theory (section 11.1). Then we explain how he tackled MCON—how he treated both MCON1 and MCON2 within a single, unified theory, effacing the distinction (sections 11.2 and 11.3). For the purposes of this book, we focus our attention on MCON2. We show that Lagrange's mechanics yields a general solution to it, though with some limitations important for our purposes. The solution rests on his most important principle—the Principle of Virtual Velocities (PVV)—along with two further principles whose content we make explicit (section 11.4). His solution has implications for philosophical mechanics; the most important is that it does not contain a solution to BODY. We show this by taking first a constructive and then a principle approach to his rational mechanics (section 11.5). Adopting the latter allows us to develop the insight further: his theory is importantly *neutral* and *open-ended* about the ontology to which it applies (sections 11.6 and 11.7). We conclude with an evaluation of his overall theory (section 11.8), including its descriptive limits.

11.2 The Principle of Virtual Velocities and "Lagrange's Principle"

Lagrange's *Mechanique* was a long time in gestation, and its origin is threefold. He created its mathematical framework (an algebraic version of the calculus of variations) in his youth, and published it as a brief paper in 1760. Its fundamental principle, or basic dynamical law, he created soon after that, and applied it to rigid-body dynamics in a 1764 prize essay on lunar motion. The chief applications of that principle—mainly to dynamics, and gravitation theory in particular—Lagrange developed over two decades at Berlin, during his tenure at the Royal Academy, where he succeeded Euler. By the time he left for Paris, in 1787, he seems to have had a first draft of *Mechanique*.[3]

---

[3] Lagrange's variational calculus, its application to dynamics, and his transition to the PVV are lucidly explained in Fraser 1983, which remains unsurpassed. Another study of Lagrange's early mechanics is Pulte 1989. Cf. also Panza 2003, and the older Galletto 1991. Caparrini 2014 surveys Lagrange's work leading up to *Mechanique*; his (forthcoming) describes the major changes from the first to the second edition. Various overviews of *Mechanique* are in Capecchi 2012, Barroso Filho 1994, and Pulte 2005.

Outwardly, the book is a diptych with two unequal parts, a statics and a dynamics. The two theories are not merely juxtaposed, however; they are held together by a fundamental premise, the Principle of Virtual Velocities. His statics and dynamics are both axiomatic-deductive in structure. The former rests on a single principle, and the latter requires a second, which we call "Lagrange's Principle."[4]

Ostensibly, each theory treats three types of object: particles, solids, and fluids. Importantly for our purposes, Lagrange leaves most of these terms undefined or equivocal (see section 11.5). For particles and solids, he studies their motions while free, and then in setups with external constraints—these latter make up the bulk of *Mechanique*, and are a significant and substantial contribution to MCON2.[5]

We explicate his rational mechanics in three steps. We begin with statics, and PVV as formulated in that context. We then move to the dynamics of unconstrained systems, expanding the scope of PVV by means of Lagrange's Principle. Finally, we extend PVV to incorporate constrained motion by introducing "equations of condition" and their associated heuristic, the method of Lagrange multipliers. This final step enabled Lagrange to treat a wide range of constrained systems, as we shall see. But first, some prerequisites on notation and basic concepts.

Lagrange's notation is nonstandard and confusing. In Part I (his statics) he uses a single operator, d, in three different regimes: timeless variation (corresponding to our δ), partial differentiation (viz. ∂), and regular differentiation d, mostly with respect to time. Context alone disambiguates how he means his operator d. In addition, he has D, for elements of integration in continuous bodies.[6] In Part II (dynamics), he introduces δ for "virtual velocities." Lastly, he has a single operator, S, for two uses: finite summation over discrete elements (as in our Σ), and integration over continuous regions (equivalent to our $\int$). Again, context alone indicates how he means it.[7] For the sake of clarity, we have disambiguated his notation wherever we could.

---

[4] We follow Papastavridis 2002 in using this label. Lagrange himself attributed the principle to d'Alembert, but his own version is importantly different.

[5] Most of the setups Lagrange treated—and for which he had developed his general heuristic—are masses with external constraints. To showcase the broad reach of his method, however, he used it to treat *internal* constraints as well: rigidity (in solids) and incompressibility (in perfect fluids).

[6] For instance, he writes the incompressibility condition as $d(DxDyDz) = 0$, with Dx etc. denoting line elements; Lagrange 1788, 442.

[7] Two lucid expositions of Lagrange's mathematical framework are Fraser 1983 (a classic study) and Dahan Dalmedico 1990.

To grasp the logic of his mechanics, it helps to start with his fundamental concepts. The key notion is force:

> In general, by "force" or "power" we understand the cause (whatever it may be) that impresses—or tends to impress—motion in the body to which we suppose that force applied. Hence, we must measure [*estimer*] that force by the quantity of motion impressed, or able to be impressed. (Lagrange 1788, 1)

Another fundamental notion is "coordinate," whereby he means a *generalized* coordinate function, not just Cartesian distance.[8] A third is "virtual velocity," a key notion that he defines twice. It is the velocity "that a balanced body is disposed to receive if the equilibrium is broken, i.e. the velocity it would actually acquire in the first instant of its motion." Equivalently, he thinks, it is the "variation, or difference" insofar as it "can result from an infinitesimally small change" induced by a force in the position of a material point.[9] Put modernly, a virtual velocity is the result of applying his variational operator δ to a coordinate x, y, r, etc. while letting time t be constant; hence the term really picks out a virtual *displacement*. Fourth, there is the "moment of a power," his term for the virtual work done by a force. He writes it as a product of the general form $mP\delta p$, where m is a mass variable, P an acceleration, and δp the "virtual velocity" of the force mP.

Lastly, there is "equation of condition," whereby he means an expression that states the constraint "given by the nature of the system."[10] Lagrange has two ways of expressing such; sometimes he writes them as equations, and sometimes he treats the expression as an analytic function (of the generalized coordinates) to which he then applies differential operators.[11] All these notions matter below, and will become clearer in context.

---

[8] "Above, we have used [Cartesian] rectangular coordinates because it simplifies calculations; but of course we can employ others just as well. Clearly, nothing in our approach requires that we use Cartesian coordinates rather than any lines or quantities that can express the positions of bodies" (Lagrange 1788, 23–4). Then he gives transformation formulas between coordinate species (e.g., Cartesian to polar and vice versa).

[9] Lagrange 1788, 11 and 13.

[10] Lagrange 1788, 45.

[11] Other synonyms he has: "finite equation" and "algebraic function." Lagrange restricts the class of admissible functions to those that can be locally approximated by a Taylor-series expansion.

(i) Principle of Virtual Velocities

In his statics, Lagrange begins with an expression he called the Principle of Virtual Velocities:

> Let a system of points or bodies drawn by arbitrary forces be in equilibrium; and give this system an arbitrary small motion whereby each point would cross an infinitely small distance (which stands for its virtual velocity). Then the sum of the forces respectively multiplied by the distance that the point (on which the force is applied) would cross (in the direction of the force) is zero. (Lagrange 1788, 10f.)

Then he gives it exact, quantitative expression:

$$P \, \delta p + Q \, \delta q + R \, \delta r + \text{etc.} = 0 \qquad (1)$$

where again P is the force and δp the corresponding virtual velocity. In our terms: if the net virtual work of the applied forces on a system vanishes, the system is balanced.

From it, he derives the equilibrium condition for the following setups. A set of free mass points under arbitrarily directed forces. A system of mass points subject to scleronomic constraints. A continuous, solid body acted on by action-at-a-distance forces. A mass point at rest on a rigid surface. An elastic string under an arbitrary distribution of external forces. A flexible string loaded with discrete masses. Three particles under rheonomic constraints. A string confined to a rigid surface while forces act on it—a "problem that is perhaps quite difficult to treat by the common principles of Mechanics, but very easy to solve by my method."[12] A rigid body under forces acting at every point. A volume of incompressible fluid, enclosed in a container or free on every side. A volume of compressible, elastic fluid.

So much for statics. The larger prize is dynamics, especially constrained motion. Lagrange's signal achievement in *Mechanique* is the incorporation of all dynamics under a single principle devised for statics. Next, we explain how he managed that.

[12] Lagrange 1788, 97.

(ii) "Lagrange's Principle"

His PVV was a principle for equilibrium—if the virtual work of the applied forces is null, the masses in that system remain at mutual *rest*—whereas dynamics treats systems in *motion*. Prima facie, then, his principle seems inapplicable beyond statics. Lagrange circumvented this obstacle by a brilliant invention: he devised a principle that "reduces" moving systems to rest, and thus makes them treatable by PVV. The basic idea is this.

Let there be a body m prevented from moving freely, be it by external obstacles fixed in space or by constraints to other bodies. Let an actual force be applied to it, in some direction D. If the body were free, it would accelerate along D by a known amount, inferable from the Second Law: $a = F/m$. However, the constraints *prevent* it from moving that way. Hence, the body's actual acceleration will be different from what the known forces and their laws entitle us to infer. It will move the body by some amount e in some direction E. Generally, $a \neq e$; that is why external constraints count as obstacles (to free motion). Now *imagine* that a *fictive* force acted on the body along E, but in the opposite direction, and equal to me. The two forces, actual and fictive, would balance each other—and the moving body would freeze in place and stay at rest. Speaking more precisely, let F be an actual force on a constrained mass m. Imagine we were able to induce on it another, *non-actual* force E such that:

$$E = -me \qquad (2)$$

In modern theory, we often call it a "reverse effective force," or REF. It follows trivially that

$$F + E = 0 \qquad (3)$$

That is, the mass would be in static equilibrium. Lastly, suppose that each force gives the mass a virtual displacement. Then

$$F \, \delta f + E \, \delta e = 0 \qquad (4)$$

This is just PVV governing the moving mass m, which is actually in motion, but has been counterfactually reduced to rest. Thus Lagrange's first great heuristic breakthrough was to allow the introduction of fictive forces that, when suitably calibrated, mentally turn a moving system into a static one. Then the moving masses become treatable by PVV, the general law of his statics: for that system in motion, the sum of each force (whether actual or fictive) times its "virtual velocity" is zero.

When he re-published his treatise, Lagrange took the opportunity to clarify his innovation—the mental device of fictive forces introduced to balance the applied forces that actually move the system:

> If we *imagine* impressing in each body the motion it is to have, but in the opposite direction, clearly the system will be reduced to rest. ... Thereby we can reduce the entire Dynamics to a single general formula. For, to apply the formula of equilibrium to the motion of a system, it is enough to *introduce* forces that come from the [actual] change in the motion of each body, which motions must be *destroyed*. (Lagrange 1811–5, volume II, 240; our emphasis)

Very rarely, modern theorists take care to remind the reader that REFs are fictive, not actual. They state Lagrange's idea as follows: "To the impressed forces add the negative accelerations of masses, as fictitious forces [*Scheinkräfte*]. Then treat the system as one in equilibrium."[13] See also Fig. 11.1.

Out of great modesty, Lagrange credited his radical innovation to his former protector, calling it "the principle of Mr. d'Alembert." Many modern writers followed him uncritically, and so "d'Alembert's Principle" has become an equivocal phrase that covers three distinct ideas.[14] To avoid confusion, we call the thought explained above "Lagrange's Principle," and state it thus:

> *Lagrange's Principle.* If reverse effective forces acted on a system of masses in motion, the masses would come to rest.

---

[13] Hamel 1949, 218.

[14] The first is from d'Alembert 1743, namely his *General Principle* (see Chapter 9). The second is from d'Alembert 1744, which d'Alembert took to be a generalization of his 1743 insight. This is most appropriately labeled "d'Alembert's Principle," and it was the inspiration for "Lagrange's Principle."

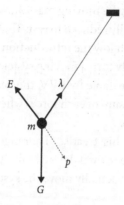

Fig. 11.1. An illustration of "Lagrange's principle" and his PVV. Here, m is a pendulum bob on a rigid rod; G is the real-impressed force on it, viz. gravity; λ is the action of the constraint induced by the rod; p is the actual momentum increment that m acquires in an instant, viz. the mass times the effective acceleration e as defined in formula (2); and E denotes a fictive, non-actual force supposed equal to p but contrary to it. "Lagrange's principle" says that E would balance the force G and the action λ. Namely, if E were actual, m would remain at rest. His PVV says that these three actions times the virtual velocities they respectively induce add up to zero.

In sum, Lagrange built his dynamics of constrained systems from two postulates, viz. PVV and "Lagrange's Principle." In *Mechanique*, this dual basis is combined into the expression[15]

$$\Sigma m(P\,\delta p + \text{etc.}) + \Sigma m\left[\left(\frac{d^2\mathbf{x}}{dt^2}\right)\delta x + \text{etc.}\right] = 0 \qquad (5)$$

From this equality, he derived the equations of motion for all the free systems solved in the book. We explain below his general approach to solutions. The setups he treated are these. A set of free mass points under central forces. The small oscillations of flexible strings. A free mass attracted by two or more forces directed to fixed centers. A system of particles interacting by action-at-a-distance forces.

These systems are all *free* in the sense that they are not subject to external constraints. In each scenario, Lagrange's chief concern is to derive

---

[15] See, for instance, Lagrange 1788, 200.

the equation of motion for that particular system. Yet, in addition, he seems interested in drawing broad conclusions too. For instance, he derives certain "general properties" of equilibrium and motion. They are what modern theory calls "first integrals of motion," viz. proof that if the net external force vanishes, then four quantities (the mass center's velocity; the system's linear momentum, angular momentum, and mechanical energy) are conserved in every closed system.[16]

11.3 Constraints: equations of condition

We have reached our chief object of interest: systems subject to external constraints. *Mechanique* exceeds all its predecessors in a key respect: it found a way to treat constraints that is *general* and *uniform*. We illustrate his insight by an example with generic import. (The notation is slightly adapted from his treatise.)

Suppose two constrained equal masses are acted on by a force each; the motion effects they undergo are as follows. The external forces induce accelerations P and R, respectively. These are the accelerations the masses *would* have acquired, were they unconstrained. Next, by "Lagrange's Principle," we must imagine that fictive REFs likewise act on the system, and induce $-d^2x/dt^2$ and $-d^2y/dt^2$, respectively: they are equal to the effective accelerations, but in the opposite direction.

Now we consider the constraints. These prevent the masses from undergoing the accelerations P and R, and so they change their motions at every instant. But these changes are not known in advance; they are not among the givens of the problem (unlike the external, applied forces, which are supposed known, or given). They *would* be so knowable, if only we knew the type and strength of the constraint forces—and the microphysical underpinnings of the constraints more broadly. However, that knowledge is generally not available, and it definitely was not at hand during this period. In the 18th century, the *only* thing known about constraints was their geometry, or "equation of condition." How to proceed?

Lagrange broke through this epistemic obstacle—our ignorance of the physics and dynamics of constraints—by reasoning as follows. Whatever their underlying physics, let the constraints produce in each mass some

---

[16] That is, he obtains the familiar conservation laws for such systems.

unknown accelerations l and m, respectively. Like any cause of motion, the constraints are associated with some "virtual velocities," say δL and δM, one for each mass. In modern terms: δL and δM are virtual displacements compatible with the constraints. These too are unknown.

But then we are no closer to really knowing how the constraints will alter the motion. What *will* yield that knowledge? Lagrange's triumph was to take two things we *do* know about the target system, and use them as stepping stones to a solution.

First, Lagrange points out that we do have *some* knowledge about the virtual displacements δL and δM. In particular, L and M designate the "equations of condition," or algebraic expressions that spell out how the coordinates *depend* on each other *because* of the constraints.[17] He takes these geometric conditions (on admissible motions) to be given by expressions of the form:[18]

$$L = 0; M = 0, \text{etc.}$$

This knowledge is what we meant above in saying that the geometry of constraints was given, or known. The condition equations state facts about the spatial-geometric configuration of the system throughout its motion.

Second, Lagrange declares that PVV *applies to the actions of constraints as well.* More precisely: each constraint can be associated with a "virtual velocity"; their combined "moments" can be added to the "moments" of the real and the imaginary forces; and still the net "moment" across the system is zero—just as PVV had it for free, *un*constrained bodies.

We have reached a crucial juncture. Lagrange has defined "moment" as the product of a virtual velocity and a force. However, generally the forces due to the constraints are *unknown*. As we noted above, they are not given. We lack the empirical facts from which to infer them before we derive the equation of motion for a system. This broad ignorance (of constraint forces) was the main stumbling block for Newton–Euler mechanics based on the Second Law. So, is the Lagrangian approach in a position to avoid Newton's and Euler's predicament?

As it happens, Lagrange was able to take this apparent weakness and turn it into a strength, as follows. (i) If the strength of a constraint force is

---

[17] The idea is that external constraints take away independence, as it were: the freedom of a coordinate to change without regard to how other coordinates change, if at all. A constraint "forces" certain coordinates to depend on others. It "conditions" their behavior.

[18] Thereby, Lagrange in effect restricts his rational mechanics to what we nowadays call *bilateral* constraints. Not all constraints are like that; see section 11.8.

unknown, then we are free to represent it by an *arbitrary* function λ, μ, ν, etc. Any function will do, provided it is analytic. The functions λ, μ, ν are the so-called *Lagrange multipliers*: scalar quantities that we may choose at will, and multiply each of them with a virtual velocity. That product has the dimension of a "moment," or work done by a force; and, being arbitrary, can stand in for unknown quantities, as constraint actions are. (ii) The direction of the unknown force is inferable from the "equation of condition" for that constraint, via its derivative dL, dM, etc. And so, the individual actions of the constraints are expressible as the "moments" (i.e., the quantities of virtual work) λdL, μdM, etc.

With this dual thought in hand, Lagrange is able to regiment the constraints under PVV, so that the full version of the principle reads:[19]

$$S\, m(X\delta x + Y\delta y + \text{etc.}) + S\, m\left[\left(\frac{d^2 x}{dt^2}\right)\delta x + \left(\frac{d^2 y}{dt^2}\right)\delta y + \text{etc.}\right]$$
$$+\, \{\lambda\delta L\ \mu dM + \text{etc.}\} = 0 \tag{6}$$

To help the reader grasp his radical innovation, Lagrange abbreviates the left-side summands above as Δ, Γ, and Λ, respectively. Then he rewrites PVV as:

$$\Delta + \Gamma + \Lambda = 0 \tag{7}$$

This is a clearer expression of the insight behind that principle: in any moving system of bodies, the virtual work of the (real) impressed forces Δ, the (imaginary) reverse effective forces Γ, and the constraint forces Λ, together vanish. *This is the statement of PVV in its most general form.* See also Fig. 11.1.

We can think of it like this. In the most general sense, constraints are any quantitative relations of dependence between the coordinates in a mechanical system. The coordinates expressing the motion of a constrained system are not free to vary independently of one another—that is what it means for a system to be constrained. We have taken the *known* information about the dependencies among the coordinates (expressed as equations of condition) and incorporated it into PVV. Lagrange explains it himself:

---

[19] See Lagrange 1788, 231 and 439.

To apply this method, . . . we will suppose that $L = 0$, $M = 0$, etc. are the condition equations that must obtain by the nature of the given problem; and that they refer to every point of the [system's] mass. We shall call them "undetermined equations of condition."

When we differentiate these equations in respect to δ, we get $\delta L = 0$, $\delta M = 0$, etc. Then we multiply the quantities δL, δM, etc. by the undetermined quantities λ, μ, etc. respectively. And then we integrate over them, which yields the formula $S\,(\lambda\delta L + \mu\delta M = \text{etc.})$. By adding it to the integral formula above [viz. Equation (1)], we obtain the general equation of equilibrium. (Lagrange 1788, 53)

By working with the equations of condition, we bypass any appeal to the forces responsible for the constraints on our system. Remarkably, incorporating constraints into PVV in this way allows him to derive equations of motion for a wide range of constrained systems. Presented synoptically, those systems are as follows. Mass points connected by massless rods or strings, but otherwise free to move in space under central forces. A system of particles moving under an initial impulse while connected by strings or rods. The small oscillations of a coupled pendulum. The small vibrations of an elastic, extended string loaded with discrete masses; and of a heavy chain swinging as it hangs from one end. Rigid-body rotation, free and forced. The motion of a rigid body in terrestrial gravity. The motion of a volume of incompressible fluid—specifically: the flow of a homogeneous fluid in a pipe of variable profile; efflux from a vertical tank; surface waves in a fluid. The "very small motions" of an elastic-compressible fluid (such as sound waves in air).

There are limitations in the types of constraints that Lagrange considered, to be sure (section 11.8). Nevertheless, *Mechanique* is a landmark work: it covers a very wide range of problems in statics and dynamics from a single principle, PVV. And, by means of a radical innovation—the method of Lagrange multipliers—it incorporates classes of constrained systems that d'Alembert and Clairaut had been unable to treat. "Though these mechanical setups are different, the steps to a solution [*la marche du calcul*] are the same. Which must count as a chief advantage of my method."[20]

This gives us the basis of Lagrange's mechanics, and a sense of the scope of systems covered. In the remainder of this chapter, we seek to better

---

[20] Lagrange 1788, 121.

understand *how* Lagrange's surprising strategy works, and the implications of this for philosophical mechanics.

11.4 Lagrange's Relaxation Postulate: the kinematics and dynamics of constraints

In addition to PVV and Lagrange's Principle, there is a third "principle" at work, of great importance for philosophical mechanics. It is an implicit but very broad assumption that covers all constrained motions. Lagrange does not name it; he gives it very little discussion, and just alludes to it as being "the spirit behind the method" of Lagrange multipliers. It took well over a century for mechanics to even begin to recognize it. Nowadays, we know it as the Relaxation Postulate, or *Befreiungsprinzip*. The pernicious dearth of axiomatized accounts of analytic mechanics makes it hard to even state it clearly. Here is an attempt:[21]

Lagrange's Relaxation Postulate
In a constrained system, the action of constraints is kinematically indistinguishable from the effect of impressed forces: if the constraints were dissolved (relaxed), the result would be additional accelerations on the now-free bodies.

The statement may seem vacuous or tautological, but is robustly informative. To see that, consider two ideas. (i) Constraints are just *geometric* or kinematic dependencies: algebraic relations between coordinates. (ii) Constraints have the same effect as *dynamical* agencies, viz. forces, on the bodies they constrain: they induce accelerations, and they do virtual work.

In building the rational mechanics of constrained systems, we start with (i) and along the way we quietly reach for (ii). But, they are distinct ideas. The Relaxation Postulate is the precise, honest acknowledgment of the need to bridge (i) and (ii). Modern mechanics regards it as a broad (non-local) postulate—an indispensable premise logically independent from the

---

[21] We rely on Hamel 1949 and Papastavridis 2002. Acknowledging the *Befreiungsprinzip* is rare, and mostly restricted to German-language treatises. In Anglophone literature, Papastavridis alone gives it the prominence it deserves. See also Neimark & Fufaev 1972.

dynamical laws. Lagrange stated it without fanfare, and with no effort to signal its foundational status:

> I must note that, in the general equation of equilibrium, we may regard the terms λdL, μdM, etc. as designating the moments of various *forces* applied to the system.
>   It follows that every condition equation is *equivalent* to one or more forces applied to the system, in the given directions.[22] Ergo, the system's state of equilibrium will be the *same*, whether we consider it as induced by these forces [equal to the Lagrange multipliers] or we suppose it to ensue from the equation of constraint. (Lagrange 1788, 47, 49; our emphasis)

We mentioned above that, by working with "equations of condition," we bypass any appeal to the forces responsible for the constraints on the system. The constraints make regions of configuration space inaccessible: they are kinematic conditions. Yet, once we think of these constraints as being realized physically, we are to treat them as part of the *dynamics* of that system.

The use of Lagrange multipliers (and their equations of condition) gives just partial information about the constraints' action. In regard to corporeal action at a point, consider two types of knowledge we could have. (i) Knowing the quantity of a vector F that, when fed into the generic formula $F = ma$, yields a determinate value for the local acceleration, **a**. (ii) Knowing the *constitutive relation*, or particular law, specific to the force behind that action: these are what one uses to replace the generic, placeholder term on the left-hand side of $F = ma$. Examples of constitutive relations are Hooke's Law for tensile stress; Newton's Law for gravitational actions; and Stokes' Law for fluid resistance on a sphere.

The method of Lagrange multipliers yields just the first kind of knowledge above. It does not tell us the particular force law giving rise to the action of the constraint. The method is powerful in part just because it allows us to set aside the need for such detailed knowledge of the forces: we use the condition equations to bracket the relevant dynamics into the kinematics, as we might say, and then solve for the motion of the system.

We saw this bracketing strategy at work in d'Alembert's *Treatise* (Chapter 9). Lagrange's *Mechanique* makes an important advance by indicating how to move beyond treating our system merely through kinematic constraints.

---

[22] Namely, in the direction of the partial derivative $\partial L/\partial q_i$.

This is a subtle, deep strand of theory-building in his book. It is a heuristic insight that allowed Lagrange to extend his theory while keeping it unified. Specifically, he took a dynamical law that applies to constrained systems, and applied it to *unconstrained* ones that are otherwise similar. The key to extending it (to free, no longer constrained systems) is Lagrange's Relaxation Postulate above. Recall, in this context "relaxation" denotes the removal of a constraint; in modern talk, the opening up of a previously closed degree of freedom.

We begin with a generic description of his approach, then we illustrate it with some examples. An apt name for his approach would be[23]

Lagrange's Relaxation Heuristic

Let Z be an arbitrary mechanical system, $L = 0$ be one of its constraint equations, and λ be its corresponding Lagrange multiplier. Now consider another system Z' exactly like Z in every respect, but with an extra degree of freedom. Namely, the degree that results by removing the constraint L above. Then the system Z' so freed will move according to the same equations as Z, except that now λ designates an *impressed force*. The kinematic and dynamical properties of Z' determine λ uniquely.

Some examples cast light on this idea. In statics, Lagrange derives the equilibrium condition for a rigid rod, with the Lagrange multiplier λ (and the accompanying equation of constraint $L = 0$) expressing the rigidity condition. Relaxing that constraint allows the body to behave as an elastic rod, which is now free to *bend*. However, bending (formerly prohibited, in the rigid body) must now be supposed to be resisted by a *force*: a "force of elasticity" that resists bending, and tends to restore the rod to its equilibrium shape.[24] In dynamics, Lagrange first derives the equation of motion for an incompressible fluid; in that setup, $L = 0$ expresses the sameness-of-volume constraint, and $\lambda \partial L$ is the resulting Lagrange multiplier. Then he *relaxes* that constraint, viz. he allows the fluid to contract or dilate—types of motion forbidden in incompressibles. But, in the fluid freed to contract, a *force* arises to resist that motion now permitted. That is internal pressure, a function of volume-change at a point.

---

[23] We base this presentation on Hamel 1917. An updated, stepwise exposition of the approach is in Papastavridis (2002, 470ff.), who illustrates it with new examples, and shows its kinship with the kinetostatic, or "rubber band" approach in modern analytic mechanics.

[24] See the details in Lagrange 1788, 87.

Unfortunately, Lagrange did not make this demarche as clear as he could have. Mostly, he suggested it:

> It follows that every condition equation is *equivalent* to one or more forces applied to the system, in the given directions.[25] . . .
>
> Conversely, these forces can replace the constraint equations that result from the nature of the given system. And so, by employing these forces, we can regard the bodies as *entirely free and devoid of constraints*. . . . That is the spirit of what the method [of Lagrange multipliers] consists in.
>
> Properly speaking, the forces above stand in [*tiennent lieu*] for the resistances that the bodies must experience in virtue of their mutual constraints; or are induced by the obstacles that, depending on the system's makeup, resist the [free] motion of bodies.
>
> Better put, those forces are nothing but the very force of these resistances [of the constraints], which must be equal and opposite to the pressures exerted on them by the bodies. As we have seen, my method gives a means to determine these forces and resistances. Which is not among its least advantages. (Lagrange 1788, 49f.; our emphasis)

He left it to readers to *infer from his use* how the Relaxation Heuristic really works. It took over a century for his posterity to grasp it.

For our purposes, it is of enormous import. In a constrained system, knowledge of the impressed forces responsible for the constraints lies outside the reach of rational mechanics, for the reasons we have just explained. Lagrange's heuristic shows how to move stepwise from a tractable constrained system to a full dynamical treatment.

In practice, this is easier said than done. For example, a conspicuous gap in his theory is the motion of deformable continua in *three* dimensions—even of simple, highly symmetric objects such as elastic bars or shells. The only three-dimensional bodies he managed to treat were rigids, even though his heuristic suggests how to move from them to deformables. It was left to Navier and Cauchy in the 1820s to take that final step. This is not to fault Lagrange, of course. Rather, it is a lesson in how difficult it is to quantify the mechanical behavior of extended bodies.

The most important take-aways for our purposes are these. *Mechanique* is a tract in *rational* mechanics—a long, connected exercise in applied differential

---

[25] Namely, in the direction of the partial derivative $\partial L/\partial q_i$.

equations. It is an *analytic* mechanics: where d'Alembert's solutions work with a mix of diagrammatic and algebraic reasoning, Lagrange's methods make no use of diagrams: "there are no figures in this book. The methods I present herein require neither construction nor geometric reasoning, just algebraic operations."[26] It is highly general, and equipped with a general heuristic for treating constrained systems—including a systematic treatment of MCON2, though with some important limits of scope—and then relaxing the constraints toward a full dynamical treatment. With this analysis of his book in hand, we turn to our next task: the significance of all this for philosophical mechanics.

11.5 Philosophical mechanics and Lagrange's *Mechanique*

Much of Lagrange's treatise is concerned with offering a theory of bodies—in equilibrium or in motion—but he says very little about what a body *is*. BODY was not his concern. But it is ours. And, as we have emphasized throughout this book, an adequate solution to BODY requires a philosophical mechanics: it demands resources beyond those of either philosophical physics or rational mechanics alone. Our task is to identify the elements of *Mechanique* relevant to BODY and to spell out their philosophical significance. To this end, we do two things. We examine Lagrange's *explicit* statements, and we also seek to uncover commitments *implicit* in the problems that he solves—we aim to recover meaning from use.[27]

Together, his overt declarations and implicit premises imply that he took a body to be any object with four properties: mobility, extension, mass, and the ability to apply forces of interaction: "Since every body is essentially heavy [*pesant*] and extended, we cannot strip them of either property without changing their nature [*les dénaturer*]."[28]

He further thinks that bodies come in two genera. But, he makes this division twice, and along different lines. One division yields solid and fluid bodies, and the distinction is broadly *dynamical*, viz. based on their specific response to applied force. His fluids come in two classes: incompressible and "elastic," or compressible:

---

[26] Lagrange 1788, vi.

[27] For more on our methodology, see Chapters 1 and 7.

[28] Lagrange 1788, 389. Caution: by "heavy" he means responsive to exogenous actions in general, not just to gravity.

In effect, the main property of fluids—the only one that distinguishes them from solids—is that all of their parts yield to the least *force*, and can move (relative to each other) most easily, no matter what the connections and mutual *actions* of these parts. . . .

As we know, fluids in general are divided into two species: incompressible, whose parts can change shape but not volume; and compressible-elastic, whose parts can change both shape and volume, and which always tend to dilate with a known *force*, commonly supposed to be some function of density. (Lagrange 1788, 122, 130; our emphasis)

Under "solids," he offers three species: rigid volumes, flexible lines (they bend but do not stretch or compress), and elastic strings. We cannot make these categories any clearer on his behalf—it will take later developments to arrive at the specificity of our modern concepts.[29] The second division is purely *geometric*, and it entails bodies that are either rigid or deformable:

In regard to mass and figure, we must regard each body as a congeries of an infinity of particles that keep the same *relative situation*, if it is solid; or can change their *situation* according to certain laws, if the body is flexible or fluid.

The solidity condition [for a body] consists in all of its points constantly keeping the same *positions and distances* relative to each other. (Lagrange 1788, 122, 130; our emphasis)

However, his two ways of carving up bodyhood are not coextensive: the first divides (i) rigids, flexibles, and elastics from (ii) fluids, both compressible and incompressible; the second divides (i) rigids from (ii) deformables, solid or fluid.

Two things to notice. First, most of this is tacit. There are no definitions of body, and no official account of body and its properties. Second, what we can extract is ambiguous. If "body" is a genus, Lagrange was unable to decide which *differentiae specificae* matter most for his mechanics. This suggests something important about Lagrange's mechanics: though overtly his tract concerns a theory of *bodies* in equilibrium and in motion, "body" is

---

[29] In modern theory, a fluid is a physical continuum that exerts very low shear stresses when it is deformed. "Solid body" is no longer a true genus.

not a robust concept therein, neither as a presupposition nor as target of his theorizing.

From the perspective of philosophical mechanics, this poses an interesting challenge. A successful philosophical mechanics goes beyond rational mechanics by specifying its objects: by telling us what the theory is *about*. As we move to assess how well a particular theory meets this requirement, we have two options: constructive and principle (see Chapter 9). The former approach to theorizing begins from the material *constitution* of the objects or systems of interest (such as bodies), and from that picture develops an account of the rules and laws that such objects or systems satisfy. The latter begins from some principles held to be very general (perhaps universal) and *independent* of the particular material constitution of the target systems. Then from those principles it draws out an account of the nature of the objects or systems (such as bodies) to which these principles apply. With these tools we can cast light on the significance of Lagrange's book for philosophical mechanics.

(i) The constructive reading of *Mechanique*

If we take a body and zoom in to the ultimate level of ontological resolution, what does Lagrange think we will see? Mass points separated by small but finite distances; tiny rigid volumes; or continuous matter of varying density?

Lagrange did not answer. Instead, he tried to formulate a rational mechanics that can be retrofitted to *any* of the three pictures of matter above. Specifically, he treats systems of discrete mass points interacting by action-at-a-distance forces or flexible strings; then rigid bodies, and finally fluids. He regards fluids as being a species of continuous matter; it is unclear whether he thinks rigid bodies are continuous, made of mass points, or a sui genesis type of body.[30] In every case, he derives the relevant equation of motion from the same one principle, PVV. So, in a robust sense, he created a rational mechanics that can be moored to *any* of the three, distinct architectures of matter. He is keenly perceptive of the differences between these matter

---

[30] The assumption that the system's component masses are tiny rigids is tacit in his treatment of a discretely-loaded string that vibrates (*Mechanique*, Part II, section V, §1); and also his later treatment of fluid resistance (edition B, Part II, section XI, §§36–7). Like Newton before him, Lagrange thinks of fluid resistance as induced by the impact of tiny *rigid* spheres on the head of a solid body moving through the fluid. That was really the default assumption until Prandtl's creation of boundary-layer theory in 1904.

theories, and he is seeking a theory that allows for them all.[31] (Admittedly, his insight has limits; see below.)

He admits that extended bodies are composites, but that clears up little. Lagrange refers to the ultimate components of body—casually and without discussion—as "points," "particles," "molecules," and "elements."[32] Only the last is clear: a Lagrange element is an infinitesimal volume dV in a physical continuum: an element of integration over continuous manifolds. He left "molecule" undefined; a pity, and a reason to be wary, because his French disciples will give "molecules" the starring role in their theory of matter (see Chapter 12). Lastly, his notions "particle" and "material point" are equivocal. The interpreter has two options. One: they are mass points. However, in an extended body, the number of mass points is *finite*, whereas Lagrange says, "Generally, when we must take into account the mass and figure of mobile bodies, we need only regard each body as a congeries [*assemblage*] of an *infinity of particles*."[33] It is unlikely that he could be confused about this simple fact. Two: they are points in a pre-theoretical notion of a *continuous* body.[34] More work is needed to decide which of these options was his considered view.

"Architecture of matter" is our umbrella term for any account that builds on the answer to a key question: what is the *size* of the mass units that make up a body? Are they zero-sized (mass points)? Finite-sized (atoms, rigid or elastic)? Infinitesimally small (volume elements)? Lagrange gave no clear answer to this question, even though his book is a mechanics of bodies and it supposes that whole-body motion emerges from the motions of its mass-bearing components. And so, his triumphant phrase—that "those who love mathematical Analysis will be delighted to see that I have turned Mechanics

[31] For instance, he knows to denote the mass of discrete particles by M, a finite number; and that of elements in continua by dM, an infinitesimal quantity. He knows to use different operators for whole-body quantities, depending on whether he regards a body as a discrete set or a true continuum. For the idea that his operator S has different meanings (viz. summation vs. integration), depending on whether masses in the target system are discrete or continuous, see Lagrange 1788, 244.

[32] For "particle" see, inter alia, 1788, 122, 330, 334, 378, 438f., 449f.); for "point" see 1788, 333, 340–50, 357–9, etc.

[33] Lagrange 1788, 330; our emphasis.

[34] Continuum mechanics *begins* with a notion of body placement, understood as a mapping from the body's points to the points of Euclidean space. A finite-time motion in this theory is defined as transplacement, viz. a mapping from one placement into another. However, when it moves to dynamics—to the study of stresses in continuous bodies—the theory leaves behind the picture of continuous bodies as sets of points. Instead, it takes the "Euler-Cut" approach: it regards them as finite volumes, and chooses arbitrary subparts in them (control volumes), which it then shrinks mentally to an infinitesimal volume element dV, so as to study its response to force and stress. For example, Lagrange seems to regard a rigid body as made up of "elements" (1788, 372, 374).

into a branch of it"—comes at great cost. From the perspective of a constructive approach, analytic mechanics clouds in obscurity its most important object.

Then let us try interpreting his theory using a principle reading instead.

(ii) The principle reading of *Mechanique*

In the landscape of Enlightenment theorizing, Lagrange's book stands out through the explicit importance it gives to principles. Like d'Alembert before him, he reflects on the principles that support his mechanics, places them in historical context, and reviews their qualities self-consciously.[35] So, approaching *Mechanique* from a principle perspective is consonant with Lagrange's own thinking. In that regard, we pursue three tasks. First, we clarify the principles and their scope. Second, we investigate their epistemic status. Third, we elucidate the relation between the kinematics and dynamics of his theory. This will enable us to clarify the extent to which the *principles* of the theory are sufficient to specify the *objects* of that theory.

As he sees it—and says explicitly—the genuine and single foundation of his theory is PVV. We know, from Chapters 7–9, that "principle" was an equivocal notion in philosophical mechanics. In Lagrange's, PVV is a principle in the strongest, most genuine sense: it is an E-principle, or *common premise* for all the equations of motion (every one of them requires it indispensably for its derivation). And, it is the privileged source of *confirmation* for all the results downstream from it. This is because of the epistemic status he ascribes to PVV, and the axiomatic-deductive structure of his theory.[36]

As it happens, that status was somewhat fluid: Lagrange was of two minds about the warrant for his PVV. In edition A, he claimed the principle is *certain* because it has four virtues. In edition B, he argued that it is *true* because it follows deductively from a necessary truth of statics. We discuss his first account now, and his second in Chapter 12.

At the outset of *Mechanique*, Lagrange declared that PVV may "be regarded as a species of axiom for Mechanics." It deserves that status, he

---

[35] Lagrange prefaces his statics and dynamics with long surveys of the various principles historically used before him. In edition B, these surveys become even more extensive. He appears keen to argue that PVV has a distinguished track record that long precedes *Mechanique*, and so his choice of principle is all the more legitimate.

[36] In effect, Lagrange's work established that any optimal solution to MCON must rest on the PVV (because it alone is compatible with the method of Lagrange multipliers).

explained, because it has a number of theoretical virtues. It is *entrenched*, or "long recognized as the fundamental principle" of equilibrium.[37] It is *simple*, in some undefined respect; presumably he means easy to grasp, because all it says is: the net virtual work is always null. It is *general*, in that it applies to every mechanical system, at rest or in motion. And, it is heuristically *fruitful*, because when we state it exactly, it becomes clear how we can apply it to other types of systems:

> We must admit that it [the PVV] has all the *simplicity* we could require of a truly fundamental principle; and we will see below how much its *generality* speaks in its favor....
> Moreover, the Principle is not just very simple in itself and very general. It has another advantage, unique and invaluable: we can translate it into a general formula that contains [the solutions to] *all the problems* [of equilibrium or of motion] that can be thought up. (Lagrange 1788, 10, 12; our emphasis)

In sum, at this stage of his career Lagrange sought certainty for his principle, not proof of truth. In singling out generality and simplicity qua distinctive marks of genuinely fundamental principles, he sides with his later mentor, d'Alembert. This changed dramatically in edition B of 1811. There he made a volte face and argued that PVV is a *theorem*, because it follows from a privileged species of true proposition; see Chapter 12 for discussion. In short, PVV is intended to be completely *general* in scope (task 1), and is taken to be *certain* (task 2).

Generality plays an important role for Lagrange in the evidential support for his theory, as we discuss below. But the desired generality has some lacunae. For the purposes of philosophical mechanics, the most serious lacuna is that his book treats only *ideal* constraints, so called. That is, constraints *assumed* to be workless: PVV requires that the combined action of constraints does not change the target system's overall momentum or energy. In reality, however, this condition does not obtain; real-world constraints do drain energy and momentum away from the bodies they constrain. The most common avenues (whereby constraints do net work) are friction, thermal effects of vibration, impact, viscosity, and similar dissipative mechanisms;

---

[37] To make his case for this, he surveys briefly how Galileo, Descartes, Wallis, and Johann Bernoulli made use of the principle in their theories; cf. Lagrange 1788, 8–12.

and these are everywhere. That stark reality poses a challenge for Lagrange. His theory does not come with an account of its own limits of representation. So, it is difficult to know *which* real-world systems of bodies it applies to, and how well it can represent their behavior. This has consequences for evidence: if the theory is to obtain empirical support, then we must know *how* to relax the workless-constraint restriction such that we can represent real-world systems. As we explained above, Lagrange had a heuristic for how to do this, but it requires us to undo exactly what made the system tractable in the first place (i.e., the bracketing of dynamics into kinematics); when this is too difficult, the evidential question remains unaddressed.

In his book PVV is stated overtly, but it is not the only principle on which his theory rests. There is also "Lagrange's Principle," although he does not state it explicitly *as a principle*. It takes analytic work to give it a precise formulation—by drawing on what Lagrange says and also what he *does* with the principle. In addition, there is the Relaxation Postulate. It too must be extracted by the reader—less from Lagrange's meager explicit statements, and more from his problem-solving practice. Having identified these three foundational pillars, the principle approach requires that we examine them for their mutual independence and consistency, and their joint sufficiency for solving the problems of mechanics. Only an axiomatic reconstruction of mechanics can discharge this task, and so we defer it here.[38]

For our purposes, we may draw two conclusions. First, Lagrange's theory has the three most important features of the principle approach: adoption—as the starting point for theorizing—of principles that are independent of the particular material constitution of the target systems; generality of the principles; and bracketing of dynamical considerations by means of a kinematic condition. In consequence, the principle approach yields a natural interpretation of *Mechanique*.

Second, we see the reason why the constructive approach fails to yield a unique ontology of objecthood for Lagrange's theory. The method he deploys is *deliberately* open-ended, in two important ways. For one, Lagrange presumed his PVV to be independent of—hence silent about—any particular choice of material constitution for the systems to which it is applied. For another, the Relaxation Heuristic tells us nothing about the *species* of forces responsible for the constraints: it asserts that, in removing the constraint, the

---

[38] Two informal axiomatic approaches are Hamel 1949 and Papastavridis 2002. We do not yet have a rigorous axiomatization of virtual-work mechanics, but Noll 1974 took the first steps toward it.

corresponding Lagrange multiplier will then designate an impressed force, but not what *species* that force is (whether gravitational, electric, and so on), nor what specific law it obeys. Since the properties of bodies, the forces they experience, and their material constitution are directly related to one another (gravitational mass, electromagnetic charge; gravitational force, electromagnetic force; point masses, line elements, etc.), we cannot know the properties of body until we know how to relax the constraints. And this is an open-ended project in Lagrange's mechanics.

11.6 Action

In Lagrange's mechanics, inter-body action comes in two species: between bodies that are free, or able to move anywhere in space; and those that are constrained, or limited in their kinematic freedoms. Both free bodies and mechanical constraints act in the same generic way: by applying *forces* on other bodies. Their effect is likewise the same: they induce virtual velocities and effective motions, viz. actual changes in generalized coordinates.

These two action species are more alike than they seem. For one, the same law governs them: the PVV, which—by allowing Lagrange multipliers—does not distinguish between these two species. For another, Lagrange first made clear, and defended, the idea that the actions of constraints *are* forces; or at least are kinematically indistinguishable from them (see section 11.4).

The true difference between these action species comes from the forces they exert. Lagrange restricts inter-body action to forces given by monogenic potentials: corporeal forces are the negative gradients of some "algebraic function of the relative distances" between them.[39] But he put no such restriction on constraint forces; in particular, rheonomic constraints clearly exceed that limitation: "nothing prevents the condition equations $L = 0$, $M = 0$, etc. from including a variable t for time as well; we just need to regard it as a constant when we vary [other variables] by means of δ," he allowed.[40]

Moreover, his treatment of Action is very general. In this respect, it exceeds all predecessors on two counts. It treats a broader spectrum of setups than any rational mechanics had done before him. And, it solves them from one common premise—PVV. Lagrange knows well the unifying power of his

[39] Lagrange 1788, 381.
[40] Lagrange 1788, 228.

approach, and draws attention to it at crucial junctures. For instance, as he sets out to treat incompressible fluids, he reviews his predecessors, praises d'Alembert, but then remarks, "because he based it on laws of equilibrium specific to fluids alone, his method makes Hydrodynamics into a separate science from the Dynamics of solid bodies."[41] In contrast, Lagrange derives the equations of fluid motion from the same one principle (of virtual work) that governs the rest of his statics and dynamics. They become "results of a single, general formula," namely, equation (6): the virtual work of the applied forces, the reverse effective forces, and the constraint forces is null.

His mechanics being based on forces is at odds with the modern but anachronistic expectation that it must have been a theory of the Lagrangian $L = T - V$ and its associated dynamical law, the Euler–Lagrange equation, ELQ. The historical Lagrange did not really think that the physical agencies T, V, and their difference were genuine and sole foundations. True, he did acknowledge ELQ as a partially alternative approach, but he did not think that it had the theoretical primacy that our modern notion of "Lagrangian" mechanics assigns it. In line with its predecessors, *Mechanique* makes good on the early modern program that mechanics is the science of forces.

Importantly, however, this is a treatment of forces within rational mechanics; it is a *mathematical* treatment. We can see the import of this by comparing Lagrange with his predecessor, d'Alembert, for whom:

We know just three ways for bodies to act on each other: either by direct impulse, as in regular collisions; or by means of some body interposed between them, and to which they are attached; or, lastly, by a mutual power of attraction, such as the Sun and the planets exert in the Newtonian system. (d'Alembert 1743, 72)

In Lagrange, there is no such appeal to physics in elaborating a taxonomy of forces. On the contrary, we find a clean separation of the *physics* by which one body acts on another from its *mathematical treatment* in rational mechanics. Specifically, to apply the mathematical machinery of Lagrange's mechanics, one need not know the physical species of force responsible for the constraints or for the motions being analyzed.

---

[41] Lagrange 1788, 437. D'Alembert's grounding assumption for his hydrodynamics was that, in any fluid at rest, internal pressure (at a point) is spherical, or equal in all directions.

11.7 Evidence

Given the vast scope and imperial ambitions of Lagrange's mechanics, a pressing question is: what evidence does he have for his novel theory? The matter is difficult, and it rewards closer study; here we can give just a synopsis of its main parts. Briefly put, he seeks to marshal evidence in two ways: *empirically* via the solutions to *particular* systems that he treats, and *mathematically* through the *generality* of the principles on which the theory is based.

(i) Evidence from empirical instances

Lagrange selects a few local consequences of PVV—one equilibrium condition and two equations of motion—that can be solved (by numeric integration) and thereby yield predictions of determinate parameter values: the shape of the earth (from the equatorial bulging ratio); the speed of sound in air; and the orbital period of the moon and sun around the earth. He regards the agreement (between his predicted values and the empirically found ones) as evidence for his equations. For instance, he derives three differential equations for a mass m attracted by two others, m' and m'', then makes two special assumptions: he lets their mutual forces be inverse-square, and continues:

> If we take m to be the mass of the earth, m' of the moon and m'' of the sun, the three equations above become those of the so-called Three-Body Problem, on which Geometers have worked a great deal recently. That the lunar and the solar orbits [around the earth] are nearly circular makes them amenable to solving by approximation. And, if you look up the papers that treat the Problem, you will see the contrivances they have devised to make that approximation as exact as possible. (Lagrange 1788, 259)

In effect, he takes that to be empirical confirmation for his equation of motion, via the special case of a "reduced" three-body problem.[42]

Two aspects deserve a note here. Prima facie, Lagrange's approach to empirical evidence resembles that of Newton and his followers, who likewise had inferred to specific parameter values for certain whole-motion

---

[42] That is the special case when one mass is much larger than the other two, and thus can be regarded as stationary.

integrals: the bulging ratio (and the associated value of g at select latitudes); the time of arrival at perihelion (for the 1759 return of Halley's Comet); and the speed of aerial sound. Very unlike Newton, however, Lagrange does not seem interested in the stronger kind of evidence to be had from systematic *discrepancies*. Namely, from the differences between predicted values (from his theory) and parameter values measured empirically. Being able to account for such discrepancies—from the very resources of the theory at issue—had yielded *strong* evidence for Newton; certainly stronger than mere loose agreement between theory and observation. Lagrange looks away from this stronger kind of empirical confirmation; when data diverges from his predicted values, he tends to speak dismissively of data:

> Hence the speed of sound [from my theory] will be 915 Paris feet per second. Experiment gives about 1,088 ft/sec, which is almost one sixth higher. But this difference [between theory and data] can only be attributed to the uncertainty of our current experimental results. (Lagrange 1788, 511)

Contrast his attitude with that of Newton, who always sought to ascertain if the discrepancies (between prediction and data) were systematic, and— when they were—to account for them by identifying their physical causes in exact, measurable terms, then by predicting (from theory) the expected size of those discrepancies.[43]

Lagrange's epistemology of mechanics has an important obscurity. Recall his statement (7) above of PVV in the source-specific form: $\Delta + \Gamma + \Lambda = 0$. If we state his principle for *free* bodies only, it reads: $\Delta + \Gamma = 0$. This implies trivially that $\Lambda = 0$: the net virtual work of the constraints vanishes.[44] But what might be his warrant for this assertion? What evidence does he have? Regrettably, he does not say; but a charitable interpreter can answer it on his behalf. Lagrange's theory is *designed* to treat *just* ideal constraints (i.e., the kind of constraints whose overall action across the system is workless): the

---

[43] The sources of confirmation (including the 1759 return of Halley's Comet) and the patterns of evidential reasoning associated with Newton's gravitation theory in the 18th century are lucidly explained by George Smith (2014). Smith first uncovered, and stressed the evidential import of, systematic discrepancies (between Newton's first-order predictions and observed orbits), and the ability of Newton's theory to trace them to physical sources (viz. further centers of gravitational attraction in the solar system).

[44] Δ is the net virtual work of the forces actually applied on the system; Γ is that of the (imaginary) reverse effective forces; and Λ that of the constraints in the system.

net virtual work due to them is zero by definition. To be sure, this evidential aspect of Lagrange's mechanics deserves a second look.

Interestingly for our purposes, the appeal to empirical evidence pertains to theorizing in physics; it takes us beyond rational mechanics and into philosophical mechanics. The second type of evidence is proper to mathematics; we turn to that now.

(ii) Evidence from mathematical generality

Lagrange puts great stock in the heuristic power of his principles: to unify theory and to solve new problems. That power drives his theory construction along two tracks. One is a recipe for instantiating the same principle for different configurations. The other is a heuristic for *extending* the theory to novel setups. Each is a form of generalization, and by these means Lagrange seeks to establish the *generality* of his principles.

The former may be called the canonical recipe. Synoptically presented, it is a series of four moves. For simplicity, we confine it to the statics of constrained systems in a plane; it is then easy to see how it works for dynamics.[45]

Step 1 Write down the Principle of Virtual Work for all the masses in the system: $P\,\delta p + $ etc. $+ \lambda\,\delta L + $ etc. $= 0$.

Step 2 The virtual velocities δp, δq, etc. are in the direction of the respective forces that induce them. Express p, q, etc. in terms of some useful set of generalized coordinates; for example, polar (R and φ). Substitute each resulting expression into δp, δq, etc. of Step 1, and work out the algebra (based on Lagrange's rules for the variational operator δ).

Step 3 Each constraint L, M is given as a dependence formula between generalized coordinates. Substitute the relevant formula into δL, δM, etc., and work out the algebra.

Step 4 For n masses in a plane, Step 1 yields $2n$ equations.[46] For k constraints, Step 3 yields k equations. Use them to eliminate

---

[45] Specifically, at Step 1 just add new terms corresponding to the reverse effective forces, and make the necessary adjustments at the subsequent steps that need them.

[46] Because two generalized coordinates are enough to describe the position of a single material point in a plane.

dependent variables from the result of Step 1. The outcome will be a set Z of $(2n-k)$ equations between $(2n-k)$ variables.

The set Z gives the sufficient conditions for the system to be in equilibrium. At this point, the mechanics problem is solved—any further steps are just problems in mathematics (e.g., integrating Z over time).

In addition, Lagrange found another heuristic—but in *dynamics* alone—which he offered as the "simplest method for obtaining the equations that determine the motion of an arbitrary system of bodies," driven by accelerative forces.[47] It too is a recipe that he stated in generic terms; we reconstruct it here. Let m be the mass of a component particle or body, q its generalized coordinates, and d denote differentiation in respect to time only.

Step 1   For that particular system, construct two functions, T and V, as follows. T is half of *vis viva*, that is, the sum over $mdq^2/2$. V is the sum of $m\Pi$, where Π is the "work function" for that system.[48]

Step 2   Now build the formula: $\Phi = d \cdot \partial T/\partial dq - \partial T/\partial q + \partial V/\partial q$

Step 3   If the masses are free, set Φ equal to zero. If they are constrained, let Φ equal the sum $\lambda \partial L/\partial dq + \mu \partial M/\partial dq$ + etc., where L, M, and the like are the equations of constraint.

Step 4   Use the dependencies expressed by $L = 0$, $M = 0$, etc. to eliminate all dependent variables in the equations obtained at Step 3. "By eliminating the unknown [quantities] λ, μ, etc., there remain as many equations as we need for the solution" of the problem.[49]

In retrospect, at Step 2 we recognize an early form of the Euler-Lagrange Equation, ELQ. His second heuristic, in effect, is a proto-version of the ELQ-based approach that nowadays we take to be *the* method of analytic mechanics. But we must caution against a misconception, or illusion of hindsight. Lagrange himself did *not* take it to be the sole recipe of his theory—not the preferred approach either. In his picture, the royal avenue to solutions was the *first* heuristic above, based on PVV. He saw the ELQ as a beneficial shortcut allowable when the dynamics of the target system permits it: namely,

---

[47] Lagrange 1788, 216.

[48] He defines Π as "an algebraic function of the distances" between the masses. Earlier, he had built it from the work done on a particle by external forces P, Q, R, etc. that are functions of the distance to it (Lagrange 1788, 375, 381). Namely, $\Pi = \int (Pdp + Qdq + Rdr + \text{etc.})$. Hence, in modern terms Π counts as a monogenic (conservative) potential.

[49] Lagrange 1788, 228.

when the interaction forces are gravity-like, namely, conservative, or given by V-type potentials; and when the forces act on point-sized particles. But, of course, not all systems are so tractable, and in that case Lagrange does not avail himself of the shortcut. For instance, his treatment of fluid motion makes no mention of ELQ.

Now that we know Lagrange's recipes for theory buildup and expansion, it is easier to grasp and appreciate his words about their value. The approach in *Mechanique*, he explained, showcases the "steady and regular workings" of his method.[50] Against this backdrop, his treatise strongly suggests an overall conclusion about its epistemic credentials. They come from the virtues of two supporting pillars, namely, its principle and the accompanying recipe. The PVV is *"simple"* in expression and very *general* in scope; the heuristic is *uniform* and *transparent*: it can be explained discursively and iterated indefinitely. It is these virtues, he implies—rather than strong support from empirical evidence—that legitimize his rational mechanics.

11.8 Assessment

Like d'Alembert before him, Lagrange touted his mechanics' vast descriptive reach as a virtue, or reason to prefer it to its competitors. He had demonstrably covered all the species of kinematic behavior that Enlightenment theory had been able to solve.

(i) Unity and scope

From our vantage point, however, his rational mechanics has yet another virtue—subtler, buried deep, but all the more valuable. Lagrange's approach yields a *complete solution* to the problem of constraints *in general*. It entails equations of motion for a system with *any* constraints, whether external or internal. More precisely, if the target object of his principles is free particles, then all Lagrange multipliers are trivially zero. If the object is free extended bodies, the multipliers denote the internal constraints specific to that body kind. And if the target is extended bodies with obstacles to their motion, the

---

[50] Lagrange 1788, vi.

multipliers pick out both classes of constraint, internal and external, that control those motions. In effect, Lagrange's approach *eliminates* the need for any principled distinction between internal and external constraints.

Admittedly, he did not stop to acknowledge his theory's unity of treatment and generality of scope. That dual merit comes to light only when we survey the long history of that theory, especially in Germany. Some figures there learned to exploit his two germinal ideas—the method of Lagrange multipliers and the relaxation of constraints—so as to cover further descriptive territory from the same coherent set of three principles. That approach culminated in Georg Hamel's comprehensive synthesis, subtitled *A unified introduction to the entire mechanics*.[51]

(ii) Lacunae

Its enormous range notwithstanding, Lagrange's rational mechanics had some real descriptive limits. It was able to treat just constraints that are holonomic, bilateral, and workless (or ideal). From a purely mathematical standpoint, constraints come in two classes: bilateral and unilateral. In turn, bilateral constraints come in two species, holonomic and anholonomic.[52] Lagrange's mechanics is fit to handle just holonomic constraints: it admits just constraints that are given by equations between coordinates (and perhaps time), which means they count as holonomic. Further, he restricted his account to constraints given by *equations* (of condition). That makes them bilateral. This reduces its scope not insignificantly. For instance, it is unable to derive the equation of motion for this simple case: a ball, at rest on a hemisphere a small distance from the pole, which starts to roll down under gravity without slipping; find its location at any time before it hits the ground. We

---

[51] See Hamel 1949, which covers—from the same three principles as *Mechanique* above—broad domains that Lagrange had been unable to treat. In particular, waves in continuous media, elasticity theory, and the dynamics of viscous fluids.

[52] Bilateral constraints are expressible as equations. Unilateral constraints are expressible as *inequalities*. Example: a particle with position vector x falls down along a hemisphere of radius r; the constraint is $| x | - r = 0$. Holonomic constraints are functions of the generalized coordinates (and perhaps time) but not of derivatives (velocity or higher order). If the equation expressing a bilateral constraint includes any generalized velocities (or higher-order derivatives), the constraint is anholonomic. Example: a sphere that rolls on a plane of length x. The constraint is the "no slip" condition for rolling, viz. $x = r\omega$, where r is the radius, and ω is the angular *velocity* of the sphere point in contact with the plane at that instant.

364 PHILOSOPHICAL MECHANICS IN THE AGE OF REASON

might try to expand the scope, but then Lagrange faces a dilemma. Either all constraint forces are monogenic, just like the applied-impressed forces. Consequence: anholonomic constraints are then not explainable as the result of *forces* acting, because Lagrange's admissible forces are not functions of generalized velocities, whereas anholonomic constraints are. Or we admit any constraint force provided it is expressible as $\lambda \partial f/\partial q$, so that anholonomic constraints are included. Consequence: Lagrange was wrong to insist that all the forces allowed in *Mechanique* are monogenic.[53]

Lastly, Lagrange treated just *ideal* constraints. We have already remarked that his formula (7) entails $\Lambda = 0$, viz. the net virtual work of the constraints vanishes across the system. That is a not insignificant limitation. In the real world, constraints are *not* workless; they drain energy away from the system, and increase its entropy. Which poses a challenge to his rational mechanics: its descriptive scope is not broad enough.

11.9 Conclusions

By mid-century, it was clear that a successful philosophical mechanics required a solution to MCON2: "Given an extended body subject to external constraints, how does it move?" In *Mechanique*, Lagrange offered an answer. Despite the limitations just discussed, he had created a single, unified theory of impressive scope, encompassing the motions of extended bodies impeded by external obstacles. In fact, it is a remarkable solution to MCON, quite generally. In order to apply the theory to physical systems, all one need do is determine the appropriate Lagrange multipliers. And herein lies the interesting challenge of moving from rational mechanics to physics. Or, in our terms, of arriving at a philosophical mechanics.

More specifically, an open question is the extent to which Lagrange's rational mechanics can be incorporated into a philosophical mechanics that yields a solution to BODY by giving a single, well-defined concept of body that is simultaneously (i) consistent with an intelligible theory of matter,

---

[53] Recall, any scleronomic constraint can be expressed as a function $f(q_1, \ldots q_n)$, viz. a dependence between coordinates alone. But, that is exactly the same condition that a potential must meet to count as monogenic. We may think of $\lambda \partial f/\partial q$ as a potential-induced force that keeps the body "glued" to the constraint surface, by requiring it to do mechanical work if it is to move free of that particular constraint.

(ii) adequate for a causal-explanatory account of the kinematic behaviors of bodies, and (iii) sufficient for the purposes of mechanics. The obstacles are significant, as we saw when we assessed his theory from a constructive and then from a principle approach. To make more progress we need to look beyond Lagrange; see Chapter 12.

12

Philosophical mechanics in the Late Enlightenment

12.1 Introduction

In the previous four chapters, our leading thread was the rational mechanics of constrained motion. In that collective project, the center of gravity was the arduous task of mathematizing the motion change at every point of an extended body with constraints. Lagrange's treatise, *Mechanique*, was the apogee of success on that count.

Throughout, Lagrange emphasized a pressing priority: the task of turning rational mechanics into a single theory, one that is *general*—it entails all the equations of motion then known; and *unified*—it derives them from one and the same basis. The basis was his Principle of Virtual Velocities (PVV) or, in modern terms, of virtual work.[1]

In the early 19th century, rational mechanics succeeded in quantifying the behavior of two new species of deformable body, which broadened its scope enormously. At the same time, alternatives emerged to Lagrange's principle approach to unification. Together, these developments had significant consequences for philosophical mechanics. More specifically, after 1800 rational mechanics managed to treat novel classes of motion phenomena: capillary effects, and the behavior of elastic solids and viscous fluids. Many of these successes came from approaches and foundations not given in *Mechanique*. For this and other reasons, a group of mathematicians objected to Lagrangian nomic unity as a goal, setting out instead to establish ontic unity, via a preferred object that they called a "molecule." However, both programs were challenged by Cauchy's work in

---

[1] In the years we survey in this chapter, the concept of mechanical work becomes explicit and gradually entrenched (see Grattan Guinness 1984), and so we abandon the 18th century language of virtual velocities in favor of the phrase "virtual work."

Philosophical Mechanics in the Age of Reason. Katherine Brading and Marius Stan, Oxford University Press.
© Oxford University Press 2023. DOI: 10.1093/oso/9780197678954.003.0012

the 1820s. His results showed that, for some very important classes of motion, the Eulerian balance-of-force approach does better than Lagrange's virtual-work principle; and that picturing matter as a deformable continuum works better than the Laplacians' discrete molecules. Ultimately, these advances and watershed changes bear on the main topic of our book—the fate of philosophical mechanics, and the chances of solving the *Problem of Bodies* (BODY).

Below, we spell out these developments and ensuing problems in sharper detail, and then draw some lessons for our main theme. We begin with a synopsis of the relevant figures and sites of research (section 12.2). We then survey the main successes of the Lagrangian nomic program in new areas of mechanics after 1800, and the problems it faced (section 12.3); and we do the same for the molecular ontic program (section 12.4). Next, we turn to Cauchy's early work in rational mechanics, where his balance-of-force approach yields an alternative basis for nomic unity, and his use of the deformable continuum challenges the molecularist approach to ontic unity (section 12.5).[2] We close by assessing the impact of these developments on philosophical mechanics after 1840 (section 12.6).[3]

12.2 Makers and spaces

The two unifying projects, ontic and nomic, were both French affairs, as was the challenger that arose to defy them subsequently. In the Late Enlightenment, the country was the world's powerhouse for mechanics.

The constructive approach relied on two leaders, Laplace and his outstanding student, Poisson, who succeeded him after 1820. They had associates in related fields—Navier in elasticity and fluid dynamics, Berthollet and Gay Lussac in chemistry, and Biot in optics. Occasionally, they received help from Haüy in crystallography and Cauchy in continuum mechanics. At the same time, this program ran into significant dissent, which blunted its appeal and curtailed its chances of success.

The principle approach originated with Lagrange. His later career in academia acquired him some very capable disciples, and also sharp but friendly

---

[2] For convenience's sake, henceforth we often use the phrase "true solid" for non-fluid bodies of three dimensions comparable in size.

[3] This chapter is based on research by MS that is ongoing, extending beyond the scope of this book.

critics (all of them young graduates from Lagrange's courses) aiming to strengthen the foundations of his program. Among these critics were Poinsot, Fourier, and Ampère.

These two programs relied on some novel logistics, thanks to happy changes in the status of exact science in France after the *ancien régime*. The most drastic innovation was to put rational mechanics in the classroom, and make it the cornerstone of elite education in science. The locus for that was the École Polytechnique, a highly competitive preparatory school for advanced education in engineering. Its founding in 1794 had two outcomes relevant here. It caused the Principle of Virtual Work to become the law of the land, as it were: the official basis for rational mechanics as taught to three generations, and codified in major textbooks. And, it resulted in the *Journal de l'École polytechnique*, a new venue for research in mechanics.

Another novel site was a private laboratory for experimental research in physics. Funded and managed by Laplace and Berthollet at Arcueil, outside Paris, it was a venue for testing empirically the theoretical work that they and some young disciples had carried out in capillary theory, physical optics, heat science, and chemistry. This mattered for the ontic program, which expected knowledge of their privileged object—the molecule, its key properties and characteristic behaviors—to come from strongly established *empirical* research in physics and chemistry.

Finally, one resource was traditional but more innovatively managed: the system of prize essay competitions at the Academy of Sciences. The innovation was that Laplace exerted more initiative over the choice of topics, all for the sake of advancing the constructive program. His overt advocacy for it, coupled with his towering prestige and political clout, led to historians calling the program "Laplacian physics."[4]

---

[4] Many other new journals for research in physics and applied mathematics were established then. The institutional landscape and publication venues at issue in this chapter have been thoroughly studied. Grattan Guinness 1990 is a comprehensive guide to them, and Gillispie 2004 is an invaluable account of the broader context for science at the time. Crosland 1967 is an older classic on research at Arcueil. Excellent surveys of the École Polytechnique in its early years are Belhoste 2003 and Gillispie 2004. Fox 1974 and 2013 are canonical presentations of "molecular" physics at the time, with an emphasis on Laplace; for Poisson and his time, the account must be supplemented with the comprehensive Arnold 1983–4.

12.3 Lagrangian nomic unification

Lagrange had pursued with great success a project to unify mechanics on d'Alembert's old terms: a formulation maximal in scope and based on very few principles. Thus the resulting unity of his mechanics was nomic, based in a privileged principle, PVV. His success lent that principle a distinguished place in French mechanics at the time:

> To give mechanics all the perfection it is capable of, the last remaining task was to combine [Lagrange's] principle (which we have just explained) with the principle of virtual velocities.[5] That is what Lagrange did. Thereby, he reduced the study of the motion of an arbitrary system of bodies to the integration of differential equations. (Laplace 1795, 307)

> Thus, now we have [Lagrange's] principle, which applies to arbitrary changes, finite or infinitesimal, that can occur in the motion of a system. By this principle, we can always reduce a case of motion to equilibrium. If we combine this principle with that of *virtual velocities*, there is no dynamical problem we cannot turn into an equation. From this vantage point, the science of *motion* and that of *equilibrium* lacks nothing. (Prony 1815, 284)

After 1800, efforts began to extend PVV to new areas not yet mathematized.

(i) Progress

All unification projects in the 1700s had been reactive, as it were: they would wait until theorists discovered further equations of motion—by whatever methods and approaches they could—and *only then* show that these newer results could be subsumed under the preferred unifying principle.[6] As the 19th century began, that reactive stance gave way to a different attitude. Theorists now turned to active unification: they used Lagrange's principles for discovery too, not just incorporation after the fact. Specifically, they relied on virtual-work approaches so as to mathematize domains not treated

---

[5] Note: what we call "Lagrange's Principle" is the statement that Lagrange himself called "the principle of Mr d'Alembert"; viz. the thesis that in a moving system the reverse effective forces exactly balance the actual-impressed forces. See Chapter 11.

[6] Recall, in *Mechanique* Lagrange derived from PVV the equations of rigid-body motion, the vibrating string, and perfect-fluid dynamics. But those equations came from d'Alembert and Euler, who had discovered them from premises and methods different from Lagrange's. See Chapter 11.

in *Mechanique*: elastostatics, capillary theory, and systems with unilateral constraints. We survey them in turn.

Elasticity

A first new area they conquered was elastostatics; in particular, the equilibrium of an elastic membrane. In the 1810s, Sophie Germain and Lagrange tackled this problem, but somewhat inconclusively.[7] It was Navier who made the real breakthrough, by deriving the equilibrium formula for a special case first, then generally.[8] He reasoned as follows. Any point in that plane is subject to *two forces* caused by the plane being deformed.[9] One is "due to tension," viz. local stretching, and the other force is "produced by the flexion," or bending of the plane. Both of these force-kinds induce virtual displacements, corresponding to "moments" (i.e., virtual work). For that point to stay at rest, Navier concludes, "the sum of these moments" must vanish:[10]

$$T\,dxdy\cdot\delta k + \varepsilon h^3 E\,dxdy\cdot\delta E = 0 \tag{1}$$

By "integrating this expression over the whole elastic plane," we obtain "the equation that expresses the equilibrium of the system." That expression then must be "treated by the methods of [Lagrange's] *Mécanique analytique*," which gives two sets of equations, one for points inside the plane, and another for points at the edges, viz. the boundary conditions.[11]

---

[7] An 1809 prize essay competition at the French Academy asked for the equilibrium condition of an elastic membrane. Sophie Germain submitted a paper in 1811, then again in 1814 with some improvements. Because she claimed to derive her formula from the "principle of moments," some scholars have thought it was Lagrange's PVV. But, it seems she meant it in Euler's older sense (of torque, or moment of a force). That is, Germain assumed equilibrium follows if the torque of the external forces balances the net torque of the membrane's "elastic force" of restoration. Spurred by her efforts, Lagrange in 1814 gave his own formula for the equilibrium, but it is unclear from what mechanical assumptions he derived it. For further discussion, see Bucciarelli & Dworsky 1980.

[8] That was the case of an elastic plate held fixed at the edges as forces press on it at arbitrary locations (e.g., as in a floor slab when people stand on it). Navier was a civil engineer.

[9] This claim is significant. Cauchy's early work in elasticity (see below) started as an effort to reduce Navier's two forces above to just one, viz. stress.

[10] See Navier 1823a, 95. T is the tension force (it stretches the plane evenly around that point), δk a virtual displacement in its direction, ε a material-specific constant (analogous to the parameter K in Hooke's Law for elastic springs), h the thickness of the plane, E the restoring force induced by the bending, and δE its corresponding virtual displacement. Navier takes the force E to be equal to the sum of the plane's curvatures in the two directions x and y; compare his result to Euler's analysis of the vibrating membrane, see Chapter 10.

[11] Navier 1823b, 178, 180. The methods he mentions are known to the reader from Chapter 11. One is a step in the Lagrange Heuristic, viz. to express the virtual displacements δk and δE in terms of Cartesian or polar coordinates. The others are purely mathematical, designed to eliminate the operator δ from the equations: by letting it commute with the operator d, then via integration by

Then he broadened his treatment for the more general case of an elastic solid (a body of comparable dimensions, not just a thin plane). To tame that type of object, again he resolved to take the "method of calculation employed in *Mécanique analytique*" (i.e., the virtual-work approach). Informally speaking, Navier's derivation went as follows. On every molecule in an elastic solid there are *internal* forces: exerted on it by all the particles nearby, hence inside the body. When the body is in the "natural state" (viz. undeformed), the internal forces vanish. When *external* forces act on the body and deform it, the internal net force on every particle becomes non-zero.[12] Moreover, every one of these forces induces a "moment," or quantity of virtual work. Following Lagrange's approach, Navier concludes: the body is in equilibrium if the "moments" of the internal forces and the external forces jointly vanish.

In exact terms, when an elastic body gets deformed, on every particle M the particles around it exert a restoring force associated with a virtual work. These count as internal forces. In addition, there are two more species: forces external to the body, acting on material particles inside it; and exogenous forces acting on its surface only.[13] These forces do virtual work on their target particle. The "general equation that expresses the equilibrium of the system" will be:[14]

$$\varepsilon \int dV \, f(\mathbf{r}) \cdot \delta F + \int dV \, X \cdot \delta x + \int dA \, X' \cdot \delta x' = 0 \qquad (2)$$

If the net virtual work of the three kinds of forces above, taken over the whole body, adds up to zero, then the elastic body will be in equilibrium.

This was a very great advance—it showcased the power of Lagrange's principle approach in a very difficult area, much harder to quantify than anything

---

parts proceed until a purely differential equation (in *d* alone) ensues. See the explanation in Lagrange (1788, 213).

[12] Because deformation changes inter-particle distances, and so it displaces every particle from the location where the net force on it was zero. In other words, Navier takes internal forces to be functions of the deformation (i.e., the *change* of relative distance), not of the mere distance.

[13] In modern continuum mechanics, Navier's taxonomy broadly corresponds to the threefold distinction between internal stresses, body forces, and surface tractions. Not long after him, Cauchy would make these distinctions rigorous; see section 12.5.

[14] See Navier 1823b, 180. In his expression, εf is the net intermolecular force, X the net applied force (external to the body, acting at that point), and X' the external force on points on the body's *surface.*

we have seen so far. Navier's breakthrough had an added benefit: it let him derive the equation of motion. In the x-direction, it reads:[15]

$$\varepsilon f(\mathbf{r}) - X = \frac{\Pi}{g} \frac{d^2 x}{dt^2} \qquad (3)$$

Namely, the internal force plus the (exogenous) body force equals the rate of change in linear momentum. Thus the motion of an extended elastic body—its motion at every point, not just its mass center—at last yielded to Lagrange's methods.

Capillary surfaces

Another major advance for the nomic unification program was due to Gauss. In rational mechanics after 1800 there were many attempts to derive the analytic formula for the shape of the meniscus surface of a fluid in a capillary tube; see below. Gauss in 1829 decided that Laplace's theory of capillarity was inadequate, and sought to replace it "from the ground up," by a "wholly different method" that proceeds "from the first principles of dynamics." Just one, in fact, namely, the "principle of virtual motions." In an incompressible fluid, he begins, at every point let there be a quantity $m\Sigma P \cdot \delta p$.[16] The fluid is in equilibrium if

$$m\,\Sigma P \cdot \delta p \; + \; m'\,\Sigma P' \cdot \delta p' \; + \; m''\,\Sigma P'' \cdot \delta p'' + \text{etc.} = 0 \qquad (4)$$

for all the "physical points" in it. That is, equilibrium obtains if the virtual work of all the forces on a fluid particle vanishes.

This part is not new, of course. Already in 1788, Lagrange had rested his entire hydrostatics on it. Gauss' innovation was to extend it by a new account of the forces acting on a fluid particle in a capillary tube. Those actions, he claimed, are "gravity" pulling downward, "attractive forces" that fluid particles exert on each other, and forces whereby those particles are

---

[15] ε, f, and X have been explained above; Π/g is Navier's representation of mass density (Π is weight per unit volume, and g the acceleration of gravity).

[16] Gauss 1833, 42f. Here, m is the particle's mass, P the net *effective* acceleration, and δp an "infinitely small motion compatible with the conditions" on the system (Gauss 1833, 43). Just as Lagrange had explained in *Mechanique*.

"attracted to some fixed points" (i.e., to the capillary tube walls). He expresses these three force kinds accordingly, and adds their respective contributions to the effective acceleration P above. The rest is just mathematical prowess. It yielded for Gauss a general solution that superseded Laplace's "incomplete theory," and did not need to start from his "principle of molecular forces."[17]

Unilateral constraints

From Chapter 11 recall that Lagrange's theory handled just bilateral constraints, viz. given by equations between dependent coordinates. Around 1827 Cournot tried to move rational mechanics beyond this limitation by a small, cautious step forward. He sought to extend Lagrange's *statics* to cases in which mass points are subject to constraints given by inequalities (e.g., $I \geq 0$ and $J \leq 0$). For that family of cases, he argued, the masses are in equilibrium when *two* conditions obtain (not just one, as in *Mechanique*). First, the virtual displacements compatible with the constraints are themselves given by inequalities:

$$\delta I \geq 0; \ \delta J \geq 0 \tag{5a}$$

Second, the virtual work of the applied forces F, G, H, etc. is *at most* null:

$$F\delta f + G\delta g + H\delta h + \text{etc.} \geq 0 \tag{5b}$$

These two formulas jointly give conditions for the masses to remain at rest.[18] Granted, this extension of PVV was rather modest. (The mechanics of unilateral constraints is difficult, and remains an area of active research.) But, it further illustrates the efforts to expand Lagrange's principle approach to the unity of mechanics.

---

[17] Gauss 1833, 42, 81.

[18] See Cournot 1827, 167f. He then illustrates his point with two examples, including the case of a mass point constrained to remain on or above a rigid plane $z = a$ parallel to the plane xy of the coordinates.

(ii) Problems of unification

A half-century after *Mechanique*, the nomic unity program had added enormous new territory. But the Lagrangian project then ran into some problems. One was internal, and it concerned the epistemology of its principle. The other was an external objection that accused the project of being explanatorily and referentially opaque. Let us inspect them more closely.

Evidential concerns

Recall from Chapter 11 a certain ambivalence about the epistemic status of PVV in *Mechanique*. There was empirical evidence for its *truth*, but it was scant, disparate, and comparatively weak. And, there were strong reasons to accept the principle as *certain*.

Lagrange's immediate posterity was unsatisfied with either. After 1796 a group of young theorists, former students of his at the École Polytechnique, made a forceful push to show that PVV is true *and* based on the strongest evidence available in rational mechanics. These critics then go on to seek that missing evidence themselves—they try to *infer* the principle from some supposedly primitive mechanical truth. Some tried to show that PVV can be inferred from the Composition of Forces (i.e., the Parallelogram Rule for statics); others argued that it follows from the Law of the Lever.[19]

Lagrange reacted to his critics rather strangely. Instead of pointing out that his principle was *certain*, he chose to agree to their terms, viz. that PVV required proof of *truth*. "I must confess that the idea [behind the principle], though in itself correct, is not clear enough for it to serve as the principle of a science whose certainty must rest on self-evidence," especially when we teach it to beginners, he fretted.[20] He just disagreed with their choices of key premise for that proof; he believed none of the candidates above really counts as a primitive truth. The proofs given so far, Lagrange objected, infer the principle from either the Parallelogram Rule or the Law of the Lever. "But, as is well known, these two are *not evident enough* that they do not require their own proof," he demurred.[21] Accordingly, in edition B he gave a proof of his own—an intended demonstration of PVV. But the proof is drawn out,

---

[19] For derivations from the Composition of Forces, see Fossombroni 1796, the first critical examination of Lagrange's warrant; and Poinsot 1806. For the Law of the Lever, see Fourier 1797. These claims help shed light on Lagrange's own attempted proof.

[20] Lagrange 1797a, 3.

[21] Lagrange 1797b, 115.

its logic opaque, and the evidence for its key premise far from transparent.[22] Whatever Lagrange thought of its strength, it did not end the debate. Years after the fact Poinsot, one of his early and astute critics, commented on the need to shore up evidentially his mechanics:

> Since Lagrange's book offered nothing clearer than the march of calculations, soon we realized that he had dispersed the obscurity hanging over the structure of Mechanics at the price of shifting that obscurity back onto the very origins of this science. (Bailhache 1975, 111)

Attempts to dispel that obscurity, by way of yet another "proof" of PVV, continued up into the 1830s. Their logical merits aside, they all assumed the target system—the thing that the Principle is about—to be a set of *discrete* mass points, free or constrained. Not continuous bodies. Which led Cournot to remark grimly:

> Consequently, in the case of *continuous* mass elements, the principle of virtual velocities holds just by induction, i.e. from particular confirmations. Hence, the foundations of the general theory that *Mécanique analytique* has so admirably developed are not yet sufficiently established a priori. We do not yet have a direct demonstration that would allow us to derive, from a *single* principle and by a *uniform* method, the laws of equilibrium for *all* systems, whether solid, fluid, etc. (Cournot 1827, 166; our emphasis)

In sum, the two editions of Lagrange's treatise are importantly different. Beneath their sameness of structure (nomic unification from PVV) lies a drastic shift in the evidential status of that principle. And yet, the epistemic credentials of his approach seemed deficient even to its followers.

---

[22] His key premise is a proposition he calls the "principle of pulleys." It is a claim about the behavior of a weight at rest hanging from a system of connected pulleys rigidly attached to pedestals fixed in space. To avoid evidential regress, the premise ought to be self-evident. It would need non-discursive a priori evidence; for instance, ostension to a diagram. But Lagrange undermines that prospect. His opposition to synthetic geometry (and its representational resources) rules out that option; recall his injunction that "analytic mechanics" has no figures to rely on. So, it is mysterious why he thinks his own proof is sound. For some discussion, mostly inconclusive, see Jacobi 1996[1847–8], Mach 1883 and Bailhache 1975.

Referential opacity

Another problem was external: a blunt reproach by figures associated with Laplace. Expressed in their terms, the objection went as follows. Recall the original rationale for the Lagrange multipliers: the forces of constraint—their physical sources and mechanisms of action—are generally unknown in advance. Only the geometry (and not the physics) of the ensuing constraints is available knowledge. The method of Lagrange multipliers exploits that geometric fact so as to infer the effect of constraints while bypassing the need for microphysical knowledge. For that reason, the Laplacians called it "abstract mechanics." To them, it seemed radically inadequate—precisely *because* it bypassed physics. So, they advocated its demise and replacement by a "physical mechanics" that incorporates the missing knowledge:

> Geometers should reprise the main questions of mechanics from this physical point of view, which corresponds to nature. So far those questions have had to be treated in a wholly abstract manner, so as to discover the general laws of equilibrium and motion. With this sort of generality and abstraction, Lagrange went as far as it is conceivable, by replacing the physical bonds of bodies with equations between the coordinates of their points. That is what *analytic* mechanics consists in. But, alongside this admirable approach, we can now advance *physical* mechanics, whose sole principle is to reduce everything to molecular actions, which transfer from point to point the action of forces, and mediate their equilibrium. . . .
>
> Generally, in applying mechanics we must take into account all the physical circumstances pertaining to the innermost nature of bodies. That need has long been felt, so as to overcome the indeterminacy behind certain questions of abstract mechanics. Such indeterminacy could in no way obtain in nature, where everything must be determinate, and admit of unique solutions. (Poisson 1828, 341f., 344)

Put in our terms, they objected that Lagrange's bracketing of dynamics into geometry left his mechanics referentially opaque: its physical content was in part occulted, and that was reason to move beyond his approach.[23] They thought they had a better way forward.

---

[23] Poisson thought that only constructive approaches could turn mechanics into a source of determinate knowledge. In his view, Lagrange multipliers do *not* give that knowledge: computing them yields a quantity (namely, $\partial\lambda/\partial q_i$) with the *dimension* of a force, but no insight into (as 17th century figures put it) the "seat" and "manner of operation" of constraint forces. In modern terms: Lagrange

12.4 Molecular ontic unification

The program we are about to explain could, and was, a plausible alternative to the Lagrangian nomic approach. But it did not begin as such—not overtly, anyway. Rather, it began as a confident project to mathematize all of physics. Namely, to use reliable results of empirical research as inputs for quantification, that is, for using the calculus of partial derivatives so as to derive those results from a physical theory of forces between certain entities they called "molecules":

> In general, all the attractive forces and repulsive forces of nature ultimately reduce to like forces [of attraction or repulsion] acting from molecule to molecule. From this supposition I had shown . . . that all capillary phenomena reduce to intermolecular attractions over imperceptible distances. . . . We have also attempted to reduce electric and magnetic phenomena to actions from molecule to molecule. . . . Lastly, the supposition of actions *ad distans* from one molecule to the next, when applied to heat, leads clearly and precisely to the true differential equations for the motion of heat in solid bodies. Thereby, this very important branch of Physics likewise enters the domain of mathematical Analysis. (Laplace 1898, 288, 290)

Though Laplace's chief aim was to reform *physics*, his program had a critical component that pertained to *mechanics* as well. The Laplacians expected all rational mechanics—all equilibrium conditions and equations of motion— to be derived from the physics of molecules. In effect, their program assumed that the unity of mechanics is really ontic. That is, mechanics is a single theory because it quantifies the possible motion-behaviors of a preferred object, namely, the molecule:

> Theorists have sought to reduce elastic phenomena as well to actions from molecule to molecule. To determine the equilibrium and motion of an elastic lamina that has been bent, we suppose that its restoring tendency at every point is inversely as the radius of curvature. But this law is a *derivative* fact, and follows from the *attractive and repulsive actions of molecules*, which are functions of distance. To show *how* the above law follows, we

multipliers are not as informative (about the microphysics of constraints) as a force law or a constitutive relation is. For related, helpful discussion, see chapters 7 and 8 in Duhem 1903.

must assume that, in an undeformed elastic body, each molecule is at rest in equilibrium under forces (attractive and repulsive) exerted on it by neighboring molecules. And, we must suppose that molecules tend to regain their equilibrium position if they become displaced by an infinitesimal amount. (Laplace 1898, 288–9)

Broadly speaking, their molecules had two features. They were discrete, not continuous—imperceptible masses separated by very small but finite distances. And, they interacted in just two generic ways (with many species), namely, by means of repulsive or attractive forces directed along the straight line between molecules.[24]

Hence, the Laplacians' commitment to ontic unity entailed that mechanics must be recast as a constructive theory: one must begin with the ontology— with an account of the preferred object, its powers and behaviors—and derive rational mechanics from it. To that end, they imposed two hard constraints on theorizing. First, appeal to molecules had to be *exact*. One was expected to start from some postulated law, or constitutive relation, for a species of molecular force. From it, one had to derive privileged quantities of some target phenomenon: the shape or height of a meniscal surface; the analytic expression for an elastic membrane in equilibrium; or the equation of motion for a vibrating string. Second, they expected the properties of molecules—the full list of their force laws and relevant parameters—to be given by *empirical physics*.

Here we must note a crucial fact. If successful, the Laplacians' ontic program would mark a watershed change in the overall project we have studied in this book: *it would relocate* BODY. The outcome of their constructive approach to theory-building would be a *physical mechanics*: a combination of rational mechanics and empirical physics. That novel project would no longer rest on philosophical physics as the notion had been understood from Descartes to Kant. It would no longer complement rational mechanics with resources supplied by a metaphysical theory of matter—an account of Nature and Action—articulated with the methods of philosophy. Instead, it would rely on the empirical methods of post-1800 mathematical physics for a picture of the micro-bodies expected to unify mechanics. Thus, Poisson's choice of words above, while polemically directed at Lagrange's theory, was much

---

[24] They were the actual parts of sensible bodies, offering the philosopher hope for recovering the bodies of our experience from this single, unified, underlying ontology.

farther reaching than that. We discuss this radical change in the Conclusions chapter of this book.

(i) Progress

The leaders of the ontic program believed, rightly or wrongly, that *all* results in rational mechanics obtained by 1800 could be recovered from facts about "molecules" and their force laws; even the mechanics of external constraints. So, they thought, in the new century their task was to *extend* mechanics to types of body and motion that 18th century work had not covered. In particular, to elastic true solids and viscous fluids.

Its initial success, and impetus for research after 1800, was in celestial mechanics. Starting from mass points interacting by direct, inverse-square gravity, Laplace had derived some impressive results in potential theory; specifically, on the local strength of attraction by spheroids. That gave him the confidence to try and use his unifying premise (of molecules interacting by central forces) at lower, sub-planetary scales. As they saw it, there were three areas then ripe for mathematics: capillary action, the motion of elastic true solids, and of viscous fluids. Accordingly, the Laplacians moved to extend their dominion over these untamed areas.

Capillary surfaces

From about 1805 onward, Laplace and others sought to infer the shape and height of the free surface that a liquid takes in a thin tube of arbitrary shape. The experimental impetus for this question was old, and other figures had been working on it (e.g., Thomas Young across the Channel). To treat it, Laplace set out to derive a formula for the total action at an arbitrary point on the bounding surface of a confined mass of fluid. Effort and skill led him to an expression for the action on a "molecule" in the fluid.[25] When that action is null, the point remains at rest. To find the meniscus shape, he just needs to integrate the formula for the action at a point.

Laplace's commitment to intermolecular forces enters this piece of theorizing at two junctures. He thinks the liquid in the tube is a mass in hydrostatic equilibrium under two forces: terrestrial gravity, which weighs

---

[25] The argument structure of Laplace's proofs is far from transparent, and secondary literature is unfortunately not much better; for an attempt at a synopsis, see Dhombres 1989, 65ff.

the fluid down; and the fluid's self-force, a two-way attraction between its molecules.[26]

Elasticity

Navier's two breakthroughs in elasticity above (see section 12.3) must count as progress for the ontic unity program as well. That is because he relied on the Laplacians' key idea about the constitution of extended bodies and their internal forces: "We regard an elastic solid body as an assemblage of matter molecules placed at extremely small distances."[27] Moreover, he explained, if one starts with molecules and their forces, the Lagrangian, virtual-work approach to the equations of motion is just one option. Another route to it is the balance-of-force approach.[28] It was this broadly Eulerian, non-Lagrangian approach that led him to his important advance below.

Fluid motion

Perhaps the greatest step forward for the ontic program in mechanics was Navier's derivation of the equation for a viscous fluid. All fluids treated in 18th century mechanics had been supposed frictionless. That assumption was known to be empirically false, but no one knew how to do better, until at last Navier broke through:

> When a fluid moves such that its molecules continually change relative distances, actions are being exerted—between these molecules, and also between them and the stationary molecules of the solid container. We must take these actions into account, if we are to know the fluid's true laws of motion. . . . My aim is to study the values of the forces due to intermolecular actions, which we must then put into the equations of fluid motion. (Navier 1821, 245f.)

[26] Admittedly, he does not know the law of these actions. However, Laplace bypasses that obstacle with a sleight of hand: he assumes that those forces are "sensible only at insensible distances"; they are given by a power-law function that decreases very rapidly (e.g., $1/r^4$ or higher). His characterization of those forces is a refrain that reliably identified his disciples at the time; they subscribed to it. See also Navier's reasoning below.

[27] Navier 1823b, 177.

[28] "Alternatively, we may consider the forces arising between a molecule M and all other molecules M' that surround it. Then we will assert that their resultant is equal to the accelerative [external] force applied at point M. Thereby, we obtain the equation that must hold at every point in the body [for equilibrium to ensue]" (Navier 1823b, 178).

His derivation relied not on PVV, but on a balance-of-force approach, in three stages. First, he focused on the state of a single "fluid molecule." As part of that, he supposed all other actions on it—the net external force and the internal pressure—to be known, or given. Second, he sought to quantify the action of viscosity in the fluid, which he called "force of adherence." Third and last, he added that action to the overall balance of force on the molecule. The resulting motion will be proportional to this net balance. As he explains, once found, the expression for viscosity "must be added, in the general equation of fluid motion, to the quantities of the forces supposed to act on the molecule."[29]

His breakthrough was at stage two above. It had two aspects. Navier took viscosity effects to be caused by an intermolecular force:

> I adopt the following principle: when two fluid molecules approach or re-cede from each other, there exists a mutual repulsion or attraction, whose strength depends on their relative speed of approach or recess. . . . by na-ture, this attraction or repulsion is a molecular force. (Navier 1821, 248)

Then he found a way to infer the net "force of adherence" on an arbitrary molecule M. Navier assumes the force is proportional to the interacting molecules' masses, their change of relative speed, and some distance func-tion such that the force decreases very rapidly above "sensible distances."[30] Next, he "composes" the intermolecular forces on M so as to find their re-sultant. Obscurely, he claims that the resultant comes out by *integration* over the region around M where viscosity is "sensible."[31] So, he concludes, after integration "there remain just the partial derivatives of the [molecule's] ve-locity, and a constant factor that is specific to the strength of intermolecular adherence in that particular fluid." In the x-direction,[32]

$$\sigma\left(3\frac{\partial^2 u}{\partial x^2}+\frac{\partial^2 u}{\partial y^2}+\frac{\partial^2 u}{\partial z^2}+2\frac{\partial^2 v}{\partial x\partial y}+2\frac{\partial^2 w}{\partial x\partial z}\right) \tag{6}$$

[29] Navier 1821, 251. By the "general equation" he means the formula (7) of Chapter 10, also known as Euler's Equation for an inviscid fluid.

[30] More precisely put, the force $V = m \cdot m' \cdot \Delta v \cdot f(r)$, where f is a power-law distance function like $f = 1/r^n$, and $n \gg 2$; v the relative speed; and r the distance.

[31] This claim is very significant, and we discuss its crucial importance below.

[32] See Navier 1821, 250f.; σ is a fluid-specific viscosity factor; u, v, and w are components of the molecule's velocity; and x, y, and z are its coordinates in the Eulerian description.

Call this quantity V. Navier's contribution to the problem of fluid motion was finding a way to quantify V; the rest came from Euler. That is, Navier added V to the equation of motion for an inviscid fluid—recall Euler's formula (7) in Chapter 10—and obtained:[33]

$$X + V - \frac{1}{\rho}\frac{\partial p}{\partial x} = \frac{\partial u}{\partial t} + \left(\frac{\partial u}{\partial x}u + \frac{\partial u}{\partial y}v + \frac{\partial u}{\partial x}w \right) \tag{7}$$

The insight behind it is: in an incompressible fluid, the net external force, the viscosity, and the local pressure gradient equal the total change in linear momentum, local and convective. It is the x-component of the Navier–Stokes equation, so called.[34]

Among the Laplacians, Poisson subsequently took up the results above, and re-derived them from somewhat different yet still broadly "molecularist" premises.[35] Notwithstanding differences in the starting premises for their individual derivations, the proponents of the molecular ontic unification program are responsible for the greatest advances in rational mechanics after 1800, as far as descriptive scope goes.

(ii) Problems of unification

The ontic agenda came to face its own problems, internal to its framework and approaches. Two in particular appear quite serious: their notion of molecule was *equivocal*; and at key junctures it became descriptively *otiose*.

Architecture

Laplacians took their molecules to be discrete: "disjoint masses" separated by very small regions of empty space, as Fourier explained. Now let us ask, what is the *geometry of mass distribution* in a molecule? It turns out that they

[33] Navier 1821, 252. X is the net external force in the x-direction, ρ the density, and p the internal pressure. The rest is as above, and also as in Chapter 10.

[34] To determine the fluid's motion completely, to Navier's formula "we must add the equation of continuity," that is, Euler's formula (6b) from Chapter 10 (Navier 1821, 252).

[35] The details are important per se, but not essential to the progress of the ontic unification program, so we forgo them here. A survey of Poisson's competition with his fellow Laplacians is Dahan Dalmedico 1992, chapter 10. There is a rich context to Navier's discovery—empirical, having to do with known departures from the inviscid-flow regime; and physical, pertaining to the Laplacian picture of molecules. For that context we again refer the reader to the admirable Darrigol 2002.

had no consensus answer. Some thought a molecule is zero-sized (i.e., a mass point). Others thought it is a deformable volume filled with mass, though very small. Yet others took it to be a rigid volume with a fixed shape, viz. an ellipsoid. They almost never stopped to ponder these differences in their individual pictures; but to the trained eye they *are* different. Cauchy and Saint-Venant endorsed the mass point:

> As to the dimensions of the atoms in which the centers of molecular force reside, we must regard them not as very small (relative to the distances between atoms), but as strictly zero. Put differently, these atoms—which are the true simples that compose matter—have no extension. (Cauchy 1868, 36)

> I think we must regard the ultimate particles of matter as being non-contiguous *points without extension*, viz. centers of action by way of repulsive and attractive forces. (Saint-Venant 1844, 8; his emphasis)

The rigid ellipsoid was Poisson's late picture of matter, caused by his need to account for the deformation of elastic solids:

> When two molecules, m and m', are not far enough from each other that their shape has no effect on their mutual action, m's action on m' is not necessarily along the line MM' [between their mass centers]. That action may not even amount to a single force. Its components will be functions of [their relative distance] r, to be sure. . . . But, they will also depend, in addition, on the angles that MM' makes with certain fixed sections inside m and m'. (Poisson 1831, 7)

Navier at times implied a molecule is finite and deformable, equipped with pores that can fill with "caloric," the fluid of heat:

> We take a solid body to be an assemblage of matter molecules at extremely small distances. These molecules exert on one another two contrary actions. One is an attractive force, proper to the molecules themselves. The other is a repulsive force due to the principle of heat. (Navier 1827, 375)

At the very least, the ontic unity group appeared split between a small fraction that argued for mass points, and a majority party that took them to be finite volumes. "Like Leucippus, Democritus and Epicurus, the majority of physicists nowadays regard atoms as small heaps of continuous, compact matter." They are "*extended* atoms," Saint-Venant noted.[36] The upshot is important: their agenda for unification had no determinate, univocal matter theory. The only tenet they shared was that molecules were discrete.

Descriptive import

Even that idea would get discarded at critical junctures. Cauchy's early success in elasticity and fluid dynamics (see below) led some Laplacians to retool their approach. To obtain the equations of equilibrium and motion in those two areas, they would start by assuming a discrete distribution of mass and velocity, which their commitment that matter is molecular required. But then right away they would average those discrete values (as contained in a small volume, say), and extrapolate the resulting averages to there being mass- and velocity quantities at *every* point inside that volume; see Fig. 12.1.

In effect, that replaces the initial assumption—discrete matter, or disjoint masses—with a continuum picture, in which mass fills that volume everywhere. In other words, to get their results, even the Laplacians had to reach for the methods of continuum modeling. That is, they ended up representing quantities at a point as elements of *continuous* extension: infinitesimal lines ds, areas dA, or volumes dV; and for whole-body quantities, they used integrals over a continuous area (its bounding surface) and over volumes inside those surfaces. For an example of that practice, look again at expression (2) above—Navier's equilibrium condition for an extended elastic solid body. He was a molecularist, and yet the condition he wrote is for a body supposed to be continuous.

In the Laplacians' mind, their replacement maneuver was just a mathematical workaround, not an abjuration of the faith. Still, it makes their theory of matter practically irrelevant. Their real tool for extending mechanics to new domains—by way of new concepts such as stress, strain, viscosity, bending moments, or elastic moduli—is a *field* theory: a mathematical picture of matter as filling space at every point in an extended body. In content, that picture is entirely equivalent to the one that continuum-matter theorists (such as Sophie Germain and Stokes) started with by design, not by accident. Ultimately then, the ontic program's overt appeal to discrete matter bits was just a ritual invocation—an empty credo belied by their mathematical

---

[36] Saint-Venant 1844, 7; his emphasis.

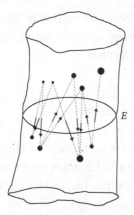 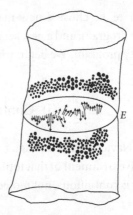

Fig. 12.1. Continuum values from discrete molecules. An intuitive indication of how Laplacians passed from molecules to mass density and internal force at a point in an extended body. Let E be a cross-section plane through a body. Left: Molecules as supposed by the Laplacian program: spatially discrete masses, exerting two-way forces on each other. At *select* points in E, the total force is non-zero, and would move a mass point (counterfactually placed there) in the direction of the resultant vector. Right: Continuous-field values—force density per unit area—at every point in the plane E, obtained by averaging the discrete values on the left (over some small area or volume). At *any* point in E, a mass placed there would be accelerated (in the direction of the average force-density vector at that point).

practice of treating matter as continuous when it mattered. And so, a *coherent* ontic unification in effect eluded their grasp.

12.5 The Cauchy package

From the early 1820s onward, both approaches to unity saw themselves defied. Initially, the challenge arose from the same quarter, driven by the same rationale, namely, the need to mathematize true solids. As it gathered steam and successes, however, the challenge came to threaten the two unification approaches separately, on their own terms. The single origin of these challenges was Cauchy's work in elasticity during the 1820s.[37] To make

---

[37] More exactly, between 1822 and 1827. The results below presuppose that matter is a deformable continuum, hence that bodies are extended volumes filled with mass everywhere. However, in terms of foundational commitments, this was a transient stage in Cauchy's work. He did not really think

progress there, he chose a new matter theory, at odds with the molecular ontic-unity program; and a new set of principles, at odds with the Lagrangian nomic-unity program. We begin with the former.

(i) Continuous matter

Soon after Navier's breakthrough paper on elastic solids,[38] Cauchy outlined an alternative treatment of that topic. Inter alia, his innovation was to assume that matter is continuous, not molecular-discrete (as Navier had supposed):

> In seeking the equations that express equilibrium conditions or laws of internal motion for solids and fluids, we may regard these bodies either as continuous masses whose density varies by insensible degrees from one point to the next; or as distinct material points, separated by very small distances. In a previous paper on fluids, I have treated them as continuous—as do the several mechanics treatises published so far. Now I take the same approach to solid bodies. (Cauchy 1890 [1828], 195)

That difference in pictures of matter is more drastic and momentous than it seems. It required Cauchy to rethink radically how matter acts by contact, and how it responds to such actions; we go over his innovations next.

Dynamics
In regard to material action, he took the Eulerian idea of internal pressure in a fluid, and generalized it so as to describe and quantify the action of internal forces in an elastic body. That is, Cauchy broadened the concept as follows. 1. In fluids, internal pressure is always normal.[39] But, in true solids it can be *oblique*. 2. In fluids, internal pressure at a point has the same value in every direction. However, in elastic solids it can have *different* values in different directions. 3. Though pressure in a solid can be oblique, its quantity is

that continuous matter was the true picture. "Cauchy's subsequent rallying to the paradigm of molecular mechanics was not a transient occurrence. The scientist persisted in that paradigm ever after that" (Dahan Dalmedico 1992, 295). The continuity of matter—in particular, of the luminiferous ether—was then a key assumption in related fields, e.g., wave optics; see Buchwald 1980.

[38] That paper was Navier 1827, which he really wrote in 1821, and presented at the Academy of Sciences in the same year.
[39] That is, perpendicular to any arbitrary surface passing through that point.

always a linear function of the normal vector at that point. See also Fig. 12.1. For his generalized idea above, later generations coined the term "stress" (Fr. *contrainte*, Ger. *Spannung*). Not yet having a word for it, Cauchy called it "pressure or tension":

> If, in a solid body (elastic or not) we let a small volume element become rigid and invariable, then on every side—and *at every point* on those sides—the element will experience a determinate pressure or tension. It is similar to the one that a fluid applies on an element in the solid body that holds it. But, there is a difference. The pressure that a stationary fluid exerts on a solid surface is *perpendicular* to it, and is everywhere independent of the solid surface's inclination [angle] relative to the planes of a coordinate system. However, the pressure or tension exerted *at a point* in a solid body—on a small area passing through the point—can be perpendicular or *oblique* to this surface. In the case of solid compression, it points inward from outside; in expansion, it points outward from inside. And, this pressure or tension *depends on the inclination* that the surface [on which it acts] has relative to the coordinate planes. (Cauchy 1889 [1822], 61; our emphasis)

Put modernly: in a deformable continuum, the stress at a point is a linear function of the normal unit vector to an arbitrary surface imagined to pass through that point; see also Fig. 12.2. In recognition of its enormous value, we call it Cauchy's Stress Principle.[40]

In the process he found that, to describe the local action of "pressure or tension," a new representational tool was needed. Specifically, to quantify the local action—on an infinitesimal part in the body, caused by the forces originating *in that body*—we must take into account the way those forces tend to stretch and shear that minimal part. In the x-direction,[41]

$$p \cos\lambda = A \cos\alpha + F \cos\beta + E \cos\gamma \qquad (8a)$$

---

[40] The phrase "Cauchy's Stress Principle" seems to come from Truesdell and Toupin 1960, 536ff. The Stress Principle is the gateway to modern continuum mechanics, and that is why we gave its genesis a close look. Dahan Dalmedico (1992, 244–9) explains the reasoning that led Cauchy to infer that stress is a function of the normal to the surface at that point.

[41] See Cauchy 1889, 68. p is the net stress vector on that point, and $\cos\lambda$ is its direction cosine, which thus projects it onto the x-axis. Cauchy's point is that p results from *three* actions: a pressure-like force A that presses or pulls on any material surface S passing through that point, and two forces F and E that tend to shear the surface in the direction of two planes perpendicular to S.

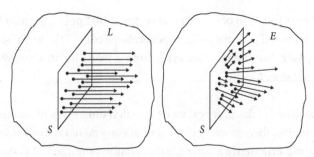

Fig. 12.2. Stress according to Cauchy. An intuitive picture of how Cauchy stress differs from hydrostatic pressure. Left: S is a plane section in an isotropic, homogeneous fluid at rest. Suppose the fluid mass on the right side of S was annihilated. Then all fluid particles on the left would be accelerated equally, in the same direction, toward L. Right: S is a plane section in an elastic body deformed by forces applied on its surface. Suppose the body mass on the right side of S was annihilated. Then the solid particles on the left would be accelerated in different directions and at different speeds, depending on how the body had been deformed.

Mutatis mutandis for the other two components. In synoptic hindsight we can grasp that, to fully represent how Cauchy's "pressure or tension" acts *at a single point in a continuous body*, we must write:

$$
\begin{matrix}
A_x & F_x & E_x \\
F_y & B_y & D_y \\
A_z & D_z & C_z
\end{matrix}
\qquad (8b)
$$

While he lacked the term, Cauchy however grasped the relevant idea: the stress is a *tensor* quantity. In mathematical structure, it differs from "Newtonian" forces, that is, the vector quantities (like gravity) that tend to displace a free particle along some curve. Rather, Cauchy stress—the action of contact forces at a point inside an extended body regarded as continuous, or filled with mass everywhere—needs *nine* different "Newtonian" actions to specify completely: three perpendicular forces, and six shearing forces .

To make his new idea intuitive, Cauchy gave it a geometric interpretation as well. He represented "pressure or tension" at a point M by means of an ellipsoid of revolution centered on it. Any radius vector in the ellipsoid measures the stress in that direction; the longest semiaxis corresponds to the maximum stress at M, and the shortest to the minimum.[42]

Kinematics

Mathematizing solid continua faced a head-on challenge: how is motion inside a continuous body to be *described*, really? The language of "material points" is vague and misleading; it seduces the reader into thinking, wrongly, that early modern kinematics was sufficient. It was not—a novel, sharper approach was needed.[43] Cauchy took the main steps toward the right answer. He considered two points X and Y in the body, very close to each other; and quantified the deformation that the *material fiber* XY can undergo.[44] He inferred that, for an infinitesimal fiber, its kinematic change (over an instant, in response to stresses) equals $\partial s/\partial x_i$, the partial derivatives of its strained length and direction s relative to its unstrained, "natural" configuration x_i.

Furthermore, just as he did with stress, Cauchy also gave a geometric interpretation of "dilation or condensation" (i.e., strain). "As I will prove, the condensations or dilations around a point, plus or minus 1, become equal (up to a sign) to the vector radii of an ellipsoid."[45]

These results turned out to matter enormously in two respects. For one, their heuristic power (to treat solids and fluids) was so great that it forced the Laplacians to reverse-engineer it from discrete matter foundations. Namely, they had to start with molecules and then find ways to represent their properties as *continuously*-distributed values, both at a point and over the whole body; again, see Fig. 12.1. For another, Cauchy's conceptual

[42] That geometric device is known as the Cauchy Stress Ellipsoid; for some details, see Dahan Dalmedico (1992, 245ff.).

[43] Here is why. Single points have just three degrees of freedom. "Points" in a continuum have an infinity of them, because they are really infinitesimal volumes dV that stresses can deform in endless different ways. This is something that a single segment—the only way to represent a displacement, before Cauchy's work—cannot adequately represent. At this juncture, our account is perforce fragmentary. The history of continuum kinematics remains mostly uncharted territory. Notes on its early stages are scattered here and there, in pieces on dynamics for continua chiefly concerned with the birth of stress concepts.

[44] In continuum mechanics, a fiber (also called a material line) of length l is a one-dimensional segment with mass values at every point in it. Hence, a fiber has a density ρ, and so its net mass is ρl. Clearly, a continuous fiber is different from two molecules at a distance l from each other. Their net mass is $m_1 + m_2$, and between them there is just empty space.

[45] Cauchy 1823, 11. Let M be a point in an elastic solid, and consider all the infinitesimal fibers that "sprout" from it, or have M in common. The action of stress on these fibers, he proved, can be represented by an ellipsoid centered at M. That ellipsoid has three semiaxes; one is the longest, and one the shortest. Those semiaxes measure the principal strains (i.e., the maximum and minimum changes of length in the fibers at M). A radius vector in that ellipsoid measures the changed length and orientation of some fiber at that location. For additional discussion, see Dahan Dalmedico 1992, 251–4.

innovation (stress and strain) fits very well with a dynamical law quite different from PVV, as we explain below. Together, his new matter theory and dynamical principle yielded a framework that challenged forcefully both approaches to unity we have examined so far.

(ii) Balance laws

Cauchy innovated in yet another respect. While his premises above challenged the molecular ontic program, his approach below threatened the virtual-work program for nomic unification. In particular, he took up and broadened Euler's old path to the dynamics of extended bodies, namely, the balance-of-force approach. Recall the generic idea behind that approach: the kinematic response of matter equals the net balance of the forces on it. Various subfields of mechanics simply adapt this genus idea to the specifics of the object type at issue. Newton had taken the first step, for the case of a free particle. Euler had applied it to a rigid body and to a volume of ideal fluid, respectively. Cauchy extended it to an elastic solid and a viscous fluid. We may distinguish two prongs to his demarche: the conceptual insight and the heuristic.

The heuristic amounts to this. To find out when an interior point remains at rest—or how it moves, if equilibrium is broken—Cauchy carved up mentally a small volume (in an elastic body) and studied two kinds of forces: those acting on its boundary surface (i.e., the "pressures or tensions" explained above); and the "external" forces, viz. those originating in sources outside the body, such as gravity. These latter forces reach deep inside the volume, and act on its mass center; they are actions at a distance. Once he had quantified the strength and direction of all these forces, he mentally shrank the volume into an infinitesimal element of integration, dV. At this juncture, Cauchy summed the two kinds of force acting on the element. That gave him the balance of forces at a point.[46]

---

[46] Many call this heuristic the "Euler-Cut" approach. However, Euler never knew to apply it in elasticity, where it matters most (for evidence, see Fraser 1991). There are some traces of him using it in fluid dynamics, but the matter is inconclusive. Cauchy made the best, clearest use of this method, thus showing its true heuristic value; see Dahan Dalmedico 1992 and Casey 1992.

The conceptual breakthrough builds on this approach. In essence, a volume element in an elastic body remains in equilibrium if the balance of forces on it vanishes:[47]

$$t + B = 0 \tag{9a}$$

In component form, the element stays at rest in the x-direction if:[48]

$$\left[\frac{\partial A}{\partial x} + \frac{\partial F}{\partial y} + \frac{\partial E}{\partial z} \right] + \rho X = 0 \tag{9b}$$

But if the net force on it is not zero, the volume element moves—in the manner specific to deformable continua. The balance-of-force approach dictates that two aspects of this process (the net action and the motion response) are equal:[49]

$$\text{Total force = Rate of linear momentum} \tag{10}$$

Inspired by Navier's success above, Cauchy's insight was that "pressure or tension" (i.e., stress) belongs on the left side above, next to gravity. This extends Euler's idea to the difficult case of deformable continuous matter. Then Cauchy produced an analogue of his law above, but for the balance of torques.[50] Because his generalization required him to devise a new

[47] In the terminology of modern continuum mechanics, t is the traction, or internal contact force, and B is the (external) body force, exerted by exogenous sources at every point inside a continuous body (e.g., gravity or magnetic force).

[48] X is the acceleration of the net external force on the material point. In square brackets is the total stress at that point relative to the Ox axis (i.e., the x-component of t above). That stress component has a "tensile" part $\partial A/\partial x$, which tends to shrink or stretch the fiber on which it acts; and two shearing components ($\partial F/\partial y$ and $\partial E/\partial z$) which turn the fiber toward the Oy and Oz axes, respectively. The other term is the x-component of the net force B external to the body. Expressed in modern form, Cauchy's result writes as $\nabla \cdot t + \rho B = 0$. See also Dahan Dalmedico 1992, 257ff.

[49] See Cauchy 1891, 343.

[50] More precisely, after 1841 Cauchy understood that, in addition to deformation, stresses also induce a rigid rotation (of the volume element, around its mass center). In consequence, Cauchy supplemented his law (10) above with an analogue for torques, or causes of change in angular momentum. That part of his mechanics has not been studied much yet, but see the quick overview in Dahan Dalmedico 1992, 254–6, and Belhoste 1991, chapter 12.

representational framework (for kinematics and dynamics), some historians think they deserve to be called Cauchy's Laws of Motion.[51]

In structure, content, and scope, balance laws are very different from virtual-work principles. The latter assert that a certain action (the virtual work of the generalized forces) vanishes across the target system. The former say that the net dynamical action (of forces and torques) equals the net kinematic change. Crucially, balance laws have the old blind spot of Newton–Euler dynamics—they have no associated recipe for solving *external* constraints.

(iii) Impact on philosophical mechanics

The alternative we presented above complicated the chances that a unified philosophical mechanics would emerge after 1840. The Cauchy package—continuous matter and his two balance laws—counts as a formidable challenge to the two unification programs that came before it.

On the one hand, it offered an alternative to Laplace's ontic approach. Cauchy did not himself believe in continuous matter; for various reasons, his loyalty was to discrete molecules.[52] And yet, his continuum-based work in mechanics escaped his orbital pull, as it were, and acquired a life of its own. In the long run, the deformable continuum became the standard basis for elasticity, fluid dynamics, and their later offshoots: plasticity; fracture, damage, and creep; mixtures and the mechanics of porous media; and the theory of Cosserat continua. In these areas, molecular foundations eventually gave way.[53]

Returning to Cauchy's historical context, to conquer elastic solids and viscous fluids, the Laplacians had to make two critical moves. One was mathematical: they started from discrete molecules, and inferred to claims about

---

[51] The term was introduced by Truesdell (1991, 182ff.). Our formula (10) expresses Cauchy's First Law in qualitative terms. Cf. Chapter 10; we might call it the Euler–Cauchy First Law. His Second Law conveys the same insight, but concerning torques and the change in angular momentum they induce.

[52] See his accessible account of them in Cauchy 1868.

[53] Nowadays they seek to regain some access through a side entrance, as it were. Since the mid-1950s there have been attempts to get concepts of stress and strain from "atomistic foundations" (i.e., from mass points). The idea is to make modern continuum mechanics explanatorily consilient with the kinetic theory of matter, which regards matter as discrete. See Murdoch 2010 for a latter-day expression of that project; incidentally, it shows how simplistic the Laplacians' workarounds at the time were.

continuous bodies. Another was conceptual: they started from a notion of mass (qua property of molecules) and along the way they quietly switched to density qua effective parameter in the equation of motion.[54]

These moves were indispensable to their work—without them, the relevant equations of motion do not follow. Or, if they follow, the Laplacians did not take the trouble to prove it. And yet, they had no clear, compelling justification for those moves. Absent that justification, the moves appear as a sleight of hand, not natural steps in a proof. Cauchy's approach requires no such costly prestidigitation. In it, *all* whole-body quantities are integrals, because a Cauchy body is a continuous volume. And, it starts from density ab initio, because the parts of his bodies (the volume elements dV) have density as a primitive feature; mass in their case is a derivative property. Thus, Cauchy's mechanics is conceptually coherent and logically impeccable to an extent that the Laplacians' ontic program could not match. As an alternative to molecular *ontic* unity, it presented the Laplacians with a major challenge.

On the other hand, the Cauchy package also offers an alternative to Lagrangian *nomic* unity: by means of balance laws. Cauchy-type breakthroughs in hydrodynamics and elasticity served to make the balance-of-force approach into an ever more credible competitor for the coveted prize—unifying mechanics from a single principle. The insight behind his formula (10) is the core of Newton–Euler–Cauchy dynamics, which unifies vast swathes of classical theory from a balance law of impressed force, rather than from a virtual-work principle. In addition to Cauchy's own results above, his two laws of motion entail all the equations of motion that 18th century figures had obtained from balance-of-force approaches: d'Alembert's wave equation (plus its extension to membranes); Euler's rigid-body dynamics; and the hydrodynamics of ideal fluids. In that respect, it can unify *nearly* as much mechanics as Lagrange was able to, but from different foundations. And, where it succeeds, Cauchy unification does *not* need to bracket the dynamics into geometry. It represents internal constraints directly and *explicitly* as effects of forces: internal forces such as elastic stresses, fluid pressure, and viscosity. It just requires a stronger mathematical framework for describing their actions. Indeed, as it turns out, the resurgence of impressed-force approaches slowly exiled virtual-work principles to areas

---

[54] For edification, consider contrastively Navier's starting assumption about the strength of viscosity V as an intermolecular force dependent on discrete *masses*, with his use of continuous *density* in the Navier–Stokes equation he inferred above.

where they have remained impregnable: the statics of structures, engineering dynamics, and the mechanics of constrained rigid bodies.[55] And so, as an alternative to Lagrangian nomic unity, achieving nomic unity via Newton–Euler–Cauchy dynamics offers strong competition.

12.6 Conclusions

In the years from Lagrange to Poisson, rational mechanics continued to make enormous advances in descriptive scope. It really was the period when mechanics came close to mathematizing the behavior of bodies as we know them—extended and deformable. Driving that leap in breadth of coverage were two competing projects that aimed to unify all of mechanics. They differed starkly in what the sought unity amounted to. For the Lagrangians, unity was nomic: mechanics is one if all the local accounts are strictly deduced from one basic principle, of virtual work. For the Laplacians, unity was ontic: mechanics counts as one theory if all local accounts—all equations of motion and equilibrium conditions—are inferable from facts about a micro-body they called a molecule.

After a period of initial élan, however, both approaches to unity came to face significant conceptual obstacles. The nomic program lacked a clear link to a picture of matter. The ontic approach began to show logical cracks in the foundations. As if to compound their respective predicaments, these approaches came to face competition from a robust outsider. Cauchy showed how to make crucial progress in expanding mechanics by means of a dual foundation—continuous matter and balance laws of force—at odds with the two unification programs above.

Their three-way competition had no clear winner. That entails a sobering fact, namely, disunity. By the mid-19th century, there was no single approach, whether ontic or nomic, that could unify *all* rational mechanics demonstrably, viz. by entailing its full gamut of differential equations. In effect, despite best efforts to achieve unity, rational mechanics had become disunified, in both its principles and its ontology.

---

[55] Again, we must restate that Newton–Euler–Cauchy approaches do *not* work for systems with *external* constraints. The virtual-work approach remains uncontested there. For historically informed glimpses into the unifying virtues of PVV in those areas, see Heyman 1996 (for statics) and Papastavridis 2002 (for analytic dynamics).

This fact leaves philosophical mechanics in a predicament when it comes to BODY. That is because, absent a single, unified mechanics (whether rational or physical, whether unified nomically or ontically), it seems there is little prospect of elaborating a single, unified concept of body, as Goal requires.[56]

[56] See Chapter 1.

Conclusions

At the beginning of the 18th century, physics was a subdiscipline of philosophy and its primary task was to solve the problem of bodies (*BODY*). By the early 1800s, this was no longer the case. Physics had become an independent discipline, and *BODY* was no longer its driving concern. In this book, we have argued that the philosophical reasons for this transformation—and its chief consequences—come into view if the 18th century is analyzed as an era of *philosophical mechanics*. That is, as an age of widespread, long-lasting, and concerted efforts to address *BODY* through the integration of *rational mechanics* into *philosophical physics*.

We built our case slowly, chapter by chapter, displaying the rich evolution of philosophical mechanics in the Age of Reason. We have already previewed our conclusions from each of these chapters (see Chapter 1), and so we will not reprise them here. Instead, looking back over the whole, we draw out some conclusions for philosophical mechanics in the Age of Reason.

1. The 18th century closed without a philosophical mechanics capable of addressing *BODY*. One reason for this was the enormous difficulty of achieving a rational mechanics for extended bodies in motion. At the start of the century, it seemed that incorporating the rules of collision should be sufficient for the purposes of philosophical mechanics (the problem of collision, or PCOL). By mid-century it was clear that much more was needed; specifically, the incorporation of a rational mechanics of constrained motion (the problem of constrained motions, or PCON). The theory of constraints developed rapidly during the later decades of that century, and the mathematical demands of this theory placed it out of reach of most philosophers at the time.

This brings us to the second reason: the *methods* of the philosophers proved inadequate by themselves for determining the nature and properties of body—or of matter in general—consistent with the demands of rational mechanics. Kant explicitly attempted the task, but with PCOL only, not PCON, in mind; and so he failed. Even Boscovich, who explicitly attempted

Philosophical Mechanics in the Age of Reason. Katherine Brading and Marius Stan, Oxford University Press.
© Oxford University Press 2023. DOI: 10.1093/oso/9780197678954.003.0013

to provide mid-century d'Alembertian rational mechanics with a physics, fell short. By the early 19th century, *empirical* rather than *philosophical* resources had become the most important source and justification for the physical commitments being integrated into rational mechanics. No longer a *philosophical mechanics*, this is the *physical mechanics* of Laplace and Poisson (see Chapter 12).

The result was a disconnect between the philosophers who continued to pursue BODY (such as Boscovich and Kant) and those such as Lagrange, Laplace, Navier, and Cauchy, whom we regard today as physicists.

2. Physics emerged at the *end* of the century as an independent discipline, with its own resources, methods, criteria for success, and also its own community of researchers. In resources, it no longer deferred to philosophy for either its ontology or its principles. For its methods, it combined those of the mathematician (in rational mechanics) with empirical methods—for the articulation and justification of both principles and ontology, and for criteria of success.

Contrary to narratives in which Newton's *Principia* is the culmination of the scientific revolution, ushering in a stable period of classical physics, in reality "physics" as we know it took another century or more to emerge. Though with hindsight the *Principia* fits the description of the new, independent physics given here, that hindsight is misleading. The *Principia* is consistent with philosophical physics *succeeding* in addressing BODY, and thereby providing (from *outside* Newton's book) an account of the bodies to which Newton's laws apply. The *Principia* sits at the *beginning* of a golden era of philosophical mechanics, when it seemed that things might go very differently, and philosophers might succeed in their task of solving BODY. It was only after a further century of struggle with BODY that physics separated itself from philosophy, and this for all the reasons we have detailed in this book.

3. Though our story ends without a solution to BODY, the problem does not go away. Bodies are all around us: from pebbles to planets, tigers to tables, and pine trees to—most importantly for many areas of philosophy—*people*. Philosophers relied on philosophical physics to provide a general account of the nature of bodies, one that could be presupposed in any other area of philosophy in which human bodies and bodily action play a role. But the task has now fragmented into two. First, as the latter chapters of our book have made clear, by the early 1800s "bodies" were no longer the presumptive objects of physics. The Problem of Bodies has become the more general

Problem of Objects (OBJECT), and the ontic disunity described in Chapter 12 shows that no solution was then in sight.[1] Second, even given a solution to OBJECT, the task remained to construct from this ontology the bodies of our experience.

The philosopher interested in BODY will have to tend to both of these tasks. As we have seen, their methods—the approaches and sources of evidence domestic to philosophy—by themselves had proven inadequate. Therein lie deep and lasting epistemological lessons: about our epistemic situation in the world, the methods whereby we may obtain knowledge of the physical world, and the kinds and limits of that knowledge. The philosopher will need to revise the Goal set out in Chapter 1 to reflect the dual task that they face. They will need the methods and results of both rational mechanics and the new physics if they are to make progress. And they will need interpretive tools: BODY is *their* problem, and no longer that of the physicist; they will need to master the details of the physics so as to unpack the philosophical moves and content as they pertain to BODY. For this, we have suggested, examining the interplay between principle and constructive approaches to theorizing provides an important means of extracting philosophical insight.

Nature and Action also require revision. The task of determining the "[1] essential properties, [2] causal powers, and [3] generic behaviors" of bodies has been transformed into the task of ascertaining [1] a set of parameters by which to specify the state of an object (or system) and [2] the dependence relations among those parameters sufficient to determine [3] changes in the state of that system, as for example in its spatiotemporal evolution. This discharges the demands of Nature, and the lesson of the 18th century is that we cannot hope to do more.[2] What we can know about the physical world depends on the physical details of us as epistemic agents, and of that world itself in which we are embedded. There is no way to go beyond these limitations, to "know more" or to somehow know other than this humanly-accessible knowledge. And so, for the purposes of OBJECT, Action is fully encompassed by the articulation of the dependence relations among the parameters; what remains is to relate this to an account of causation among the bodies of our experience.

---

[1] See Chapter 1 for the explanation of OBJECT.

[2] Notice that in addition to the constructive route for addressing Nature we now have the principle route, and this opens the way to domain-specific, effective theories for which there is no underlying constructive theory (but which nevertheless should not be thought of as "instrumental" just because they are domain-specific).

Clearly, in the wake of the extraordinary developments of the 18th century, there is much for the philosopher to do as they try to develop a philosophical mechanics. But these challenges lie ahead, in the 19th, 20th, and 21st centuries. As for the 18th century, the Age of Reason, it was remarkable for the evolving relationships between philosophy, physics, and mechanics. And for the project we have called philosophical mechanics, it was a golden era.

Bibliography

Aiton, E. 1972. *The Vortex Theory of Planetary Motions*. New York: Elsevier.

Alexander, H. G., ed. 1970. *The Leibniz–Clarke Correspondence*. Manchester University Press.

Anonymous. 1688. Review of I. Newton, *Philosophiae Naturalis Principia Mathematica*. *Journal des Sçavans* 16: 237–8.

Antman, S. 1980. The equations for large vibrations of strings. *American Mathematical Monthly* 87: 359–70.

Arnold, D. 1983–4. The *Mécanique Physique* of Siméon Denis Poisson. *Archive for History of Exact Sciences* 28: 243–87, 343–67; 29: 37–94.

Bailhache, P. 1975. Introduction. *L. Poinsot: La théorie générale de l'équilibre et du mouvement des systèmes*. Paris: Vrin.

Barroso Filho, W. 1994. *La mécanique de Lagrange: principes et méthodes*. Paris: Karthala.

Baumeister, Chr. 1747. *Elementa philosophiae recentioris*. Leipzig.

Beeson, D. 1992. *Maupertuis: An Intellectual Biography*. Oxford: Voltaire Foundation.

Belhoste, B. 1991. *Augustin-Louis Cauchy*. Springer.

Belhoste, B. 2003. *La formation d'une technocratie: L'École Polytechnique et ses élèves de la Révolution au Second Empire*. Paris: Belin.

Bernoulli, D. 1738. *Hydrodynamica*. Strasbourg.

Bernoulli, D. 1746. Nouveau problème de mécanique. *Mémoires de l'académie des sciences* 1: 54–70.

Bernoulli, Jakob. 1691. Demonstratio centri oscillationis, ex natura Vectis. *Acta Eruditorum* July: 317–21.

Bernoulli, Jakob. 1703. Démonstration générale du centre de balancement ou d'oscillation, tirée de la nature du Levier. *Histoire de l'Académie des sciences de Paris*, 78–83.

Bernoulli, Johann. 1715. De centro turbinationis inventa nova. *Acta Eruditorum* June: 242–57.

Bernoulli, Johann. 1717. Nouvelle théorie du centre d'oscillation. *Histoire de l'Académie Royale des Sciences*, 208–30.

Bernoulli, Johann. 1727. *Discours sur les loix de la communication du mouvement*. Paris.

Bernoulli, Johann. 1742a. Hydraulica. *Opera omnia* 4: 387–493. Lausanne.

Bernoulli, Johann. 1742b. Meditationes de chordis vibrantibus. *Opera omnia* 3: 198–210. Lausanne.

Bernoulli, Johann. 1742c. Problema statico-dynamicum. *Opera omnia* 4: 332–41. Lausanne.

Bernoulli, Johann. 1746. Meditatio de natura centri oscillationis. *Opera omnia* 2: 168–86. Lausanne.

Bertoloni Meli, D. 1993. *Equivalence and Priority: Newton vs Leibniz*. Oxford: Clarendon.

Besterman, Th. 1959. *Les lettres de la Marquise du Châtelet*, 2 vols. Geneva: Institut et Musée Voltaire.

Biener, Z. 2018. Newton's 'Regulae philosophandi.' *The Oxford Handbook of Newton*, edited by Chr. Smeenk and E. Schliesser. Oxford University Press. https://doi.org/10.1093/oxfordhb/9780199930418.013.4

Biener, Z., and E. Schliesser. 2017. The certainty, modality, and grounding of Newton's laws. *The Monist* 100: 311–25.

Biener, Z., and C. Smeenk. 2012. Cotes's queries: Newton's empiricism and conceptions of matter. *Interpreting Newton: Critical Essays*, edited by A. Janiak and E. Schliesser, 105–37. Cambridge University Press.

Boscovich, R. 1745. *De viribus vivis*. Rome.

Boscovich, R. 1922. *A Theory of Natural Philosophy*. Translated by J. M. Child. Chicago: Open Court.

Boss, V. 1972. *Newton and Russia*. Harvard University Press.

Brading, K. 2012. Newton's law-constitutive approach to bodies: a response to Descartes. *Interpreting Newton: Critical Essays*, edited by A. Janiak and E. Schliesser, 13–32. Cambridge University Press.

Brading, K. 2013. Three principles of unity in Newton. *Studies in History and Philosophy of Science* 44: 408–15.

Brading, K. 2018. Newton on body. *The Oxford Handbook of Newton*, edited by C. Smeenk and E. Schliesser. Oxford University Press. https://academic.oup.com/edited-volume/34749/chapter-abstract/296599802?redirectedFrom=fulltext&login=false

Brading, K. 2019. *Emilie Du Châtelet and the Foundations of Physical Science*. New York: Routledge.

Brading, K. forthcoming. Newton's *Principia* and philosophical mechanics. *Theory, Evidence, Data: Themes from George E. Smith*, edited by M. Stan and Chr. Smeenk. Springer.

Brading, K., and M. Stan. 2021. How physics flew the philosophers' nest. *Studies in History and Philosophy of Science* 88: 312–20.

Breidert, W. 1983. Leonhard Euler und die Philosophie. *Leonhard Euler, 1707–1783: Beiträge zu Leben und Werk*, edited by J. Burckhardt, E. Fellmann, and W. Habicht, 447–58. Basel: Birkhäuser.

Broman, T. 2012. Metaphysics for an enlightened public: the controversy over monads in Germany, 1746–1748. *Isis* 103: 1–23.

Brown, H. 2005. *Physical Relativity: Space–Time Structure from a Dynamical Perspective*. Oxford University Press.

Brown, H., and P. Pooley. 2006. Minkowski space–time: a glorious non-entity. *The Ontology of Spacetime*, edited by D. Dieks, 1: 67–89. Elsevier.

Brunet, P. 1952. *La vie et l'œuvre de Clairaut*. Paris: Presses Universitaires de France.

Bub, J. 2000. Quantum mechanics as a principle theory. *Studies in History and Philosophy of Modern Physics* 31: 75–94.

Bucciarelli, L., and N. Dworsky. 1980. *Sophie Germain: An Essay in the History of the Theory of Elasticity*. Dordrecht: D. Reidel.

Buchwald, J. Z. 1980. Optics and the theory of the punctiform ether. *Archive for History of Exact Sciences* 21: 245–78.

Calero, J. S. 2008. *The Genesis of Fluid Mechanics 1640–1780*. Translated by J. H. Watson. Springer.

Calinger, R. 1969. The Newtonian–Wolffian controversy: 1740–1759. *Journal of the History of Ideas* 30: 319–30.

Calinger, R. 1976. Euler's 'Letters to a Princess of Germany.' *Archive for History of Exact Sciences* 15: 211–33.

Calinger, R. 2016. *Leonhard Euler: Mathematical Genius in the Enlightenment*. Princeton University Press.

Cannon, J., and S. Dostrovsky. 1981. *The Evolution of Dynamics*. Springer.

Caparrini, S. 2014. The history of the *Méchanique analitique*. *Lettera Matematica* 2: 47–54.

Caparrini, S. forthcoming. Remarks on *Mecanique Analytique*. *Theory, Evidence, Data: Themes from George E. Smith*, edited by M. Stan and Chr. Smeenk. Springer.

Capecchi, D. 2012. *History of Virtual Work Laws*. Basel: Birkhäuser.

Carré, L. 1706. Des loix du mouvement. *Histoire de l'Académie Royale des Sciences*, 442–61.

Casey, J. 1992. The principle of rigidification. *Archive for History of Exact Sciences* 43: 329–83.

Cauchy, A. L. 1823. Recherches sur l'équilibre et le mouvement intérieur des corps solides ou fluides, élastique ou non élastique. *Bulletin de la Société Philomatique*, 9–13.

Cauchy, A. L. 1868. *Sept leçons de physique générale*. Edited by F. N. Moigno. Paris.

Cauchy, A. L. 1889. De la pression ou tension dans un corps solide. [1822] *Oeuvres complètes d'Augustin Cauchy*, 2:7: 60–78. Paris.

Cauchy, A. L. 1890. Sur les équations qui expriment les conditions d'équilibre ou les lois du mouvement intérieur d'un corps solide élastique ou non élastique. [1828] *Oeuvres complètes d'Augustin Cauchy*, 2:2: 195–226. Paris.

Cauchy, A. L. 1891. Sur l'équilibre et le mouvement intérieur des corps considérés comme des masses continues. [1829] *Oeuvres complètes d'Augustin Cauchy*, 2:9: 342–72. Paris.

Chalmers, A. 2017. *One Hundred Years of Pressure: Hydrostatics from Stevin to Newton*. Springer.

Child, J. M. 1922. Introduction to *A Theory of Natural Philosophy* by R. Boscovich, xi–xviii. Chicago: Open Court.

Clairaut, A. 1739. Solution de quelques problèmes de dynamique. *Histoire de l'Académie royale des sciences*, 1–22. Paris.

Clairaut, A. 1745. Sur quelques principes qui donnent la solution d'un grand nombre de problèmes de dynamique. *Histoire de l'Académie royale des sciences*, 1–52. Paris.

Clatterbaugh, K. 1999. *The Causation Debate in Modern Philosophy, 1637–1739*. New York: Routledge.

Condillac, E. 1749. *Traité des sistêmes*. The Hague.

Cournot, A. 1827. Extension du principe des vitesses virtuelles au cas où les conditions des liaisons sont exprimées par inégalités. *Bulletin des sciences mathématiques, astronomiques, physiques et chimiques* 8: 165–70.

Crosland, M. 1967. *The Society of Arcueil*. Harvard University Press.

Crousaz, J. P. 1721. Dissertation sur les causes du ressort. Bordeaux.

Crousaz, J. P. 1752. Discours sur le principe, la nature, et la communication du mouvement. *Recueil des pièces qui ont remporté les prix de l'Académie royale des sciences*. Paris.

Dahan Dalmedico, A. 1990. Le formalisme variationnel dans les travaux de Lagrange. *La Mécanique analytique de Lagrange et son héritage*, edited by A. Lichnerowicz, 81–106. Turin: Accademia delle Scienze di Torino.

Dahan Dalmedico, A. 1992. *Mathématisations: Augustin-Louis Cauchy et l'École Française*. Paris: Editions du Choix.

d'Alembert, J. 1743. *Traité de dynamique*. Paris.

d'Alembert, J. 1744. *Traité de l'équilibre et du mouvement des fluides*. Paris.

d'Alembert, J. 1747. Recherches sur la courbe que forme une corde tenduë mise en vibration. *Histoire de l'Académie royale des sciences et des belles-lettres de Berlin*, 214–19. Berlin.

d'Alembert, J. 1749. *Recherches sur la précession des équinoxes, et sur la nutation de l'axe de la terre, dans le systême Newtonien*. Paris.

d'Alembert, J. 1754. Corps. *Encyclopédie ou Dictionnaire raisonné des sciences, des arts et des métiers* 4: 261–63. Paris.

d'Alembert, J. 1759. *Essai sur les éléments de philosophie*. Paris.

d'Alembert, J. 1995. *Preliminary Discourse to the Encyclopedia of Diderot*. Translated by R. N. Schwab. University of Chicago Press.

Darrigol, O. 2002. Between hydrodynamics and elasticity theory: the first five births of the Navier-Stokes Equation. *Archive for History of Exact Science* 56: 95–150.

Darrigol, O. 2005. *Worlds of Flow*. Oxford University Press.

Darrigol, O., and U. Frisch. 2008. From Newton's mechanics to Euler's equations. *Physica D: Nonlinear Phenomena* 237: 1855–69.

Descartes, R. 1637. *Discours de la methode pour bien conduire sa raison, & chercher la verité dans les sciences*. Leiden.

Descartes, R. 1991. *Principles of Philosophy*. Translated by V. R. Miller and R. P. Miller. Dordrecht: Kluwer.

Detlefsen, K. 2019. Du Châtelet and Descartes on the role of hypothesis and metaphysics in science. *Feminist History of Philosophy: The Recovery and Evaluation of Women's Philosophical Thought*, edited by E. O'Neill and M. Lascano, 97–128. Springer.

Dhombres, J. 1989. La théorie de la capillarité selon Laplace. *Revue d'histoire des sciences* 42: 43–77.

Downing, L. 2012. Maupertuis on attraction as an inherent property of matter. *Interpreting Newton: Critical Essays*, edited by A. Janiak and E. Schliesser, 280–98. Cambridge University Press.

Du Châtelet, E. 1740. *Institutions de physique*. Paris. Translated into English as *Foundations of Physics* (in her 2009, 2014, 2018b).

Du Châtelet, E. 1958. *Les lettres de la Marquise Du Châtelet*. Edited and translated by T. Besterman, 2 vols. Geneva: Institut et Musée Voltaire.

Du Châtelet, E. 2009. *Selected Philosophical and Scientific Writings*. Edited and translated by I. Bour and J. Zinsser. University of Chicago Press.

Du Châtelet, E. 2014. On the divisibility and subtlety of matter. *Philosophy, Science, and History: A Guide and Reader*. Edited and translated by L. Patton, 332–42. New York: Routledge.

Du Châtelet, E. 2018a. *La correspondance*, edited by U. Kölving and A. Brown, vol. 1. Ferney-Voltaire, France: Centre international d'étude du XVIIIe siècle.

Du Châtelet, E. 2018b. *Foundations of Physics*. Translated by K. Brading et al. Available at www.kbrading.org.

Ducheyne, S. 2012. 'The Main Business of Natural Philosophy': Isaac Newton's Natural-Philosophical Methodology. Springer.

Ducheyne, S. 2015. Petrus van Musschenbroek and Newton's *vera stabilisque Philosophandi methodus*. *Berichte zur Wissenschaftsgeschichte* 38: 279–304.

Ducheyne, S., and P. Present. 2020. Pieter van Musschenbroek on laws of nature. *British Journal for the History of Science* 50: 637–56.

Ducheyne, S., and J. van Besouw. 2017. Newton and the Dutch 'Newtonians': 1713–1750. *The Oxford Handbook of Newton*, edited by Chr. Smeenk and E. Schliesser.

Oxford University Press. https://academic.oup.com/edited-volume/34749/chap
ter-abstract/296601089?redirectedFrom=fulltext&login=false#:~:text=Newton%20
and%20the%20Dutch%20%E2%80%9CNewtonians%E2%80%9D%3A%201
713%E2%80%931750%20%7C%20The,particular%20on%20three%20strands%20
in%20the%20reception%20of

Ducheyne, S., and J. van Besouw. 2022. 's Gravesande's foundations for mechanics. *Encyclopedia of Early Modern Philosophy and the Sciences*, edited by Ch. Wolfe and D. Jalobeanu, 1–8. Springer.

Duhem, P. 1903. *L'évolution de la mécanique*. Paris: Joanin.

Einstein, A. 1982. What is the theory of relativity? [1919] *Ideas and Opinions*, 2nd ed., 227–31. New York: Crown Publishers.

Eneström, G. 1905. Der Briefwechsel zwischen Leonhard Euler und Johann I Bernoulli. *Bibliotheca Mathematica* 6: 16–83.

Erxleben, J. P. 1772. *Anfangsgründe der Naturlehre*. Göttingen.

Euler, L. 1736. *Mechanica, sive scientia motus*. St. Petersburg.

Euler, L. 1745. Dissertation sur la meilleure construction du cabestan. *Pièces qui ont remporté le prix de l'Académie royale des sciences en 1741*, 29–87. Paris.

Euler, L. 1746a. De motu corporum in superficiebus mobilibus. *Opuscula varii argumenti*, 1:1–136. Berlin.

Euler, L. 1746b. Recherches physiques sur la nature des moindres parties de la matière. *Opuscula varii argumenti*, 1:287–300. Berlin.

Euler, L. 1746c. De la force de percussion et de sa véritable mesure. *Mémoires de l'académie des sciences de Berlin* 1: 21–53.

Euler, L. 1746d. Enodatio quaestionis utrum materiae facultas cogitandi tribui possit necne. *Opuscula varii argumenti*, 1:277–86. Berlin.

Euler, L. 1748. Sur la vibration des cordes. *Histoire de l'Académie royale des sciences et des belles-lettres de Berlin*: 69–85.

Euler, L. 1752a. Découverte d'un nouveau principe de mécanique. *Mémoires de l'académie des sciences de Berlin* 6: 185–217.

Euler, L. 1752b. Recherches sur l'origine des forces. *Mémoires de l'académie des sciences de Berlin* 6: 419–47.

Euler, L. 1757. Principes généraux du mouvement des fluides. *Mémoires de l'académie des sciences de Berlin* 11: 274–315.

Euler, L. 1761. Principia motus fluidorum. *Novi commentarii academiae scientiarum Petropolitanae* 6: 271–311.

Euler, L. 1765a. Recherches sur la connoissance mécanique des corps. *Mémoires de l'académie des sciences de Berlin* 14: 131–53.

Euler, L. 1765b. Du mouvement de rotation des corps solides autour d'un axe variable. *Mémoires de l'académie des sciences de Berlin* 14: 154–93.

Euler, L. 1766. De motu vibratorio tympanarum. *Novi commentarii academiae scientiarum Petropolitanae* 10: 243–60.

Euler, L. 1771. Genuina principia doctrinae de statu aequilibrii et motu corporum tam perfecte flexibilium quam elasticorum. *Novi commentarii academiae scientiarum imperialis Petropolitanae* 15: 381–413.

Euler, L. 1802. *Letters of Euler on Different Subjects in Physics and Philosophy. Addressed to a German Princess*, vols. 1 and 2, 2nd ed. Translated by H. Hunter. London.

Euler, L. 1862. Anleitung zur Naturlehre. *Opera postuma*, edited by P. H. Fuss and N. Fuss, 2:449–560. St. Petersburg. Also in *Euleri Opera Omnia*, 3:1:16–180.

Euler, L. 1960. *Lettres à une princesse d'Allemagne sur divers sujets de physique et de philosophie*. [1760] *Euleri Opera Omnia*, series 3, vols. 11 and 12. Basel: Birkhäuser.

Feigenbaum, L. 1985. Brook Taylor and the method of increments. *Archive for History of Exact Sciences* 34: 1–140.

Firode, A. 2001. *La dynamique de d'Alembert*. Paris: Fides.

Fleckenstein, J. O. 1949. *Johann und Jakob Bernoulli*. Basel: Birkhäuser.

Flores, F. 1999. Einstein's theory of theories and types of theoretical explanation. *International Studies in the Philosophy of Science* 13: 123–34.

Fossombroni, V. 1796. *Memoria sul principio delle velocità virtuali*. Florence.

Fourier, J. 1797. Mémoire sur la statique. *Journal de l'École Polytechnique* 5: 20–61.

Fox, R. 1974. The rise and fall of Laplacian physics. *Historical Studies in the Physical Sciences* 4: 89–136.

Fox, R. 2013. Laplace and the physics of short-range forces. *Oxford Handbook of the History of Physics*, edited by J. Z. Buchwald and R. Fox, 406–31. Oxford University Press.

Fox, C. 2016. The Newtonian Equivalence Principle: how the relativity of acceleration led Newton to the equivalence of inertial and gravitational mass. *Philosophy of Science* 83: 1027–38.

Fraser, C. 1983. Lagrange's early contributions to the principles and methods of mechanics. *Archive for History of Exact Sciences* 28: 197–241.

Fraser, C. 1985. D'Alembert's Principle: the original formulation and application in Jean d'Alembert's *Traité de dynamique*. *Centaurus* 28: 31–61, 145–59.

Fraser, C. 1991. Mathematical technique and physical conception in Euler's investigation of the elastica. *Centaurus* 34: 211–46.

Fraser, D. 2008. The fate of 'particles' in quantum field theories with interactions. *Studies in History and Philosophy of Modern Physics* 39: 841–59.

Freudenthal, G. 2002. Perpetuum mobile: the Leibniz–Papin controversy. *Studies in History and Philosophy of Science* 33: 573–637.

Friedman, M. 2013. *Kant's Construction of Nature*. Cambridge University Press.

Galilei, G. 1974. *Two New Sciences*. Translated by S. Drake. University of Wisconsin Press.

Galletto, D. 1991. Lagrange e la *Mécanique Analytique*. *Memorie dell'Istituto Lombardo*, 29: 77–179. Milan: Istituto Lombardo di Scienze e Lettere.

Garber, D. 1992. *Descartes' Metaphysical Physics*. University of Chicago Press.

Garber, D. 2009. *Leibniz: Body, Substance, Monad*. Oxford University Press.

Gaukroger, S. 1982. The metaphysics of impenetrability: Euler's conception of force. *British Journal for the History of Science* 15: 132–54.

Gauss, C. 1833. Principia generalia theoriae figurae fluidorum in statu æquilibrii. *Commentationes Societatis Regiae Gottingensis* 7: 39–88.

Gilbaud, A. and C. Schmit. 2017. A historiographical overview of the current state of research into Jean Le Rond D'Alembert. *Centaurus* 59: 251–62.

Gillispie, C. 2004. *Science and Polity in France: The Revolutionary and Napoleonic Years*. Princeton University Press.

Gouhier, H. 1926. *La vocation de Malebranche*. Paris: Vrin.

Grattan Guinness, I. 1984. Work for the workers: advances in engineering mechanics and instruction in France, 1800–1830. *Annals of Science* 41: 1–33.

Grattan Guinness, I. 1990. *Convolutions in French Mathematics, 1800–1840*, vols. I–II. Springer.

Grigorian, A. T., and V. Kirsanov. 2007. *Letters to a German Princess* and Euler's physics. *Euler and Modern Science*, edited by N. Bogolyubov, G. Mikhailov, and A. Yushkevich, 307–16. Providence: Mathematical Society of America.

Grimberg, G. 1998. D'Alembert et les équations aux dérivées partielles en hydrodynamique. PhD diss., Université Paris 7.

Grimsley, R. 1963. *Jean d'Alembert*. Oxford: Clarendon Press.

Guicciardini, N. 2009. *Isaac Newton on Mathematical Method and Certainty*. MIT Press.

Guzzardi, L. 2020. *Ruggiero Boscovich's Theory of Natural Philosophy*. Springer.

Hagengruber, R., ed. 2012. *Emilie du Châtelet between Newton and Leibniz*. Springer.

Hahn, R. 1971. *The Anatomy of a Scientific Institution: The Paris Academy of Sciences, 1666–1803*. University of California Press.

Hamel, G. 1917. Über ein Prinzip der Befreiung bei Lagrange. *Jahresberichte der Deutschen Mathematiker-Vereinigung* 25: 60–7.

Hamel, G. 1949. *Theoretische Mechanik*. Springer.

Hankins, Th. 1970. *Jean d'Alembert: Science and the Enlightenment*. Oxford: Clarendon Press.

Hanov, M. Ch. 1762. *Philosophia naturalis, sive physica dogmatica*, vol. 1. Halle.

Harper, W. 2012. *Isaac Newton's Scientific Method*. Oxford University Press.

Hepburn, B., and Z. Biener. 2022. Mechanics in Newton's wake. *Cambridge History of Philosophy of the Scientific Revolution*, edited by D. M. Miller and D. Jalobeanu, 293–311. Cambridge University Press.

Hermann, J. 1716. *Phoronomia*. Amsterdam.

Heyman, J. 1996. *Elements of the Theory of Structures*. Cambridge University Press.

Holden, T. 2004. *The Architecture of Matter*. Oxford University Press.

Howard, D. 2005. 'And I shall not mingle conjectures with certainties': Einstein on the principle theories–constructive theories distinction. Invited lecture delivered at the conference *2005: The Centenary of Einstein's Annus Mirabilis*. London: British Academy.

Howard, N. C. 2003. Christiaan Huygens: the construction of texts and audiences. Phd. diss., Indiana University, Bloomington.

Hult, J. 1985. Eulers *Briefe an eine deutsche Prinzessin. Festakt und Wissenschaftliche Konferenz aus Anlass des 200. Todestages von Leonhard Euler*, edited by W. Engel, 83–90. Berlin: Akademie-Verlag.

Huygens, Chr. 1673. *Horologium oscillatorium*. Paris: F. Muguet.

Huygens, Chr. 1889–1950. *Oeuvres complètes de Christiaan Huygens*, 22 vols. The Hague: Société Hollandaise des Sciences.

Huygens, Chr. 1912. *Treatise on Light*. Translated by S. P. Thompson. London: Macmillan.

Huygens, Chr. 1977. The motion of colliding bodies. Translated by R. Blackwell. *Isis* 68: 574–97.

Huygens, Chr. 1986. *The Pendulum Clock, or Geometrical Demonstrations Concerning the Motion of Pendula as Applied to Clocks*. Translated by R. Blackwell. Iowa State University Press.

Iltis, C. 1967. *The controversy over living force, Leibniz to d'Alembert*. PhD diss., University of Wisconsin-Madison.

Iltis, C. 1970. D'Alembert and the *vis viva* controversy. *Studies in History and Philosophy of Science* 1: 135–44.

Jacobi, C. G. J. 1996. *Vorlesungen über analytische Mechanik* [1847-8]. Edited by H. Pulte. Braunschweig: Vieweg.

Jacquier, Fr. 1785. *Institutiones philosophicae*, vol. 4. Venice.

Jalobeanu, D. 2011. The Cartesians of the Royal Society: the debate over the nature of collisions (1668–1671). *Vanishing Matter and the Laws of Motion*, edited by D. Jalobeanu and P. Anstey, 103–29. New York: Routledge.

Jalobeanu, D., and P. Anstey, eds. 2011. *Vanishing Matter and the Laws of Motion*. New York: Routledge.

Jammer, M. 1957. *Concepts of Force*. Harvard University Press.

Janiak, A. 2008. *Newton as Philosopher*. Cambridge University Press.

Janik, L. G. 1982. Searching for the metaphysics of science: the structure and composition of Madame Du Châtelet's *Institutions de physique*, 1737-1740. *Studies on Voltaire and the Eighteenth Century* 201: 85–113.

Janssen, M. 2009. Drawing the line between kinematics and dynamics in Special Relativity. *Studies in History and Philosophy of Modern Physics* 40: 26–52.

Kant, I. 1905. Neuer Lehrbegriff der Bewegung und Ruhe. [1758] *Kant's Gesammelte Schriften*, edited by the Prussian Academy of Sciences, 2:13–26. Berlin: G. Reimer.

Kant, I. 1910. Monadologia physica. [1756] *Kant's Gesammelte Schriften*, edited by the Prussian Academy of Sciences, 1:473–88. Berlin: G. Reimer.

Kant, I. 1911. Metaphysische Anfangsgründe der Naturwissenschaft. [1786] *Kant's Gesammelte Schriften*, edited by the Prussian Academy of Sciences, 4:465–566. Berlin: G. Reimer.

Kant, I. 1992. *Theoretical Philosophy 1755–1770*. Edited and translated by D. Walford and R. Meerbote. The Cambridge Edition of the Works of Immanuel Kant. Cambridge University Press.

Kant, I. 2004. *Metaphysical Foundations of Natural Science*. Edited and translated by M. Friedman. Cambridge Texts in the History of Philosophy. Cambridge University Press.

Kant, I. 2012. *Natural Philosophy*. Edited by E. Watkins. The Cambridge Edition of the Works of Immanuel Kant. Cambridge University Press.

Keill, J. 1698. *An Examination of Dr. Burnet's Theory of the Earth*. Oxford.

Keill, J. 1702. *Introductio ad veram physicam seu lectiones physicae*. London: Thomas Bennett.

Keill, J. 1726. *An Introduction to Natural Philosophy*. 2nd ed. London.

Lagrange, J. L. 1788. *Mechanique analytique*. Paris: Veuve Desaint.

Lagrange, J. L. 1797a. *Théorie des fonctions analytiques*. Paris.

Lagrange, J. L. 1797b. Sur le principe des vitesses virtuelles. *Journal de l'École Polytechnique* 5: 115–8.

Lagrange, J. L. 1811-5. *Mécanique analytique*, 2 vols. Paris.

Lamé, G. 1852. *Leçons sur la théorie mathématique de l'élasticité des corps solides*. Paris: Bachelier.

Laplace, P. S. 1795. *Exposition du système du monde*, vol. 1. Paris.

Laplace, P. S. 1898. Mémoire sur les mouvements de la lumière dans les milieux diaphanes. [1810] *Oeuvres complètes de Laplace*, 12:267–300. Paris: Gauthier-Villars.

Leduc, C. 2015. La métaphysique de la nature à l'Académie de Berlin. *Philosophiques* 42: 11–30.

Leduc, C., and D. Dumouchel, eds. 2015. *Philosophiques*, vol. 42, issue 1.

Leibniz, G. W. 1686. Brevis demonstratio erroris memorabilis Cartesii et aliorum circa legem naturae. *Acta Eruditorum* August: 161–3. French translation in *Nouvelles de la République des Lettres*: 996–9.

Leibniz, G. W. 1687. Réplique de M[onsieur] L[eibniz] a Monsieur l'Abbé D[e] C[atelan] . . ., touchant ce qu'on a dit Monsieur Descartes que Dieu conserve toujours dans la nature la même quantité de mouvement. *Nouvelles de la République des Lettres* February: 131–45.

Leibniz, G. W. 1695. Specimen dynamicum. *Acta Eruditorum* April: 145–57. English translation in *Philosophical Papers and Letters*, edited by L. Loemker, 435–51. Kluwer.

Leibniz, G. W. 1860. *Mathematische Schriften*, vol. 6. Edited by C. I. Gerhardt. Halle.

Leibniz, G. W. 1875. *Philosophische Schriften*, vol. 1. Edited by C. I. Gerhardt. Berlin.

Leibniz, G. W. 1956. *Philosophical Papers and Letters*. Edited and translated by L. Loemker. University of Chicago Press.

Lennon, T. M. 1993. *The Battle of the Gods and Giants: The Legacies of Descartes and Gassendi, 1655–1715*. Princeton University Press.

Le Ru, V. 2015. La notion de temps chez d'Alembert: une notion mathématique? *Les mathématiques et l'expérience*, edited by E. Barbin and J. P. Cléro, 131–47. Paris: Hermann.

Le Ru, V. 2021. D'Alembert ou comment penser les grandes questions métaphysiques à partir de son principe de dynamique. *Les chemins du scepticisme en mathématiques*, edited by J. P. Cléro, 253–66. Paris: Hermann.

Mach, E. 1883. *Mechanik in ihrer Entwickelung*. Wien.

MacLaurin, C. 1724. Démonstration des loix du choc des corps. *Pièce qui a remporté le prix de l'Académie royale des sciences proposé pour l'année 1724*. Paris: Jombert.

Malebranche, N. 1958–70. *Oeuvres complètes*. Edited by A. Robinet, 21 vols. Paris: Vrin.

[Malebranche, N.] 1692. *Des loix de la communication du mouvement, par l'auteur de la Recherche de la Verité*. Paris: Pralard.

Maltese, G. 1992a. *La storia di F=MA*. Firenze: Olschki.

Maltese, G. 1992b. Brook Taylor and Johann Bernoulli on the vibrating string. *Physis* 29: 703–44.

Maltese, G. 1993. Toward the rise of the modern science of motion: the transition from synthetic to analytic mechanics. *History of Physics in the 19th and 20th Centuries*, edited by F. Bevilacqua, 51–67. Bologna: SIF.

Maupertuis, P. L. M. 1732. *Discours sur les différentes figures des astres*. Paris.

Maupertuis, P. L. M. 1734. Dissertation on the different figures of the celestial bodies. Translated by J. Keill in *An Examination of Dr. Burnet's Theory of the Earth*, 1–67. Oxford and London.

Maupertuis, P. L. M. 1740. Loix de repos des corps. *Histoire de l'Académie royale des sciences*, 170–7. Paris.

Maupertuis, P. L. M. 1744. Accord des différentes loix de la nature qui avoient jusqu'ici paru incompatibles. *Histoire de l'Académie royale des sciences*, 417–26. Paris.

Maupertuis, P. L. M. 1746. Les loix de mouvement et du repos, déduites d'un principe de métaphysique. *Histoire de l'Académie des sciences et belles lettres de Berlin*, 267–94. Berlin.

Mazière, P. 1727. Les loix du choc des corps à ressort parfait ou imparfait. *Pièce qui a remporté le prix de l'Académie royale des sciences proposé pour l'année 1726*. Paris: Jombert.

McDonough, J. 2020. Not dead yet: teleology and the 'Scientific Revolution.' *Teleology: A History*, edited by Jeffrey K. McDonough, 150–79. Oxford University Press.

McMullin, E. 1978. *Newton on Matter and Activity*. University of Notre Dame Press.

Meier, G. 1765. *Metaphysik*. 2nd ed., vol. 2. Halle.

Mouy, P. 1927. *Les lois du choc des corps d'après Malebranche*. Paris: Vrin.

Murdoch, A. I. 2010. On molecular modeling and continuum concepts. *Journal of Elasticity* 100: 33–61.

Murray, G., W. Harper, and C. Wilson. 2011. Huygens, Wren, Wallis, and Newton on rules of impact and reflection. *Vanishing Matter and the Laws of Motion*, edited by D. Jalobeanu and P. Anstey, 153–92. New York: Routledge.

Musschenbroek, P. 1726. *Elementa physicae*. Leiden.

Musschenbroek, P. 1744. *The Elements of Natural Philosophy: Chiefly Intended for the Use of Students in Universities*, vol. 1. Translated by J. Colson. London.

Nagel, F. 2005. Jacob Hermann—Skizze einer Biographie. *Gelehrte aus Basel an der St. Petersburger Akademie der Wissenschaften des 18. Jahrhunderts*, edited by F. Nagel and A. Verdun, 55–76. Aachen: Shaker.

Nagel, F., and A. Verdun, eds. 2005. *'Geschickte Leute, die was praestiren können.' Gelehrte aus Basel an der St. Petersburger Akademie der Wissenschaften des 18. Jahrhunderts*. Aachen: Shaker.

Nakata, R. 1994. Joseph Privat de Molières: reconciler between Cartesianism and Newtonianism in collision theory. *Historia Scientiarum* 3: 201–13.

Nakata, R. 2002. The general principles for resolving mechanical problems in d'Alembert, Clairaut and Euler. *Historia Scientiarum* 12: 18–42.

Navier, H. 1821. Sur les lois des mouvements des fluides, en ayant régard à l'adhésion des molécules. *Annales de Chimie et de Physique* 19: 244–60.

Navier, H. 1823a. Extrait des recherches sur la flexion des plans élastiques. *Bulletin de la Société Philomathique de Paris*, 92–102.

Navier, H. 1823b. Sur les lois de l'équilibre et du mouvement des corps solides élastiques. *Bulletin de la Société Philomathique de Paris*, 177–81.

Navier, H. 1827. Sur les lois de l'équilibre et du mouvement des corps solides élastiques. *Mémoires de l'Académie Royale des Sciences* 7: 375–93.

Neimark, J., and N. Fufaev. 1972. *Dynamics of Nonholonomic Systems*. Providence: American Mathematical Society.

Newton, I. 1687. *Philosophiae Naturalis Principia Mathematica*. London.

Newton, I. 1713. *Philosophiae Naturalis Principia Mathematica*. Edited by R. Cotes, 2nd ed. London.

Newton, I. *Principes mathématiques de la philosophie naturelle*. Translated by E. Du Châtelet. Paris.

Newton, I. 1999. *The Principia*. Edited and translated by I. B. Cohen, with A. Whitman. University of California Press.

Newton, I. 2014. *Philosophical Writings*. Edited by A. Janiak, 2nd ed. Cambridge University Press.

Ney, A. 2020. *The World in the Wave Function*. Oxford University Press.

Noll, W. 1974. La mécanique classique, basée sur un axiome d'objectivité. *The Foundations of Mechanics and Thermodynamics*, 135–44. Springer.

Oldenburg, H. 1968. *The Correspondence of Henry Oldenburg*. Vol. V: *1668–1669*. Edited and translated by A. R. Hall and M. Boas Hall. University of Wisconsin Press.

Ostertag, H. 1910. *Der philosophische Gehalt des Wolff-Manteuffelschen Briefwechsels*. Leipzig: Quelle & Mayer.

Ott, W. 2009. *Causation and Laws of Nature in Early Modern Philosophy*. Oxford University Press.

Ott, W., and L. Patton, eds. 2018. *Laws of Nature*. Oxford University Press.

Panza, M. 2003. The origins of analytic mechanics in the 18th century. *A History of Analysis*, edited by H. N. Jahnke, 137–53. Providence: American Mathematical Society.

Papastavridis, J. 2002. *Analytical Mechanics*. Oxford University Press.

Pemberton, H. 1728. *A View of Isaac Newton's Philosophy*. London.

Pepe, L. 2008. Il giovane Lagrange e i fondamenti dell'analisi. *Sfogliando la* Méchanique Analitique, edited by G. A. Sacchi Landriani and A. Giorgilli, 37–49. Milan: LED.

[Pfautz, C.] 1688. Review of *Philosophiae Naturalis Principia Mathematica* by Isaac Newton. *Acta Eruditorum* June: 304–12.

Planck, M. 1919. *Einführung in die Mechanik deformierbarer Körper*. Leipzig: S. Hirzel.

Poinsot, L. 1806. Théorie générale de l'équilibre et du mouvement des systèmes. *Journal de l'École Polytechnique* 13: 182–241.

Poisson, S. D. 1811. *Traité de mécanique*. 1st ed. Paris.

Poisson, S. D. 1828. Mémoire sur l'équilibre et le mouvement des corps élastiques. *Annales de Chimie et de Physique* 37: 337–54.

Poisson, S. D. 1831. *Nouvelle théorie de l'action capillaire*. Paris.

Pourciau, B. 2006. Newton's interpretation of Newton's Second Law. *Archive for History of Exact Sciences* 60: 157–207.

Prony, R. 1800. *Mécanique philosophique*. Paris.

Prony, R. 1815. *Leçons de mécanique analytique*. Paris.

Pulte, H. 1989. *Das Prinzip der kleinsten Wirkung und die Kraftkonzeptionen der rationalen Mechanik*. Stuttgart: Franz Steiner.

Pulte, H. 2005. Lagrange, *Méchanique analitique*, first edition. *Landmark Writings in Western Mathematics 1640–1940*, edited by I. Grattan-Guinness, 208–24. Elsevier.

Pyle, A. 2003. *Malebranche*. New York: Routledge.

Ravetz, J. 1961. The representation of physical quantities in eighteenth-century mathematical physics. *Isis* 52: 7–20.

Reichenberger, A. 2012. Emilie du Châtelet's *Institutions* in the context of the vis viva controversy. *Emilie du Châtelet between Newton and Leibniz*, edited by R. Hagengruber, 157–71. Springer.

Robinet, A. 1955. *Malebranche et Leibniz: Relations personnelles*. Paris: Vrin.

Robinet, A. 1970. *Malebranche de l'Académie des sciences. L'œuvre scientifique, 1674–1715*. Paris: Vrin.

Rohault, J. 1671. *Traité de physique*. Paris.

Rohault, J. 1697. *Physica*. Edited and translated by S. Clarke. London.

Rohault, J. 1723. *System of Natural Philosophy*. Edited and translated by J. Clarke. London.

Saint-Venant, A. B. de. 1844. *Mémoire sur la question de savoir s'il existe des masses continues, et sur la nature probable des dernières particules des corps*. Paris.

Saulmon, M. 1721. Du choc des corps dont le ressort est parfait. *Histoire de l'Académie royale des sciences*, 126–45.

Saulmon, M. 1723. Sur le choc des corps à ressort. *Histoire de l'Académie royale des sciences*, 101–7.

Schmit, C. 2015. Les dynamiques de Jean-Jacques Dortous de Mairan. *Revue d'Histoire des Sciences* 68: 281–309.

Schmit, C. 2017. Beginnings of a new science. D'Alembert's *Traité de dynamique* and the French Royal Academy of Sciences around 1740. *Centaurus* 59: 285–99.

Schmit, C. 2020. *La philosophie naturelle de Malebranche au XVIIIe siècle*. Paris: Classiques Garnier.

Scott, W. 1970. *The Conflict between Atomism and Conservation Theory, 1644 to 1860.* New York: American Elsevier.

's Gravesande, W. J. 1720. *Mathematical Elements of Physics Prov'd by Experiments: Being an Introduction to Sir Isaac Newton's Philosophy.* Translated by J. Keill. London.

Shapin, S. 1998. *The Scientific Revolution.* University of Chicago Press.

Smith, G. E. 2002. The methodology of the *Principia. Cambridge Companion to Newton*, edited by I. B. Cohen and G. E. Smith, 138–73. Cambridge University Press.

Smith, G. E. 2008. Newton's *Philosophiae Naturalis Principia Mathematica. Stanford Encyclopedia of Philosophy*, edited by E. Zalta. Winter ed. 2008. https://plato.stanford.edu/archives/win2008/entries/newton-principia/.

Smith, G. E. 2014. Closing the loop: testing Newtonian gravity, then and now. *Newton and Empiricism*, edited by Z. Biener and E. Schliesser, 262–352. Oxford University Press.

Smith, G. E. 2019. Newton's numerator in 1685: a year of gestation. *Studies in History and Philosophy of Modern Physics* 68: 163–77.

Smith, S. 2007. Continuous bodies, impenetrability, and contact interactions. *British Journal for the Philosophy of Science* 58: 503–38.

Smith, S. 2013. Kant's picture of monads in the *Physical Monadology. Studies in the History and Philosophy of Science* 44: 102–11.

Stan, M. 2009. Kant's early theory of motion. *The Leibniz Review* 19: 29–61.

Stan, M. 2017. Newton's concepts of force among the Leibnizians. *Reading Newton in Early Modern Europe*, edited by M. Feingold and E. Boran, 244–89. Brill.

Stan, M. 2022. From metaphysical principles to dynamical laws. *Cambridge History of Philosophy of the Scientific Revolution*, edited by D. M. Miller and D. Jalobeanu, 387–405. Cambridge University Press.

Stan, M. forthcoming. *Kant's Natural Philosophy.* Cambridge University Press.

Taylor, B. 1713. De motu nervi tensi. *Philosophical Transactions*, 26–32.

Taylor, B. 1717. *Methodus incrementorum.* London.

Terrall, M. 2002. *The Man Who Flattened the Earth.* University of Chicago Press.

Truesdell, C. 1954. *Rational Fluid Mechanics, 1687–1765.* Zurich: Orell Füssli.

Truesdell, C. 1960. *The Rational Mechanics of Flexible or Elastic Bodies, 1638–1788.* Zurich: Orell Füssli.

Truesdell, C. 1968. The creation and unfolding of the concept of stress. *Essays in the History of Mechanics*, 184–238. Springer.

Truesdell, C. 1968b. *Essays in the History of Mechanics.* Springer.

Truesdell, C. 1991. *A First Course in Rational Continuum Mechanics.* 2nd ed. Boston: Academic Press.

Truesdell, C., and R. Toupin. 1960. The classical field theories. *Principles of Classical Mechanics and Field Theory*, edited by S. Flügge, 226–858. Springer.

van Besouw, J. 2020. 's Gravesande's philosophical trajectory: 'between' Leibniz and Newton. *Intellectual History Review* 30: 615–40.

van Inwagen, P. 1990. *Material Beings.* Cornell University Press.

Veldman, M. n.d. Mathematizing metaphysics: the case of the Principle of Least Action. Unpublished manuscript.

Verdun, A. 2015. *Leonhard Eulers Arbeiten zur Himmelsmechanik.* Springer.

Verdun, A. forthcoming. *Leonhard Euler's Principle of Angular Momentum.* Springer.

Vilain, C. 2000. La question du centre d'oscillation de 1660 à 1690; de 1703 à 1743. *Physis* 37: 21–51, 439–66.

Villaggio, P. 2008. Introduction. *Die Werke von Johann I und Nikolaus II Bernoulli*, 6:27–241. Basel: Birkhäuser.

von Harnack, A. 1900. *Geschichte der königlich Preussischen Akademie der Wissenschaften zu Berlin*, vol. 1A. Berlin.

Watkins, E. 1997. The laws of motion from Newton to Kant. *Perspectives on Science* 5: 311–48.

Watkins, E. 2005. *Kant and the Metaphysics of Causality*. Cambridge University Press.

Watkins, E. 2006. On the necessity and nature of simples: Leibniz, Wolff, Baumgarten, and the pre-Critical Kant. *Oxford Studies in Early Modern Philosophy* 3: 261–314.

Watkins, E. 2019. *Kant on Laws*. Cambridge University Press.

Westfall, R. S. 1971. *Force in Newton's Physics*. Elsevier.

Whyte, L. L. 1961. Boscovich's atomism. *Roger Joseph Boscovich. Studies of His Life and Work*, edited by L. L. Whyte, 102–26. London: Allen & Unwin.

Wilson, C. 1987. D'Alembert versus Euler on the precession of the equinoxes and the mechanics of rigid bodies. *Archive for History of Exact Sciences* 37: 233–73.

Wilson, M. 2009. Determinism and the mystery of the missing physics. *British Journal for the Philosophy of Science* 60: 173–93.

Wilson, M. 2018. *Physics Avoidance*. Oxford University Press.

Winter, E. 1956. *Die Registres der Berliner Akademie der Wissenschaften 1746–1766*. Berlin: Akademie Verlag.

Wolff, Chr. 1731. *Cosmologia generalis*. Frankfurt.

Wolff, Chr. 1740. *Philosophia rationalis, sive logica*. 3rd ed. Frankfurt, 1740.

Index

For the benefit of digital users, indexed terms that span two pages (e.g., 52–53) may, on occasion, appear on only one of those pages.

Figures are indicated by f following the page number

Action, 2, 98, 103, 119, 132, 169, 180, 219–20, 226, 231–32, 254–55, 356–57
  bodily action, 10, 11, 104–5, 127–28, 130, 133, 143, 149–50, 153, 265, 267–68, 397–98
  capillary action, 379
  contact action, 3, 23–24, 32, 47, 56, 106, 143–45, 170, 176, 177, 181, 183–84, 202, 204, 210, 211, 213, 296–97
  continued action, 164–65
  corporeal action, 85, 210–11, 346
  God's action, 79, 80–81, 126
  mutual action, 90, 160, 164–66, 211, 383
  Nature, 2–3, 10, 98, 104–5, 106, 122, 124, 128–30, 149–50, 151, 158, 177, 180, 199, 254–55, 261–62, 265, 378–79, 398
  philosophical action, 18
  physical action, 41, 89
action-at-a-distance, 56, 95, 121, 129–30, 171, 177, 185, 211–13, 337
  Newtonian, 144–45, 210
  as unintelligible, 144–45, 210
Age of Reason, 1, 223, 396, 399
Ampère, André-Marie, 335
analytic mechanics, 24, 238, 254, 333, 345, 348–49, 352–53, 361–62, 376
ancien régime, 222–23, 368
Ancient Greece, 244–45
Archimedes, 241, 255
Aristotle, 19–20, 89n.41
atomism, 104, 134–35, 141, 199, 205

Basel School, 232–33
Baumeister, Fr., 102
Berkeley, George, 21

Berlin (Germany), 220–22
Berlin Academy of Science, 137, 221–22
  French, as official language, 221
  monad dispute, 154
Bernoulli, Daniel, 232–33, 295
  Leibniz problem, solving of, 215
  methodus indirecta, 318–19
Bernoulli, Jakob, 227–28, 232, 240, 241, 243–44, 256
  compound pendulum, 241, 255–56, 259–60
  internal constraints, as workless, 260
  Lever Principle, 241, 242–43
  new dynamical law, 241
  statics, principle of, 255–56
Bernoulli, Johann, 29, 51–52, 55, 56, 57, 145, 146, 228, 232, 233, 234, 235, 302, 307
  center-of-oscillation problems, 256–57
  elastic recoil, model of, 56
  gurges, notion of, 320
  hard bodies, 50–51
  methodus directa, 319, 320
  pressure, as action, 320
  statics, 256
  vibrating strings, 236–37
Berthollet, Claude Louis, 367, 368
bodies in motion, 8, 78, 109, 149, 202, 214, 275, 276
  mathematizing of, 308
  rational mechanics of, 279–80, 396
  bodily action, 10, 11, 104–5, 127–28, 130, 133, 149–50, 153, 265, 267–68, 397–98
body kinds, 38, 48, 49–50, 326–27, 330
body types, 84, 87, 90, 326–27

Boscovich, Roger Joseph, 23, 129, 139,
    168, 169–70, 179–81, 184–85, 187,
    188, 189–90, 192, 197, 199–200, 211,
    288, 397
  acceleration field, 195
  Boscovich force, 185–87, 186f, 195–
    96, 197
  collisions, treatment of, 190, 191–92
  d'Alembert, comparison with,
    294, 396–97
  divisibility of extension, 175
  elastic bodies, 189
  elastic collision, 191
  elusive mass, 191, 193
  extended bodies, 171, 174–75, 176, 177
  force of inertia, 194–95, 196, 197, 198
  forces, as causes, 179
  hard bodies, 189
  impenetrability of bodies, 176
  points under forces, 185
  quantity of motion, 187, 190, 197–98
  soft bodies, 188–89, 190–91
botany, 12
Bradley, James, 310n.14
Brief Demonstration, A . . . (Leibniz), 34,
    63, 68, 74–75, 81
Britain, 220
British Isles, 61–62, 63, 64

calculus, 232, 248–49, 258, 301
  differential, 29
  improving of, 215
  invention of, 64–65
  new, 233
  Newton, 235
  partial derivatives, 319–20, 377
  variations, 334
Carré, Louis, 39, 52–53
Cartesian method, 29, 107, 108
Cartesian physics, 10, 107–8, 124
Cartesians, 63–65, 70, 74, 82–83, 95–96,
    100, 105, 110, 112
  criticism of, 67–68
  system-building, tendency toward, 110
Catelan, Father, 37
Cauchy, Augustin-Louis, 24–25, 305–6,
    327, 348, 367, 386, 397
  action of internal forces, 386–87

balance-of-force approach, 390, 393–94
balance of torques, 391–92
Cauchy package, 392, 393–94
Cauchy's Laws of Motion, 306, 391–92
continuous matter, 392, 394
continuous volume, 393
discrete molecules, 392
Euler-Cauchy laws of continuum
    mechanics, 259, 305
mass point, 382–83
strain, 389
stress, 388f, 389
stress, tensor quality of, 388
tension, 386–87, 388–89
causation, 21, 101, 119, 121, 122–23,
    125, 398
  debate over, 124–26, 127, 128–29
  metaphysics of, 126–28, 129, 179–80
  primary, 20–21
  secondary, 20–21, 124, 125–26, 127, 129
  substance, 1, 20–21, 23, 97, 98, 124, 127,
    128–29, 131
celestial mechanics, 20, 233, 306, 309, 379
centrifugal force, 55, 57, 78
Clairaut, Alexis, 52, 227–28, 233, 245–46,
    247, 254–55, 257–58, 260, 344
  centroid approach, 250
  general principle of accelerative forces,
    247–49, 248f
  rigid constraints, 249, 260
  theory building, 254
Clarke, John, 99
Clarke, Samuel, 64
Clatterbaugh, K., 126
collision theory, 3–4, 8, 10, 22–23, 170, 204
  BODY, addressing of, 132, 169
  Boscovich, 192
  causal-explanatory basis, 183–84
  Descartes' physics, foundational for, 5
  Hermann, 84, 87, 88
  Leibniz, 82–83
  Malebranche, 27–28, 29, 30–31, 32–33,
    34, 36, 37–38, 41, 42, 44, 47, 58, 61,
    68, 74–75
  natural philosophy, foundational for,
    5, 52, 95
  philosophical mechanics of collision, 5
  principle approach, 31

problem of collisions (PCOL),
    addressing of, 129–30, 229
Wolff, 89, 90
composition of motion, 111–12, 269, 279–
    80, 292, 294
compound pendulum, 24, 231, 232, 234,
    238, 239–40, 259–60, 261–62, 281,
    282, 286
    Bernoulli's Lever Principle, 241, 242–
        45, 242f
    dual constraints of, 238–39, 261
    Huygens Formula, 240–41, 244, 255–56
    linear, 282, 283, 284
concurrentism, 125–26
Condorcet, Marquis de, 222–23
conservation of *vis viva*, 107–8, 123, 149,
    180–81, 217, 247–48, 257, 295, 304
    See also *vis viva*
constraints
    external, 15, 24, 224, 225, 231, 232,
        240, 254–55, 260, 261, 300, 331–32,
        333, 335, 338, 340–41, 362–63, 364,
        379, 392
    internal, 14, 24, 224, 231, 244–45, 259–
        60, 261, 299–300, 301, 305–6, 308–9,
        322, 324, 326, 330, 331–32, 333, 362–
        63, 393–94
contact action, 3, 23–24, 32, 47, 56, 106,
    143–44, 170, 181, 183–84, 202, 204,
    211, 213, 296–97
    action at a distance, 176, 177
    intelligibility of, 144–45, 210
Continuity Equation, 322, 329
continuous matter, 141, 142–43, 205, 351–
    52, 392, 394
    deformable, 24–25, 391–92
continuum mechanics, 352n.34, 367,
    389n.44, 392n.53
    Euler-Cut approach, 259
Cosserat continua, 392
Cotes, R., 108–9
Crousaz, Jean-Pierre de, 53–54, 55, 56,
    100, 103, 149–50

d'Alembert, Jean Le Rond, 52, 227–28,
    236, 240, 244–45, 259–60, 264, 267,
    270–71, 272, 273–74, 275, 276–77,
    284–87, 288, 289, 290–91, 295,

297–98, 310–11, 312, 320, 346–47,
    348–49, 369
composition of motion, 279–80,
    292, 294
d'Alembert's Principle, 227, 319, 339
equilibrium, 277–80, 288
force of inertia, 271, 272, 288
fourteen problems, 269, 281
General Principle, 257–58, 277, 280, 281,
    282, 284, 290, 291, 294, 296–97, 304
inertia, 269, 279–80, 294
Lever Principle, version of, 275–76
mass, notion of, 288, 291–93
Parallelogram Rule, 273
principle approach, 293
problem of collisions (PCOL), 282
problem of constrained motion
    (PCON), 282
Theory (Boscovich), comparison to, 294
Treatise on Dynamics, 24, 25–26,
    263, 297
virtual motions, 276, 280
wave equation, 315f, 316, 393–94
d-contact, 211–13
deductive nomic unity, 220
Democritus, 383–84
Descartes, René, 3–5, 8, 21–23, 29, 32–33,
    34, 37, 41, 47–48, 74–75, 104, 134,
    143, 201–2, 204, 270, 272, 378–79
angle of incidence, 277
Cartesian circle, 20
Cartesian dualism, 20, 125
Cartesian skepticism, 20
divisible matter, 133–34, 199
doctrine of method, 29
force of rest, 29
hard bodies, 30, 32, 41, 49, 67–68
laws of nature, 219
mass qua measure, 59
Principles of Philosophy, 13, 27–28, 30,
    45–46, 95–96, 230
quantity of matter, 141, 204
rules of collision, 65
theory of collisions, 7
theory of matter, 4
de l'Hôpital, marquis de, 29, 232
Discours (J. Bernoulli), 56
Discourse on Method (Descartes), 106

Du Châtelet, Émilie, 122–23, 148–49
  bodies, account of, 154
  bodies, as phenomenal, 136
  collision process, account of, 152, 153
  Foundations of Physics, 25–26, 106, 122,
    136, 148, 153, 154
  free will, 147, 148, 149
  Newtonianism, early advocacy of, 122
  as plenist, 141
  primitive force and derivative force,
    distinction between, 151
  PSR, adoption of, 151
  vis viva debate, 145
dynamical laws, 12, 88, 181, 182, 183,
    218n.22, 219, 225–26, 225n.36, 238,
    305–6, 345–46
  See also Euler's First Law
dynamical mass, 193–94, 195–96, 197–98,
    207, 209
  Newtonian, 191–92, 194–95, 206

École Polytechnique, 368, 374
Einstein, Albert, 25–26, 265
  constructive and principle approaches,
    distinction between, 265n.5, 268
  Lorentz transformations, 266
  special relativity (SR), 266
elastic collision, 32–33, 54–55, 87, 88, 94,
    165, 191
elastic rebound, 44–45, 47, 52–53, 54–55,
    56, 66–67, 69–70
elasticity, 42, 53–54, 66–68, 72, 73–74, 80–
    81, 105, 163, 164–67, 188, 276–77,
    306, 327–28, 347, 367, 370, 380, 384,
    385–86, 392, 393–94
  geometric, 72–73
  kinematic, 72–73
  perfect, 71–72
  without, 276–77, 279
Elements of Natural Philosophy
  (Musschenbroek), 70, 99
elusive mass, 57, 61, 69–70, 141, 191,
    193, 204
empirical evidence, 45, 70, 107–8, 130, 146,
    149, 255, 266, 267, 358–59, 360, 362, 374
empirical physics, 378–79
England, 54–55, 64–65, 99
  See also Britain

Enlightenment, 21–22, 210, 225–26,
    227–28, 232–33, 246, 251–52, 302–3,
    353, 362
  Late, 367
  Long, 302
Epicurus, 383–84
Euler, Leonhard, 1, 23, 24, 25–26, 97, 132,
    133, 135, 137, 139–40, 154, 156, 169,
    170, 202, 221–22, 228–29, 232–33,
    240, 301, 304, 306, 316, 319–20, 323,
    328–29, 330, 331–32, 333, 334, 382
  architecture of matter, 328
  causes of changes in bodies, 217–18
  collisions, 155–56, 159, 160–63, 164,
    165, 166, 167–68
  continuity, 140
  Continuity Equation, 322, 329
  deformable bodies, 165, 327
  dynamical equation, 313
  dynamical mass, 207–8
  elastic and inelastic bodies, difference
    between, 163–64, 165–66
  E-principle, 311, 324
  Euler-Cauchy laws, 259, 305–6
  Euler Cut, 174n.11, 259n.48, 352n.34,
    390n.46
  Euler Heuristic, 228, 305, 324, 325–
    26, 331
  Euler's First Law, 303, 324, 325, 327–28,
    329, 331
  Euler-Lagrange equation (ELQ),
    259n.48, 357, 361–62
  Euler Heuristic, 228, 305, 324, 326, 331
  Euler's Equation, 313, 323
  extended bodies, 139, 140, 215, 216,
    309, 311–12
  extension, 138, 139, 140, 155
  first principles, 326
  flexible bodies, 318
  fluid motion, 319–20, 321–22, 323
  force, origin of, 158, 159
  frictionless bodies, 318
  gurges, 320
  impenetrability, 159, 160, 164, 165–66
  inertia, 156–57, 159–60, 208–9
  inertia and force, 158
  inertial force, 208
  as intellectual giant, 138n.14, 220

kinematic equation, 313
moment of inertia, 209, 310
monadic theory, rejection of, 138–39
motus turbinatorius, 316
Newton-Euler-Cauchy
    dynamics, 393–94
Newton-Euler dynamics, 392
Newton-Euler mechanics, 342
piecemeal approach of, 331
principle of sufficient reason
    (PSR), 157–58
problem of constraints, 307–9
property of bodies, 155, 157, 158
quantity of matter, 218
rigid bodies, 309, 310–11, 312, 313,
    314, 390
true essence of bodies, 154
vibrating string, 316–17
Europe, 3, 61–62, 63, 64, 210, 221–
    22, 232–33
Evidence, 3, 10, 98, 105, 106, 108, 112–14,
    115, 119, 121, 122, 124, 131, 132,
    149, 151, 158, 169, 231–32, 261–62,
    265, 296
    empirical evidence, 45, 70, 107–8, 130,
        146, 149, 255, 266, 267, 358–59, 360,
        362, 374
    observational evidence, 107, 119
    Principle, 10, 98, 106, 108, 113–14, 115,
        119, 121, 122, 124, 131, 132, 149, 151,
        169, 261–62, 265, 296
explanatory ontic unity, 220
extended bodies, 14, 109, 133–34, 135,
    138, 171, 177–78, 187–88, 199–200,
    302, 305–6
    balance-of-force approach, 390
    as composites, 352
    constitution of, 380
    with constraints, 223–24, 231–32,
        299, 331
    d'Alembert, 285–86
    as derivative, 176
    divisibility of, 136–37, 139, 140, 177
    dynamical laws, 327–28
    elements of matter, 328
    evolution formula, 300
    as free particles, 308
    global expression, 300

internal constraints, 24, 231,
    333, 362–63
Lex Secunda (Newton), as application
    of, 306
local expression, 300
mathematizing of, 299, 300, 307–8, 331
MCONI, 224, 225
MCON2, 224, 225
mechanical behavior of, 348
motions of, 207, 213–14, 215, 217, 226,
    232, 297, 300, 307–8, 324, 327, 329,
    331, 362–63, 364, 396
physical monads, 174
as point particles, 207–8
rational mechanics of, 228, 229, 305,
    307, 323, 396
rotational inertia, 209
simples, 136, 137, 175
theory of motion, 204, 209
extended body with constraints,
    257, 366
extended matter, 175, 231–32, 249,
    261, 327
extension, 4, 5, 29, 30–31, 75–76, 90,
    112–13, 133–34, 139, 170, 244–45,
    248–49, 250, 349, 383, 393–94
    atomist approach, 138
    of bodies, 136–37, 138, 141, 151, 158,
        174–75, 286
    Boscovich, 174–75, 191–92
    Cartesian, 37, 134
    as causally inert, 90
    constraints, 287
    continuity, 140
    continuous, 384
    corporeal, 171
    d'Alembert, 270–71, 287
    divisibility of, 136, 137, 139, 171,
        173, 175
    Du Châtelet, 150–51, 158
    elastic, 105
    Euler, 138, 139, 140, 155–56
    geometric, 133–34, 136–37
    impenetrability, 142–43, 177, 270–140
    inertia, 159
    as infinitely divisible, 104, 136–37,
        138, 139
    mathematical, 135

extension (*cont.*)
    matter, as sole essential attribute of, 104
    motive force, 150–51
    notion of body, 134
    physical, 135, 136–37, 138, 172, 175
    Principles of Sufficient Reason
        (PSR), 150
    Principle of Virtual Velocities
        (PVV), 373
    as property of bodies, 133, 138, 139,
        140, 141, 155
    quantity of matter, 141, 142
    resisting force, 151
    simple, 95
external constraints, 15, 24, 224, 231,
    232, 240, 254–55, 260, 300, 331–32,
    333, 335, 338, 340–41, 362–63, 364,
    379, 392
    as kinematic possibilities, 225, 261

fluid dynamics, 320–21, 367, 384, 392
Fontenelle, Bernard Le Bovier de, 53
force of inertia, 84–85, 90, 105, 155–56,
    185, 186–87, 194–95, 196, 197,
    198, 204, 207–8, 209, 269, 271, 272,
    288, 294
Foundations of Physics, 25–26, 106, 122,
    136, 148, 153, 154
Fourier, Jean-Baptiste Joseph, 367–
    68, 382–83
France, 22–23, 27–28, 34, 37, 47–48, 49,
    54–55, 57, 59, 61–62, 95, 122, 222–23,
    233, 368
Frederick II, 221–22, 233
French Academy of Sciences, 27, 222–
    23, 368

Galileo, 82–83, 232–33, 237–38
    Galilean relativity, 57, 60–61–, 69–70
    law of free fall, 76, 78, 237n.13
Gauss, Carl Friedrich, 222, 372–73
Gay-Lussac, Joseph Louis, 367
geometric contact (g-contact), 210–13
geometry, 6–7, 35, 76, 87, 110, 134, 138,
    171, 172, 218, 246, 251, 290–91, 306,
    327, 328, 329, 331, 341, 376, 393–94
    of constraints, 342, 376
    differential, 228n.40
    of mass distribution, 382–83

synthetic, 6, 375n.22
    of a system, 225–26
Germain, Sophie, 370, 384–85
Germany, 86, 102, 222, 363
Goal, 2, 3, 22, 23, 131, 395, 398
gravity, 7–8, 20, 66, 78, 83, 105, 106, 112–
    13, 195–96, 228, 241, 243–44, 247,
    256–57, 310–11, 322, 361–62, 363–
    64, 372–73, 390, 391–92
    center of, 41, 84, 187–88, 193, 197, 215,
        233, 281, 286, 293, 308, 313, 328–
        29, 366
    inverse-square, 379
    Newtonian, 168, 388
    terrestrial, 344, 379–80
gurges, 319, 320

Halley's Comet, 358–59, 359n.43
Haüy, René Just, 367
Hermann, Jakob, 82–83, 84, 85–86, 87,
    221, 232–33
    actant bodies, 83
    active versus passive, 86
    elastic collision, 87
    force of inertia, 84–85
    inelastic collision, 84
    inert bodies, 83
    material approach of, 88
    oblique impact, 86
    vis visa, 83, 85–86
Hooke's Law, 94, 327–28, 346
H-principle, 258, 260
Hume, David, 1, 21
    causation, 121, 124, 126
    epistemic lessons, 124
Huygens, Christiaan, 8–9, 27, 31, 33, 63,
    78–79, 83, 87, 88, 108, 209, 221, 234,
    257, 299
    compound pendulum, 240
    Huygens Formula, 240–41, 244, 255–57
    Torricelli-Huygens principle, 304
Hydraulica (Johann Bernoulli), 319
Hydrodynamica (D. Bernoulli), 320
hydrostatics, 56, 372–73

impenetrability, 30–31, 66, 67–68, 77, 104,
    133, 156, 165, 166, 211
    of bodies, 159, 160, 164–66, 176,
        211, 270

collisions, 159–61, 164, 165–67, 176
contact action, 143, 176
continuity, 185
dynamical conception of, 176
elasticity, division between, 165–66
as essential property of bodies, 159
extension, 142–43, 155, 177, 270
of fluid, 325–26
force of inertia, 155–56
geometric conception of, 270
Imperial Academy, 233
inertia, 77, 156, 159, 208
acceleration, measured by, 209
D'Almbert, 271, 272, 276–77, 279–80
division of bodies, as inert and
actant, 83
force of, 84–85, 90, 105, 155–57, 158,
159–60, 185, 186–88, 194–95, 196,
197, 198, 204, 207–8, 209, 269, 271,
272, 288, 294
law of, 90
linear motion, 209
of matter, 84, 85
moment of inertia, 209, 301, 310
Newton, 59, 271
principle of, 276–77, 278, 279
rotational, 209
Introduction to Natural Philosophy
(Euler), 139
Introduction to Natural Philosophy, An
(Keill), 64–65

Jacquier, Fr., 102–3
Jalobeanu, Dana, 9
Jammer, M., 176–77
Journal de École polytechnique
(journal), 368

Kant, Immanuel, 1, 23, 97, 129, 139, 145,
168, 169–70, 171, 172, 175, 176, 180–
81, 199–200, 221, 378–79, 396–97
apparent motion, 183
on bodies, 172, 199
body, notion of, 184
causal knowledge, 178, 179–80, 183
collisions, treatment of, 181
contact action, 211
Critical Turn, 21, 182n.30
d-contact, 211

divisibility, 173
dynamical laws, 182
equal-force bodies, 182–83
extended bodies, 171, 176, 177–78
first causes, 172, 178–79
force of motion, 183–84
geometry, 136–72
impact, explanation of, 182
matter theory, 213
monads, 173
notion of mass, 183–84
physical monads, 174, 176, 177–78
true motion, 181–82
Keill, John, 64–65, 70, 73–74, 99, 100, 102,
107, 109, 110, 111, 112, 121, 149–50
axioms, 111–12, 119–20
causal account, of collision process, 66
collisions, 65, 66
constructive approach, 68–70
elastic collisions, 66–67
hard bodies, with rebound, rejection
of, 67–68
as Newtonian, 68–69
Oxford lectures, 74–75, 108–9, 110
post-collision behavior, 69
Kelvin-Voigt equation, 327–28
Kepler, Johannes, 84–85
kinematic behavior, 40, 45, 87, 95–96,
276–77, 288, 362
kinematic constraints, 15, 16, 225–26,
239–40, 346–47
kinematics, 41, 72, 87, 219–20, 250, 267,
289–90, 297–98, 327, 329, 346, 353,
354–55, 389, 391–92
of collision, 84, 90
of free fall, 88f
knowledge claim, 247–48, 258
Kuhn, Thomas, 25–26

Lagrange, Joseph-Louis, 24–25, 97, 222,
227–28, 240, 246, 249, 259–60, 294,
301, 327, 333, 336, 341–42, 351, 353,
358, 362, 364–65, 372–73, 375, 376,
393–94, 397
architecture of matter, 352–53
bilateral and unilateral
constraints, 363–64
bodies, approach to, 349, 350–51
critics, reaction to, 374–75

Lagrange, Joseph-Louis (*cont.*)
  d'Alembert, comparison to, 357
  deformable continua, 348
  dynamics, 361–62
  elasticity, 370
  empirical evidence, 358–59
  equations of condition, 336, 342, 346
  Euler-Lagrange Equation (ELQ),
    259n.48, 357, 361–62
  extended bodies, as composites, 352
  fictive forces, 339
  fluids, as incompressible and
    compressible, 349–50
  generality of principles, 354–55, 360
  ideal constraints, 364
  inter-body action, 356
  Lagrange Heuristic, 331–32,
    370–71n.11
  Lagrange multipliers, 24, 259, 260, 330,
    335, 342–43, 344, 345, 346, 348, 356,
    362–63, 364, 376
  Lagrange's Principle, 335, 339–40, 340*f*,
    341, 345, 355, 366–67, 371–72, 373
  Mechanique analytique, 190, 228–29,
    333, 334, 335, 337, 340, 341, 344,
    346–47, 348–49, 353–54, 355, 357,
    362, 364, 366–67, 369–70, 373, 374
  principle approach, 355, 367–68
  Principle of Virtual Velocities (PVV),
    337, 338, 339, 342–44, 345, 351–52,
    353–54, 355–57, 358, 359–60, 361–
    62, 366, 369, 374–75
  Principle of Virtual Work, 260, 360
  proof of truth, 374–75
  Relaxation Heuristic, 347–48, 355–56
  Relaxation Postulate, 15, 345–46, 355
  reverse effective force (REF), 338
  solids, 350
  three-body problem, 358
Lamé, Gabriel, 306
Lange, Marc, 45n.29
Laplace, Pierre-Simon, 367, 369, 376, 377–
    78, 379, 396–97
  capillarity theory, 372
  celestial mechanics, 306
  incomplete theory, 372–73
  intermolecular forces, commitment
    to, 379–80

Laplacian physics, 368
  ontic approach, 392
Laplacian School, 377, 379, 382, 384
  abstract mechanics, 376
  continuously-distributed
    values, 389–90
  continuum modeling, 384
  discrete molecules, 382–83, 392–93
  equation of motion, 392–93
  extended bodies, 380
  field theory, 384–85
  nomic unity, 393–94
  ontic unity, 378–79, 393, 394
Law of Continuity, 3, 34–35, 36–37, 41, 77,
    82, 106, 123
laws of motion, 6, 9, 32, 77, 89, 91, 148,
    152, 263–64, 267–68, 289, 380
  Cauchy, 306, 391–92, 393–94
  constructive approach, 11
  Euler, 227
  general rule, 52–53, 215, 229, 258
  Newton, 1, 8, 20, 66, 152, 153, 217,
    219, 261
  principle approach, 11, 268
Leibniz, Gottfried Wilhelm, 1, 23, 27–28,
    31, 32–33, 34, 57, 75, 83, 88–89, 95,
    105, 127, 136, 137, 143, 151, 170,
    177–78, 181, 183, 232–33, 234
  active force, 104–5, 130, 143–44
  calculus, version of, 232
  Clarke, correspondence between, 59–
    60, 63, 64, 74–75, 79–82, 106
  collisions, 78–79, 81, 82–83, 130
  collision theory, causal-explanatory
    basis for, 183–84
  continuity, 185
  dead and living force, distinction
    between, 77–78, 143–44
  derivative force, 77
  dynamics, 78
  extended bodies, 135
  force, notion of, 76–77
  God's action, 79–81
  hard bodies, denial of, 52
  Law of Continuity, 34–35, 36–37
  Malebranche, criticism of, 34–35, 36–
    37, 51–52, 64, 74–75
  passive force, 104–5, 130, 143–44

primitive force, 77
Principles of Sufficient Reason
    (PSR), 76, 77
quantity of motion and quantity of
    force, distinction between, 81–82, 95
solicitation, notion of, 77–78
theory of force, 75
vis viva controversy, 77, 78, 80, 82
Leibnizians, 63–64, 95–96, 100, 105, 143–
    44, 176, 183
Leibniz-Malebranche agenda, 181, 183
Leibniz-Clarke Correspondence, 59–60,
    63–64, 79, 82, 106
    God's action, 79, 80
Le Seur, Thomas, 102
Letters to a German Princess (Euler), 138
Leucippus, 383–84
Locke, John, 21
Lois de la communication des mouvements
    (Malebranche), 36
Lorentz transformations
    dynamical commitments, 267
    as kinematic condition, 267
    special relativity (SR), 266

Mach, Ernst, 25, 215, 303, 305–6
Mach-Kuhn perspective, 25–26
MacLaurin, Colin, 49, 50, 51–52
    "Axioms and Principles," 49–50
Mairan, Jean-Jacques Dortous de, 145, 146
Malebranche, Nicolas, 1–28, 29, 30, 35,
    48, 52–53, 55, 57, 63–64, 65, 66–70,
    71–72, 73–74, 75
    as Cartesian, 27
    Cartesian mistakes, perpetuation
        of, 27–28
    causal-explanatory account, of
        collisions, 32, 47–48
    collision theory, 30–31, 32–33, 35–36,
        37, 38, 47–48, 51–52, 57, 58, 59, 60, 61
    Descartes' doctrine of method,
        adoption of, 29
    elastic body collisions, 43–47
    extension, as essential property of
        bodies, 30–31
    Galilean relativity, 57, 60–61, 69–70
    hard body collision theory, 38–41, 48,
        49, 50, 51, 52, 61

kinematic behavior, 40, 45
Leibniz-Malebranche agenda, 181, 183
Leibniz's criticism of, 34, 36–37, 41
M-mass, 58–59
mature collision theory, 28, 37–38, 41,
    47, 61, 74–75
mental gluing, 94
new French physics, leading
    authority in, 29
occasionalism of, 20–21, 27, 38,
    127, 149–50
soft body collision theory, 41, 42
true speed, 59–60
voluntarism, 40, 45
Mariotte, Edme, 257
mass
    of a body, 58, 142, 162, 165, 193–95,
        208, 209
    concept of, 210
    conservation of mass, 329
    as derivative property, 393
    dynamical, 193–96, 197–98, 206, 207, 209
    as elusive concept, 210
    elusive, 57, 61, 69–70, 141, 193, 204
    fluid, 228
    geometric, 193–94, 205–6, 208
    gravitational, 355–56
    as hard-won concept, 210
    inertial, 59, 143–44, 197, 327–28
    kinematic, 194–95, 197, 198, 205–6
    linear, 209
    mass center, 83, 86, 162–63, 181–82,
        255–56, 313, 340–41, 372, 383, 390
    mass density, 305, 321–22, 329
    mass distribution, 301, 328, 382–83
    mass flow, 301, 329
    M-mass, 58–59
    mass points, 207, 211–13, 249–50, 281,
        305, 306, 324, 328–29, 337, 340, 344,
        351–53, 373, 375, 379, 382–84
    mass variable, 336
    moment of inertia, 209, 301, 310
    Newtonian, 69–70, 95, 191–92, 193,
        197–98, 204, 207–8, 388
    Newton's second law of motion, 206
    as primary, 293
    quantity of matter, 69–70, 206, 207–8,
        209, 218, 279, 292–93

mass (*cont.*)
    quantity of motion, 292, 293
    qua measure, 59
    rigid bits of, 252–53
    speed, 146–47, 274, 285, 292–93
    as term, 204, 205–6, 291–92
    as theoretical constructs, 293
    true speed, 61, 183
    velocity, 146, 190, 292, 384
    virtual displacements, 341–42
    volumetric, 205
material properties, 10, 251, 327–28
Mathematical elements of physics . . . ('s
    Gravesande), 70
matter theory, 1, 3–5, 10, 11, 12, 29–31, 32,
    47–48, 52–53, 61–62, 66–67, 74, 76,
    90, 104, 105, 199, 213, 217–18, 219–
    20, 225–26, 229, 234–35, 251–53,
    295, 330, 385–86, 389–90
    Boscovich, 174–75
    Cartesian, 22–23, 134, 149–50
    Châtelet, 265–66
    continuous, 205
    Malebranche, 30, 44–45, 47
    material properties, 251
    rules of collision, 93–94, 95–96
    single, 229
    Taylor, 254
    theory of bodies, 266
    types of, 133–34
    univocal, 383–84
    Wolff, 89–90
Maupertuis, Pierre Louis, 1, 114, 120,
    122–24, 144–46, 147, 148, 149, 210,
    221–22, 233
    principle of least action (PLA), 123,
    145, 227
Mazière, Father, 55
mechanics of constrained motion
    (MCON), 202, 223–24, 299, 300, 304,
    333–34, 364
    solving, 24, 260, 301, 306
    See also MCON1; MCON2; theory of
    constrained motions
mechanics of constrained motion
    (MCON1), 14, 16, 224, 225, 226,
    227–28, 240, 249–50, 251, 254, 261,
    262, 299–300, 302–3, 327, 328,
    331, 334
    internal constraints, 14, 24, 224, 299–
    300, 333
    inviscid fluids, 323
    rigid-bodies, 225, 313
    solutions to, 15, 16, 224, 226, 304, 313,
    314, 323, 324, 325, 326, 330
    as term, 331
    vibrating strings, 314
mechanics of constrained motion (MCON2),
    15, 16, 224, 225, 226, 227–28, 231, 254–
    55, 261, 262, 333–34, 348–49
    external constraints, 15, 224, 240,
    300, 333
    solutions to, 16, 224, 249, 364
Mechanica (Euler), 14, 207–8, 228–29, 308
Mechanique analytique (Lagrange), 228–
    29, 333, 340, 357, 364, 366–67, 369–
    70, 373, 374
    constraints, general and uniform, 341
    explicit statements, 349
    as ground-breaking, 227–28
    implicit commitments, 349
    kinematic constraints, 346–47
    as landmark work, 344
    MCON2, contribution to, 335
    origin of, 334
    principle reading of, 353, 355
    Principle of Virtual Velocities (PVV),
    353–54, 362, 374
    rational mechanics, contribution to, 334
    statics, 337
    as tract in rational mechanics, 348–49
    unified theory of, 364
mechanical philosophy, 5, 6, 125, 144–45
mechanics. *See* analytic mechanics;
    celestial mechanics; continuum
    mechanics; mechanics of constrained
    motion (MCON, MCON1,
    MCON2); philosophical mechanics;
    physical mechanics; practical
    mechanics; rational mechanics;
    universal mechanics
Meditationes (Johann Bernoulli), 255
*Metaphysical Foundations of Natural
    Science* (Kant), 184
metaphysics, 1, 34, 48, 60, 61–62, 89, 100,
    101, 102, 125, 135, 148, 171, 172,
    177–78, 204, 220–21, 222
    of body, 29–30, 37, 220

BODY, at intersection of, 22
of causation, 125, 126–28, 129, 179
first causes, 179
geometry, union with, 172
of material nature, 184
of matter, 90, 220–21
physics, division between, 101
of science, 19
scientific, 22
of substance and causation, 20–21, 101,
124, 128–29, 131
Methodus (Taylor), 235, 253, 255
molecules, 352, 377, 378, 379–80,
381, 383
attractive and repulsive forces, 377–
78, 383
as continuously distributed
values, 389–90
discrete, 366–67, 378, 382–84, 392–93
fluid motion, 380, 381
as otiose, 382
monads
dispute over, 137–38
Du Châtelet, 136–37
Euler, 154, 170
Kant, 173
Leibniz, 20–21, 135
monad theory, 141
physical, 174, 177–78, 207
prize competitions, 137
rejection of, 138
Motus nervi (Taylor), 235, 252, 255
Musschenbroek, Pieter van, 1, 70–71,
72–73, 74, 99–100, 101, 103, 107,
112–13, 114, 116, 117, 118, 119, 120,
136–37, 171
atomism, 135
elasticity, 72, 73–74
epistemic access, 117, 118, 121
hard bodies, 71–72, 73
hypotheses, 116
inductive reasoning, 116, 119
soft bodies, 72, 73

natural philosophy, 1, 7–8, 12, 42, 45–46,
48, 57, 63, 64–65, 74–75, 82, 99, 115,
116–17, 131, 134, 157, 171
BODY problem, at heart of, 132,
167, 198–99

Boscovich, 189–90
Cartesian, 61–62, 63, 95
collision theory, foundations of, 5, 8,
13, 16–17, 23, 27, 28, 47–48, 52, 61–
62, 95
Enlightenment, 21–22
Euler, 157–58, 165–66, 217–18
geometric mass, 205
monads, relevance of, 137
Newtonian, 122
problem of collisions (PCOL), as
foundational problem, 129–30
physics, as used interchangeably, 6, 99,
102, 103
Natural Philosophy (Euler), 138–39, 157
Nature, 2, 10, 85, 99, 103, 104, 107, 110,
114, 120, 125, 128–29, 130, 133, 149–
50, 151, 155, 169, 177, 180, 189, 199,
219–20, 231–32, 288, 296
Action, 2–3, 98, 103, 104–5, 106, 122,
124, 129–30, 132, 149, 158, 180, 226–
55, 261–62, 296–97, 378–79, 398
bodily action, 10, 265
Nature, Action, Evidence, and Principle
(NAEP), 2, 3, 11, 23, 97, 106, 124
Navier, Claude-Louis, 327, 348, 367, 370,
371, 381, 383, 397
continuous matter, 386
equation of motion, 382
Navier-Stokes equation, 326, 382
viscosity effects, 381
viscous fluid, 380
Neile, William, 10, 32, 78–79, 183–84
Newton-Euler-Cauchy dynamics, 393–94
Newton-Euler dynamics, 392
Newton, Isaac, 3–4, 5, 8, 19–20, 21–22, 23,
25–26, 31, 63, 64–65, 74–75, 81, 113,
116, 143, 166–67, 171, 185, 187, 208,
214, 215, 221, 223–24, 227–28, 232–
33, 234, 238, 258, 261–62, 270–71,
292–93, 299, 304, 318, 331, 358–59
action-at-a-distance, 95
calculus of fluxions, 235
first law of motion, 143, 194–95
first rule of reasoning, 107–8,
120, 122–23
force of inertia, 204, 271
free particle, 304, 388, 390
gravitation theory, 257

Newton, Isaac (*cont.*)
  gravity, 7–8
  impressed force, 143–44
  inductive reasoning, 115–16, 117
  inherent force, 143–44, 204
  Lex Secunda, 303, 304
  mass, concept of, 141, 207–8, 210
  mass, as term, 204
  Newton-Euler-Cauchy dynamics, 393–94
  Newton-Euler dynamics, 392
  Newton-Euler mechanics, 342
  point particle mechanics, 193
  Rules of Reasoning, 112, 122
  second law of motion, 305–6
  second rule of reasoning, 120
  theory of universal gravitation, 20, 121
  third law of motion, 160, 187, 190
  three laws, 215
  universal mechanics, 6–7
Newtonians, 63–65, 70, 74, 82–83, 95–96,
    98, 100, 104, 105, 106, 107, 108, 115,
    116, 119, 124, 135, 143–44, 167, 176,
    178, 265–66
Newtonian physics, 124, 142
nomic unity, 12, 24–25, 219, 220, 263–64,
    295, 297–98, 367, 374
  deductive nomic unity, 220
  Lagrangian nomic unity, 366–67, 385–
    86, 393–94
  ontic disunity, based on, 326
nutation
  terrestrial, 218n.22, 310n.14

observational evidence, 107, 119n.39
occasionalism, 125–26, 127, 179
  criticism of, 149–50, 150n.45
  Malebranchean, 20–21, 145n.34, 149–
    50, 150n.45
  modulo, 27, 38
Oldenburg, Henry, 8–9, 78–79
On Living Forces (Boscovich), 180–81
ontic disunity, 326, 330, 397–98
ontic unity, 12, 24–25, 220–21, 263–64,
    295, 297–98, 366–67, 378, 380, 383–
    84, 385–86
  explanatory, 220
  molecular, 393
  in physical theory, 12

Opticks (Newton), 80, 81

Papin, Denis
  Leibniz, controversy with,
    73n.18, 80n.29
Parallelogram Rule, 218n.23, 236, 247–
    48, 248f, 252, 254, 257, 273–74,
    273f, 374–75
Paris Academy of Sciences, 222–23, 233
  prize competitions concerning rules of
    collision, 27–28, 57, 61
particle physics, 177
Pemberton, Henry, 73–74, 107, 112, 115,
    116–17, 119, 122–23
Peter the Great, 233
Pfautz, Christoph, 215n.18
phenomena, 25, 78–79, 113–14, 116, 179
  capillary, 377
  causal-explanatory account, 122–23
  celestial, 233
  elastic, 377–78
  of extended bodies, 136
  gravitational, 115, 121, 195–96
  by induction, 111
  motion, 366–67
  natural, 47–48
  of nature, 171, 178–79
  observed, 10, 119
  true causes of, 115
  ultimate causes, 115
philosophical mechanics, 3–4, 5, 8, 10, 11,
    13, 20, 23, 24–26, 54–55, 89, 97, 98,
    130–31, 133, 153, 154, 166, 170, 180,
    201–2, 213–14, 218, 220, 222–23,
    231–32, 238, 249, 251, 254–55, 257,
    262, 263–64, 266, 268–69, 279, 287–
    88, 295, 296, 297, 299, 300, 302–3,
    323, 331, 333, 334, 344–45, 348–49,
    360, 367, 395, 396
  of bodies, 10
  of bodily action, 153
  BODY, 203, 214, 226, 333, 349, 364–65,
    366–67, 396
  Boscovich, 180, 184–85, 190, 192, 198,
    199–200
  of collisions, 5, 10, 13, 63, 64, 74, 88–89,
    95–96, 132, 133, 151, 160–61, 166,
    167–68, 169, 170, 180, 198

conclusions for, 396–99
Euler, 221–22, 326, 328, 329
of extended matter, as elusive, 261
golden age, 397
of impact, 7, 74, 91, 164
Kant, 180, 183–84, 199–200
Lagrange, 351, 353, 354–55, 364–65
matter theory, 94, 219–20, 229
MCON2, 364
nomic unity, 295, 326
Principal of Virtual Velocities (PVV), 353, 354–55
as term of art, 3
virtual-work approach, 261–62
philosophical physics, 7–8, 11, 13–14, 16–17, 98, 101–2, 104–5, 115, 122, 130, 135, 154, 167, 169, 177, 179, 203, 204, 214, 218, 231, 289, 349, 378–79, 397–98
of bodily action, 133
BODY, addressing of, 103, 130–31, 198–99, 209, 217, 333, 397
collisions, foundational role of, 132
constructive approach, 266
of impact, 167–68, 228–29
matter theory, 229
rational mechanics, divide between, 23–24, 201, 333
rational mechanics, integration of, 25–26, 130–31, 132, 180, 266, 396
philosophical theory of matter, 13–14
philosophy, 1, 3–4, 5, 7, 11, 12, 17–19, 21–22, 23–25, 64, 98, 100, 108–9, 110, 115, 132, 133, 177–78, 198–200, 202, 218, 222–23, 228, 254–55, 299, 378–79, 397–98, 399
Cartesian, 79
Descartes, 29
experimental, 111
French, 27
Leibniz, 79, 82
modern, 19, 97, 125
moral philosophy, 1, 100
Newton, 73–74, 99
physics, as sub-discipline of, 6, 23, 25, 97, 98, 99, 100–2, 103, 229, 230, 266, 295, 396
political philosophy, 1

theoretical, 102
transcendental, 172
See also mechanical philosophy; natural philosophy
Phoronomia (Hermann), 64, 82–83, 84, 88, 92, 228–29, 232–33
physical influx, 20–21, 125–26
physical mechanics, 376, 378–79, 396–97
physical monadology, 171, 174, 176
Physical Monadology (Kant), 171, 181
physical monads, 174, 177–78, 207
physical theory, 12, 50, 377
nomic unity, 12
ontic unity, 12
physics, 1, 11, 12, 13–14, 17, 19, 20, 22, 23, 25, 45–46, 61–62, 79, 90, 98, 101, 102, 103, 106, 108, 111–12, 117, 118–19, 120, 121, 122–23, 126–27, 128, 135, 140, 143, 148, 158, 165–66, 167, 168, 177, 180, 203, 218, 219, 228, 229–30, 254–55, 263–64, 265, 288, 376, 399
avoidance, 225–26
bodies, 397–98
BODY, 129–30
Boscovich, 169–70, 177–78, 179, 180, 184–85, 188, 190–91, 193, 198, 199
Cartesian, 10, 107–8, 110, 122, 124, 134
classical, 397
of collisions, 133, 160–61, 164, 166–67, 170
complete, 160–61
of constraints, 261–62
constructive approach, 268
D'Alembert, 277, 294, 396–97
Descartes, 4, 5, 143, 201
eighteenth century, 99, 101–2, 157, 179–80, 266, 294, 295
eighteenth century v. modern, 7, 98
empirical, 378–79
Euler, 157, 160–61, 167, 217–18
experimental research, 368
extended, 169
extended bodies, 133–34, 172, 177, 331
French, 29
of impact, 159–60, 167–68
impenetrability, 161, 166, 175–76
as independent discipline, 25, 264, 396, 397

physics (*cont.*)
  Kant, 169–70, 172, 177–79, 180, 199
  Lagrange, 341–42, 357
  of Laplacian, 24–25, 368, 377
  Leibniz, 127, 143–44
  mathematical, 378–79
  mathematical treatment, 357, 377
  Maypertuis, 122–23
  mechanics, , 7, 19–20, 133, 137, 154,
    156, 166–43, 185
  metaphysics, 101
  modern, 6, 19–20
  of molecules, 377–78
  natural philosophy, interchanged with,
    6, 99, 103, 132n.2
  new physics, 170, 303, 398
  Newtonian, 115, 121–22, 124, 142,
    143–44, 210
  non-foundationalist approach to, 140
  as "normal science," 25
  particle physics, 177
  philosophy, as sub–discipline of, 23, 25,
    97, 98, 99, 100, 101–2, 103, 169n.2,
    177–78, 295, 396
  principle of least action (PLA), 122–23
  Principles of Sufficient Reason (PSR), 150
  as qualitative, 7
  rational mechanics, 7, 8, 11–12, 133,
    199–200, 203, 204, 225–26, 360, 364
  as study of bodies, 19–20, 100–1, 102,
    133, 199
  as term, 6, 7, 98, 132n.2
  terrestrial, 20
  theoretical, 295
  transformation of, 177–78, 199
  underlying, 341–42
  Wolff, 89
Planck, Max, 302, 315
plenism, 205
plenum, 27, 43, 49, 141, 142–43
Poinsot, Louis, 367–68, 374–75
Poisson, Siméon Denis, 228n.40, 367, 376,
  376–77n.23, 382, 383, 394, 396–97
practical mechanics, 6–7
pre-established harmony, 20–21, 125–
  26, 127
Principia (Newton), 3–4, 6, 7–8, 20, 25,
  63, 73–74, 102, 108–9, 112, 115, 122,
  140, 143–44, 155, 166–67, 204, 210,
  214, 219, 223–24, 238, 246, 271, 304,
  333, 397
Principle, 3, 11, 105, 106, 113–14
  auxiliary principle, 247n.34
  conservation principle, 187
  dynamical principle, 236, 331, 389–90
  equilibrium principle, 292–93
  equivalence principle, 195–96
  Evidence, 10, 98, 106, 108, 115, 119,
    121–22, 124, 132, 149, 169, 261–62,
    265, 296
  general Principle of order, 34–35
  heuristic principle, 257–58
  Principle approach (bodies), 11, 268
  Principle approach to collisions, 31
  Principle approach (objects), 12
  relativity principle, 49–50, 266, 267, 274
Principle of Least Action (PLA), 122–23,
  145, 227, 304
Principles of Philosophy (Descartes), 3–4,
  27–28, 45–46, 230
Principles of Sufficient Reason (PSR), 76,
  77, 82, 90, 106, 107–8, 113–14, 122,
  144, 145, 150–51, 157, 158
Principle of Virtual Velocities (PVV), 344,
  351–52, 355–57, 358, 359–60, 361–
  62, 366, 369, 373, 374
  Composition of Forces, 374
  constraints, 342, 344, 354–55
  as E-principle, 353
  equations of condition, 335
  fictive forces, 339
  Lagrange multipliers, 356
  Lagrange's Principle, 335, 338, 339, 340,
    342, 343–44, 345, 353–54, 355, 375,
    381, 389–90
  as neutral and open-ended, 334
  proof of truth, 374–75
  as theorem, 354
  unconstrained bodies, 342
  virtual velocity, 339
Principle of Virtual Work, 259, 260, 340*f*,
  360, 368
Problem of Bodies (BODY), 1–2, 97, 132,
  231, 263, 396
problem of collisions (PCOL), 13–14, 16,
  22, 23, 64, 95, 96, 97, 151, 177, 180–
  81, 183–84, 198–99, 201–2, 226, 227,
  229, 231, 328, 331, 396–97

BODY, as necessary for solving, 97, 98, 130–31, 132, 167–68, 175–76
BODY, solutions to, 131, 180
Boscovich, 190, 199
d'Alembert, 282, 297
importance of, 130
as investigative tool, 130
natural philosophy, as foundational problem in, 129–30
problem of constrained motion (PCON), relationships between, 16–17
solution to, 95–96
solving of, 129–30
solving of, as necessary, 13, 129–30
problem of constrained motion (PCON), 13, 16, 22, 23–24, 201–2, 223, 226, 231, 249, 251, 257, 281, 300, 396–97
BODY, as solution to, 16–17, 229, 261, 299
compound pendulum, 231, 232, 261
d'Alembert, 282, 297, 298
Euler, 308–9
extended bodies, 232, 245
external constraints, 232
implications for, 231
MCON1 and MCON2, integration of, 227, 261
problem of collisions (PCOL), relationships between, 16–17
rational mechanics, contributions from, 231–32
vibrating string, 231, 261
See also rational mechanics of constrained motion
properties of bodies, 10, 68–69, 95–96, 103, 112–14, 138, 143, 154, 155, 157, 167, 180, 230, 265–66, 297–98, 355–56
causes of change in bodies, 158, 217–18
essential, 125
general, 185, 204
generic, 128–29
observable, 115
Prussia, 222

quantity of matter, 69–70, 133, 141, 142, 167, 193–94, 204, 206, 207–8, 209, 218, 292–93
volume of body, 205

rational mechanics, 3–4, 5, 7–8, 10, 11–12, 13–14, 16, 19–20, 24, 90, 133, 170, 184, 185, 199, 202, 204, 207, 213, 214, 220–21, 222, 227, 228, 229, 231, 233, 238, 251–52, 259, 261–62, 263, 266, 268, 278, 302–3, 304, 307–8, 318, 328, 331–32, 345–46, 351, 360, 373, 374, 378, 379, 382, 394, 397, 398
advanced research in, 233
BODY, relevant to, 201, 202, 203, 209, 213–14, 218, 219–20, 223, 225, 226, 227, 233, 261, 297, 299, 333, 349, 364–65
capillary effects, 372
Cauchy, 367
compound pendulum, 234, 238
of constrained motion, 217, 225–26, 254–55, 257, 294, 299, 333, 366, 396
d'Alembert, contribution to, 279–80, 285, 287–88, 291, 294, 296, 297, 357, 396–97
deformable body, 366–67
disunified, 394
elite education, cornerstone of, 368
equations of motion, center stage of, 302
Euler, 154, 209, 218, 307, 308–9, 313, 320, 323, 324, 326, 327, 330
of extended bodies, 24, 217, 219, 228, 231–32, 305, 323, 324, 396
of external constraints, 333
of frictionless fluids, 318
generality, 258
growth in, 229
heuristic predicament of, 246
H-principle, 258
innovation and development, as powerhouse of, 227
of internal constraints, 308–9
of inviscid fluids, 320
Lagrange, 335, 348, 351–52, 356–57, 362–65, 366–67
Laplacians, 377, 378–79
Lever Principle, 259
in Long Enlightenment, 302
mass in, 207, 209, 210
as mathematical and exact, 6–7
mathematical treatment, 357
MCON1, solving of, 228
Mechanique (Lagrange), contribution to, 334, 348–49, 364, 366–67

rational mechanics (*cont.*)
   nomic unity, favoring of, 219, 220
   philosophical physics, divide between,
     23–24, 201
   philosophical physics, integration with,
     25–26, 130–31, 132, 169, 180, 199–
     200, 396
   problem of constrained motion
     (PCON), 261
   problem of constraints, 261
   rise of, 16–17
   single matter theory, 229
   single theory of matter, 330, 366–
     67, 394
   Taylor and Bernoulli, 234–35, 251
   as term, 6
   theory of constraints, 201–2
   Treatise (d'Alembert), 264, 271, 287–88,
     295, 297
   of vibrating string, 234, 236, 252–
     53, 262
   See also mechanics; PCON
rebound, 9, 10, 32–33, 35–36, 40, 43–44,
     47, 50, 55, 66, 73–74, 88, 90, 105, 147,
     164, 296–97
   after collision, 40
   elastic, 44–45, 47, 52–55, 56, 66–67, 69–
     70, 87, 92–93, 193n.61
   failure to, 72
   as k-elastic bodies, 43, 49, 50–51
   k-hard bodies, 49, 67–68
   kinematic, 67, 69, 72–73, 83, 95–96
   perfect, 49, 94
   physical reason of, 53–54
   reason for, 164–65
   stiffness, 42
relationism, 181–82
Relaxation Heuristic, 347, 348, 355–56
Relaxation Postulate, 345–47, 355
Rohault, Jacques, 99, 100, 101, 111–12
Royal Academy of Paris, 103
Royal Academy of Sciences, 222–23
Royal Society of London, 32–33

Saint-Venant, Jean Claude, 382–84
Saulmon, Nicolas, 53–54, 55, 56
Scott, W., 176
Search after Truth (Malebranche),
     27–28, 29

secondary causation, 20–21, 124, 125–26,
     127, 129
's Gravesande, W. J., 70, 73–74, 99, 107,
     112, 113, 117, 118, 171
special relativity (SR), 266
Specimen Dynamicum (Leibniz), 75, 77, 82
Stokes, George G., 384–85
Stokes' Law, 346
St. Petersburg (Russia), 54–55, 220–21,
     233, 307
System of Natural Philosophy (Rohault)

Taylor, Brook, 234, 235, 236, 240, 251,
     252–54, 255, 256, 257, 314, 316
   equivalent pendulum, 249–50
   multiple derivation, 255
   reverse conditional, 235–36
   shape, 250–51
   theories of matter, 253*f*, 254
Taylor and Bernoulli papers, 255–56
   descriptive incompleteness, 234
   equivalent pendulum, 249–50
   frequency of moving string, 235
   harmonic condition, 235
   heuristic silence, 234
   multiple derivation, 255
   vibrating string, 255
theoretical physics, 295
Theory of Natural Philosophy, A (Bosovich)
   Treatise (d'Alembert), comparison to, 294
theory of constrained motions
     (MCONI), 14, 16
   See also MCON; MCON1; MCON2
theory of matter, 2, 10, 11, 31, 37–38, 91,
     104, 206, 217, 219–20, 221–22, 330,
     352, 384–85
   continuum, 211
   Descartes, 4, 5
   explicit, 11, 13–14, 265–66
   intelligible, 23, 131, 135, 364–65
   kinetic, 392n.53
   Malebranche, 32, 47
   metaphysical, 378–79
   philosophical mechanics of collision, 5,
     10, 13, 89
   underlying, 69–70, 71–72, 177–78,
     234, 254
   underlying matter theory, 10, 32, 47–48,
     66–67, 104, 330

Torricelli, Evangelista, 318
Torricelli-Huygens principle, 87n.38, 304
Toupin, R., 387n.40
Treatise on Dynamics (d'Alembert), 24,
    25–26, 263, 297
Treatise on Light (Huygens), 108
true motion, 181–82, 183
Truesdell, Clifford, 251, 303, 304n.8,
    387n.40, 392n.51
    gurges, 319n.43

underdetermination problem, 108
universal mechanics, 6–7

Varignon, Pierre, 318
vibrating string, 232, 234–35, 236–37,
    255–56, 261, 262, 314, 378
    compound pendulum, 24, 231, 234,
    240, 261
    mass-point pendulum, 249–50
vibration theory, 249–50, 308–9
virtual displacement, 336, 338, 370n.10

virtual-work approach, 261–62, 371, 380
vis viva, 34, 36–37, 41, 75, 77, 78, 79, 82,
    83, 85–86, 88, 94, 95, 107–8, 123, 146,
    148, 153, 183, 217, 237, 268–69, 361
    as controversy, 68, 74–75, 79
    debate over, 143–44, 145, 180–81
    God's action, issue of, 80
    principle of conservation of, 149,
    180–81, 247–48, 257, 295, 304
    See also conservation of *vis viva*
Voltaire, 122
voluntarism, 40, 45, 113–14

Wallis, John, 8–9, 10, 27, 33
Wolff, Christian, 1, 64, 86, 89, 90–94,
    100–1, 102, 137, 154, 156, 160,
    174, 221–22
    elastic collision, 94
    g-contact, 210–11
Wren, Christopher, 8–9, 27, 33, 78–79

Young, Thomas, 379